Stationary Stochastic Processes

Theory and Applications

CHAPMAN & HALL/CRC
Texts in Statistical Science Series

Series Editors
Francesca Dominici, *Harvard School of Public Health, USA*
Julian J. Faraway, *University of Bath, UK*
Martin Tanner, *Northwestern University, USA*
Jim Zidek, *University of British Columbia, Canada*

Analysis of Failure and Survival Data
P. J. Smith

The Analysis of Time Series — An Introduction, Sixth Edition
C. Chatfield

Applied Bayesian Forecasting and Time Series Analysis
A. Pole, M. West, and J. Harrison

Applied Categorical and Count Data Analysis
W. Tang, H. He, and X.M. Tu

Applied Nonparametric Statistical Methods, Fourth Edition
P. Sprent and N.C. Smeeton

Applied Statistics — Handbook of GENSTAT Analysis
E.J. Snell and H. Simpson

Applied Statistics — Principles and Examples
D.R. Cox and E.J. Snell

Applied Stochastic Modelling, Second Edition
B.J.T. Morgan

Bayesian Data Analysis, Second Edition
A. Gelman, J.B. Carlin, H.S. Stern, and D.B. Rubin

Bayesian Ideas and Data Analysis: An Introduction for Scientists and Statisticians
R. Christensen, W. Johnson, A. Branscum, and T.E. Hanson

Bayesian Methods for Data Analysis, Third Edition
B.P. Carlin and T.A. Louis

Beyond ANOVA — Basics of Applied Statistics
R.G. Miller, Jr.

The BUGS Book: A Practical Introduction to Bayesian Analysis
D. Lunn, C. Jackson, N. Best, A. Thomas, and D. Spiegelhalter

A Course in Categorical Data Analysis
T. Leonard

A Course in Large Sample Theory
T.S. Ferguson

Data Driven Statistical Methods
P. Sprent

Decision Analysis — A Bayesian Approach
J.Q. Smith

Design and Analysis of Experiments with SAS
J. Lawson

Elementary Applications of Probability Theory, Second Edition
H.C. Tuckwell

Elements of Simulation
B.J.T. Morgan

Epidemiology — Study Design and Data Analysis, Second Edition
M. Woodward

Essential Statistics, Fourth Edition
D.A.G. Rees

Exercises and Solutions in Biostatistical Theory
L.L. Kupper, B.H. Neelon, and S.M. O'Brien

Extending the Linear Model with R — Generalized Linear, Mixed Effects and Nonparametric Regression Models
J.J. Faraway

A First Course in Linear Model Theory
N. Ravishanker and D.K. Dey

Generalized Additive Models: An Introduction with R
S. Wood

Generalized Linear Mixed Models: Modern Concepts, Methods and Applications
W. W. Stroup

Graphics for Statistics and Data Analysis with R
K.J. Keen

Interpreting Data — A First Course in Statistics
A.J.B. Anderson

Introduction to General and Generalized Linear Models
H. Madsen and P. Thyregod

An Introduction to Generalized Linear Models, Third Edition
A.J. Dobson and A.G. Barnett

Introduction to Multivariate Analysis
C. Chatfield and A.J. Collins

Introduction to Optimization Methods and Their Applications in Statistics
B.S. Everitt

Introduction to Probability with R
K. Baclawski

Introduction to Randomized Controlled Clinical Trials, Second Edition
J.N.S. Matthews

Introduction to Statistical Inference and Its Applications with R
M.W. Trosset

Introduction to Statistical Limit Theory
A.M. Polansky

Introduction to Statistical Methods for Clinical Trials
T.D. Cook and D.L. DeMets

Introduction to the Theory of Statistical Inference
H. Liero and S. Zwanzig

Large Sample Methods in Statistics
P.K. Sen and J. da Motta Singer

Linear Models with R
J.J. Faraway

Logistic Regression Models
J.M. Hilbe

Markov Chain Monte Carlo — Stochastic Simulation for Bayesian Inference, Second Edition
D. Gamerman and H.F. Lopes

Mathematical Statistics
K. Knight

Modeling and Analysis of Stochastic Systems, Second Edition
V.G. Kulkarni

Modelling Binary Data, Second Edition
D. Collett

Modelling Survival Data in Medical Research, Second Edition
D. Collett

Multivariate Analysis of Variance and Repeated Measures — A Practical Approach for Behavioural Scientists
D.J. Hand and C.C. Taylor

Multivariate Statistics — A Practical Approach
B. Flury and H. Riedwyl

Multivariate Survival Analysis and Competing Risks
M. Crowder

Pólya Urn Models
H. Mahmoud

Practical Data Analysis for Designed Experiments
B.S. Yandell

Practical Longitudinal Data Analysis
D.J. Hand and M. Crowder

Practical Multivariate Analysis, Fifth Edition
A. Afifi, S. May, and V.A. Clark

Practical Statistics for Medical Research
D.G. Altman

A Primer on Linear Models
J.F. Monahan

Principles of Uncertainty
J.B. Kadane

Probability — Methods and Measurement
A. O'Hagan

Problem Solving — A Statistician's Guide, Second Edition
C. Chatfield

Randomization, Bootstrap and Monte Carlo Methods in Biology, Third Edition
B.F.J. Manly

Readings in Decision Analysis
S. French

Sampling Methodologies with Applications
P.S.R.S. Rao

Stationary Stochastic Processes: Theory and Applications
G. Lindgren

Statistical Analysis of Reliability Data
M.J. Crowder, A.C. Kimber, T.J. Sweeting, and R.L. Smith

Statistical Methods for Spatial Data Analysis
O. Schabenberger and C.A. Gotway

Statistical Methods for SPC and TQM
D. Bissell

Statistical Methods in Agriculture and Experimental Biology, Second Edition
R. Mead, R.N. Curnow, and A.M. Hasted

Statistical Process Control — Theory and Practice, Third Edition
G.B. Wetherill and D.W. Brown

Statistical Theory, Fourth Edition
B.W. Lindgren

Statistics for Accountants
S. Letchford

Statistics for Epidemiology
N.P. Jewell

Statistics for Technology — A Course in Applied Statistics, Third Edition
C. Chatfield

Statistics in Engineering — A Practical Approach
A.V. Metcalfe

Statistics in Research and Development, Second Edition
R. Caulcutt

Stochastic Processes: An Introduction, Second Edition
P.W. Jones and P. Smith

Survival Analysis Using S — Analysis of Time-to-Event Data
M. Tableman and J.S. Kim

The Theory of Linear Models
B. Jørgensen

Time Series Analysis
H. Madsen

Time Series: Modeling, Computation, and Inference
R. Prado and M. West

Texts in Statistical Science

Stationary Stochastic Processes

Theory and Applications

Georg Lindgren

CRC Press
Taylor & Francis Group
Boca Raton London New York

CRC Press is an imprint of the
Taylor & Francis Group an **informa** business
A CHAPMAN & HALL BOOK

CRC Press
Taylor & Francis Group
6000 Broken Sound Parkway NW, Suite 300
Boca Raton, FL 33487-2742

Printed in the United States of America on acid-free paper
Version Date: 20121121

International Standard Book Number: 978-1-4665-5779-6 (Hardback)

Library of Congress Cataloging-in-Publication Data

Lindgren, Georg, 1940-
Stationary stochastic processes : theory and applications / Georg Lindgren.
pages cm. -- (Chapman & Hall/CRC texts in statistical science)
Includes bibliographical references and index.
ISBN 978-1-4665-5779-6 (hardback)
1. Stationary processes. 2. Stochastic analysis. I. Title.

QA274.3.L56 2013
519.2'2--dc23 2012028123

Visit the Taylor & Francis Web site at
http://www.taylorandfrancis.com

and the CRC Press Web site at
http://www.crcpress.com

Statistics is a science in my opinion, and it is no more a branch of mathematics than are physics, chemistry and economics; for if its methods fail the test of experience – not the test of logic – they are discarded.

John W. Tukey (1953)

Contents

List of figures

Preface

This book has grown out of my own experiences as teacher and researcher at a department in mathematical statistics with responsibilities both to an engineering and a science community. The spirit of the text reflects those double responsibilities.

The background

The book *Stationary and Related Stochastic Processes* [35] appeared in 1967. Written by Harald Cramér and M.R. Leadbetter, it drastically changed the life of PhD students in mathematical statistics with an interest in stochastic processes and their applications, as well as that of students in many other fields of science and engineering. Through that book, they got access to tools and results for stationary stochastic processes that until then had been available only in rather advanced mathematical textbooks, or through specialized statistical journals. The impact of the book can be judged from the fact that still, after almost fifty years, it is a standard reference to stationary processes in PhD theses and research articles.

Even if many of the more specialized results in the Cramér-Leadbetter book have been superseded by more general results, and simpler proofs have been found for some of the statements, the general attitude in the book makes it enjoyable reading both for the student and for the teacher. It will remain a definite source of reference for many standard results on sample function and crossings properties of continuous time processes, in particular in the Gaussian case. Unfortunately, the book only appeared in a first edition, and it was out of print for many years. The Dover reprint from 2004 filled that gap.

Another book, that at its time created a link between the mathematical theory of stationary processes and their engineering use, is the book *An Introduction to the Theory of Stationary Random Functions*, by A.M. Yaglom from 1962, [125]. The idea of bridging the gap between a strict mathematical treatment and the many practically useful results in the theory of stationary processes is even more explicitly expressed in the later, two-volume work by the same author, *Correlation Theory of Stationary and Related Random*

Functions from 1987, [126]. That work, with its extensive literature survey, presents a rather complete overview of correlation and spectrum theory for stationary processes at its time, with lots of extensions, references, and historical notices.

The third influential book is *Probability*, by Leo Breiman [19], also long out of print until it was reprinted in 1992. In the preface of that book, the author mentions the two-armed character of probability theory. The right arm is rigorous mathematics, based on measure theory; the left arm sets probability in action with gambling, motions of physical particles, and, now, almost fifty years later, remote sensing interpretation and systems control.

This book is for the left handed student, who wants to learn what the right handed side of stationary process theory has to offer, without going very far into the mathematical details! It is my hope that also the right handed probabilist could appreciate some of the fascination that lies in the potential of probabilistic thinking in engineering and science. The Preface to Yaglom's [126], Volume II, expresses these thoughts in a judicious way.

This text has grown out from a series of PhD courses on *Stationary stochastic processes*, which have been given at the Department of Mathematical Statistics at Lund University. Previous versions of the text have also been used in Stockholm, Trondheim, and Umeå. The audiences have consisted of Masters and PhD students in Mathematics and Statistics, but also to a large extent of PhD students in Automatic control, Electrical engineering and Signal processing, Image analysis, and Economics. As far as possible, I have tried to incorporate the links between these concrete topics and the theoretical tools developed.

I am grateful to three decades of PhD students who have followed the courses in Lund and elsewhere. Their comments and suggestions have removed many unclear and obscure points and statements. Their demands have also had the result that the text has grown, with new topics along with the increasing interest in advanced statistical tools in applications.

The text

The text is intended for a "second course" in stationary processes, and the material has been chosen to give a fairly broad overview of the theory behind widely scattered applications in engineering and science.

The reader I have in mind has typically some experience with stochastic processes and has felt an urge to know more about "what it really is" and "why." As a "first course" in stationary processes the reader could consult "episode one" [82], an introductory text for scientists and engineers that has been used for a long time in Lund for students in electrical and computer engineering, and also for students in physics an economics.

The emphasis in the present text is on second order properties and Fourier methods. Neither time series analysis nor Markov assumptions play any prominent role.

Since the days of the three texts mentioned in the introduction, applications of second order stationary processes have expanded enormously. The basic theory remains, but the applications, in particular new measurement techniques, have inspired much new theory. This is certainly true for spatial and spatio-temporal models, counting processes in space and time, extremes and crossings, and, not least, on computational aspects.

The aim is to provide a reasonably condensed presentation of sample function properties, spectral representations for stationary processes, and fields, including a portion of stationary point processes. One of the most important applications of second order process theory is found in linear filtering theory. The material on linear filters is divided into two chapters, one with a general technique of how to use the spectral representations and one on three important applications.

One of the most difficult topics to explain in elementary courses on stationary processes is ergodicity. I have tried to give both a mathematical and a statistical view with the main aim to couple mathematics and probability to statistics and measurements.

A chapter on crossing problems and extremes with an introduction to some of its applications has grown out from my own research interests.

Some knowledge of the mathematical foundations of probability helps while reading the text; I have included some of it in a probability appendix, together with the existence and basic convergence properties, as well as some Hilbert space concepts. There is also an appendix on how to simulate stationary stochastic processes by spectral methods and the FFT algorithm.

Some suggestions for reading

The reader who wants to design a course on parts of the text may make use of the following selections:

- For a general and concrete level:
  - Section 1.1 (Section 1.2), Sections 1.3–1.5
  - Section 2.1
  - Sections 3.2–3.4
  - Chapter 4 except Section 4.4

  - Section 5.2
  - Sections 7.1.1, 7.2.1
  - Section 8.1.1
- For specialized and concrete levels:
  - Section 3.5
  - Section 4.4
  - Sections 5.1 and 5.3
  - Chapter 7
  - Chapter 8
  - Appendix B
- For a more abstract level one could add:
  - Section 1.2, Appendix A, plus a book on general probability theory
  - Sections 2.2–2.5
  - Section 3.1
  - Chapter 6

The website http://www.crcpress.com/product/isbn/97814665577796 contains corrections and updates, as well as worked applications.

MATLAB® and Simulink are registered trademarks of The Mathworks, Inc. For product information, please contact:

The Math Works, Inc.
3 Apple Hill Drive
Natick, MA 01760-2098 USA
Tel: 508-647-7000
Fax: 508-647-7001
E-mail: info@mathworks.com
Web: www.mathworks.com

Acknowledgments

To all those students who have been confronted with early versions of this text, as well as the colleagues who have read, lectured from, and commented on the text, I express my great gratitude. I have felt as we have been fellow students of the many facets of the topic. I mention in particular Anastassia Baxevani, Rikard Berthilsson, Klas Bogsjö, Jonas Brunskog, Halfdan Grage, Peter Gustafsson, Oskar Hagberg, Lars Holst, Ulla Holst, Henrik Hult, Pär Johannesson, Timo Koski, Torgny Lindström, Karl-Ola Lundberg, Ulla Machado, Anders Malmberg, Dan Mattsson, Sebastian Rasmus, Bengt Ringnér, Tord Rikte, Igor Rychlik, Jesper Rydén, Mikael Signahl, Eva Sjö, Martin Sköld, Martin Svensson, Jörg Wegener, Linda Werner, and Sofia Åberg.

During the preparation of the final text I have had fruitful discussions with David Bolin, Finn Lindgren, Johan Lindström, Jimmy Olsson, Krzysztof Podgórski, Magnus Wiktorsson; thank you!

But it all goes back to Harald Cramér and Ross Leadbetter!

Georg Lindgren
Lund

List of notations

Roman letters

A, A_k	random amplitudes
$A_{\mathbf{s}}(\mathbf{v})$	set of functions $y(s)$ with $y(s_j) \leq v_j$
$A_{xy}(\omega) = \lvert f_{xy}(\omega)\rvert$	cross-amplitude spectrum
$A(z)$	generating polynomial
$A \Delta B$	$(A - B) \cup (B - A)$, difference between sets
a_T, b_T	inverse scale, and location, parameter in asymptotic extreme value distribution
$BB_T(t)$	Brownian bridge
$\mathscr{B}, \mathscr{B}_n, \mathscr{B}_\infty, \mathscr{B}_T$	Borel fields in $\mathbb{R}, \mathbb{R}^n, \mathbb{R}^\infty, \mathbb{R}^T$
C	generic constant
$C(x_1, x_2)$	$E((x_1 - m_1)\overline{(x_2 - m_2)})$, covariance of complex variables
$C(z)$	generating polynomial
$\mathbb{C}$	space of continuous functions
$c_{xy}(\omega)$	co-spectrum density
$D(u_n), D'(u_n)$	mixing and non-clustering conditions in asymptotic extreme value theory
$d(\mathbf{s}, \mathbf{t})$	$\sqrt{E(\lvert x(\mathbf{s}) - x(\mathbf{t})\rvert^2)}$, distance measure
$\mathrm{d}Z(\omega)$	differential spectral component
$\mathrm{d}\widetilde{Z}(\omega) = \frac{\mathrm{d}Z(\omega)}{f(\omega)}$	normalized differential spectral component
$E(x\lvert y), E(x\lvert\mathscr{A})$	conditional expectation given y or given $\mathscr{A}$
$e(\mathbf{s}, \mathbf{t})$	Euclidian distance
$F_{t_1 \cdots t_n}(b_1, \ldots, b_n)$	n-dimensional distribution function for $x(t_1), \ldots, x(t_n)$
$\mathbf{F} = \{F_{\mathbf{t}^n}\}$	family of finite-dimensional distribution functions
$F(\omega), F_x(\omega)$	spectral distribution function
F^{ac}, F^d, F^s	absolutely continuous, discrete, and singular part of distribution function
$\widetilde{F}(\omega)$	$(F(\omega) + F(\omega - 0))/2$, midpoint at spectral jump
$F_{xy}(\omega), f_{xy}(\omega)$	cross-spectral distribution and density functions
$\mathscr{F}_0, \mathscr{F}$	field and σ-field of events

f	friction coefficient
$f_{x\|y=v}(u)$	conditional density function
$f(\omega), f_x(\omega)$	spectral density functions for stationary process
$f_x^{(d)}(\omega)$	spectral density after sampling with sampling distance d
$f_x^{\text{time}}(\omega)$	time wave spectrum
$f_x^{\text{space}}(\kappa)$	space wave spectrum
$f_{PM}(\omega), f_J(\omega)$	Pierson-Moskowitz and JONSWAP spectral densities
$f(\omega,\theta)$	$f_x^{\text{time}}(\omega)g(\omega,\theta)$, spectral density for random field with directional spreading
$f_c(\omega)$	$\frac{\lambda}{2\pi}+f(\omega)$, Bartlett spectrum for counting process
$f_n^{\text{per}}(\omega)$	periodogram computed from data
g	constant of gravity
$(g,h)_{\mathscr{H}(F)}$	$\int g(\omega)\overline{h(\omega)}\,\mathrm{d}F(\omega)$, inner product in Hilbert space $\mathscr{H}(F)$
$g(\omega)$	$\int e^{-i\omega u}h(u)\,\mathrm{d}u$ frequency response function
$g(\omega,\theta)$	directional spreading function
H	Hurst parameter, index of fractional Brownian motion
$H(\varepsilon)$	entropy
h	water depth
$\mathscr{H}, \mathscr{H}(\Omega)$	Hilbert space of random variable with finite variance
$\mathscr{H}(F)$	Hilbert space of functions with $\int \|g(\omega)\|^2\,\mathrm{d}F(\omega) < \infty$
$\mathscr{H}(x)$	$\mathscr{S}(x(s); s\in T)$, Hilbert space generated by $x(t), t\in T$
$\mathscr{H}(x,t)$	$\mathscr{S}(x(s); s\le t)$, Hilbert space generated by $x(s), s\le t$
$h(u)$	impulse response function
$\mathscr{I}$	field of unions of a finite number of intervals, or σ-field of invariant sets
$J_m(t)$	Bessel function of the first kind of order m
$K_\nu(t)$	Bessel function of the second kind of order ν
$\mathscr{M}_a^b$	σ-field generated by $x(t), a\le t\le b$
N	Avogadro's number
$N(t)$	Poisson process
$\widetilde{N}(I)$	normalized point process of crossings
$N_I(x,u), N_I^+(x,u)$	number of u-crossings/upcrossings by $x(t), t\in I$
$\mathbb{N}$	the set of natural numbers
P	general probability measure
$P^{hw}(A)$	horizontal window conditioned probability after level crossing

$P^u(A)$	distribution of Slepian model for process after u-upcrossing
$P_j^{\max}(A)$	distributions for Slepian models after local maximum, $j=1,2$
$P_x, P_y, \ldots$	probability measure for processes $x(t), y(t), \ldots$
$P_x^{(n)}, P_y^{(n)}, \ldots$	n-dimensional probability measure for $x(t), y(t), \ldots$
$p^u(z)$	Rayleigh density for derivative at level upcrossing in Gaussian process
$q_{xy}(\omega)$	quadrature spectrum
R	Boltzmann's constant
$R(t)$	envelope process
$\mathbb{R}, \mathbb{R}^n, \mathbb{R}^\infty, \mathbb{R}^T$	real line, n-dimensional real space, space of infinite real sequences, space of real function defined on T
$r(s,t)$	$C(x(s),x(t))$ covariance function
$r(t)$	$C(x(s+t),x(s))$ covariance function for stationary process or field
$r_{xy}(t)$	cross-covariance between $x(s+t)$ and $y(s)$
$r_N^*(t)$	covariance estimate based on the periodogram
$r_{\Delta_1}(s,t), r_{\Delta_2}(s,t)$	covariance functions for Gaussian residual in Slepian models after local maximum
$r_\kappa(s,t)$	covariance function for Gaussian residual in Slepian model after level upcrossing
$\mathscr{S}(L)$	space spanned by random variables in L
$\mathscr{S}(x,\varepsilon)$	ε-surrounding around point or function x
$\mathrm{sinc}(x) = \frac{\sin x}{x}$	sinc function
T	parameter space, often $\mathbb{R}$ or $\mathbb{R}^+$, or absolute temperature
$T(\omega)$	transformation of a probability space
$T_2 = 2\pi\sqrt{\omega_0/\omega_2}$	mean period
$t_n^{(k)} = k/2^n$	dyadic number
$(u,v)_{\mathscr{H}(x)}$	$E(u\overline{v})$, inner product in Hilbert space $\mathscr{H}(x)$
$V_f(I)$	total variation of $f(t), t \in I$
$w(t)$	Wiener process
$w'(t)$	Gaussian white noise
$w(\tau)$	covariance intensity function
$w_c(\tau) = \lambda\delta_0(\tau) + w(\tau)$	complete covariance intensity function
$\widehat{x}(t)$	Hilbert transform of $\{x(t), t \in \mathbb{R}\}$
$\mathbb{Z}$	the set of integer numbers
$Z(\omega), Z_j(\omega)$	spectral processes

Greek letters

$\alpha = \omega_2/\sqrt{\omega_0\omega_4}$	spectral width parameter
γ	peakedness parameter in JONSWAP spectrum
$\gamma(\mathbf{u},\mathbf{v})$	semi-variogram
$\Delta F_\omega, \Delta F_k$	spectral distribution jump at ω and ω_k
$\Delta Z_\omega, \Delta Z_k$	spectral process jump at ω and ω_k
δ_{jk}	Kronecker delta
$\delta(t), \delta_0(t), \delta_\tau(t)$	Dirac delta functions
$\varepsilon = \sqrt{1-\alpha^2}$	spectral width parameter
$\zeta = c/(2\sqrt{mk})$	relative damping in linear oscillator
θ	wave direction in random field
κ	space frequency, wave number
κ_k	spectral moment in space
$\kappa^2_{xy}(\omega)$	squared coherence spectrum
λ	intensity in point process
λ_k	eigenvalue to a covariance operator
$\mu^+ = \frac{1}{\pi}\sqrt{\omega_2/\omega_0}$	mean frequency
$\mu(u)$	mean number of u-crossings per time unit
$\mu^+(u)$	mean number of u-upcrossings per time unit
$\mu_{\min} = \mu_{\max}$	mean number of local extremes per time unit
$\xi_u(t)$	Slepian model for process after u-upcrossing
$\xi_1^{\max}(t), \xi_2^{\max}(t)$	Slepian models for process after local maximum
$\rho^2 = \omega_1^2/\omega_0\omega_2$	spectral width parameter
$\rho_{xy}(\omega)$	in-phase correlation in vector processes
$\widetilde{\rho}_{xy}(\omega)$	out-of-phase correlation in vector processes
$\mathbf{\Sigma}$	covariance matrix
$\sigma^2/2\pi$	constant spectral density of white noise
$\sigma(x)$	σ-field generated by random variable
$\Phi_{xy}(\omega)$	phase spectrum
ϕ, ϕ_k	random phases
$\phi_k(t)$	eigenfunction to a covariance operator
ϕ, Φ	standard normal density and distribution functions
$\varphi(s), \varphi(x \mid y)$	characteristic function and conditional expectation
χ_A	indicator function for event A
ψ	phase parameter in linear oscillator
$\psi_\varepsilon(t)$	characteristic function
Ω	sample space
$(\Omega, \mathscr{F})$	measurable space, sample space with σ-field $\mathscr{F}$
$(\Omega, \mathscr{F}, P)$	probability space with probability measure P

ω	frequency or element in sample space
$\omega_0 = \sqrt{k/m}$	resonance frequency in linear oscillator
ω_2	$-r''(0) = V(x(t))$, second spectral moment
ω_k	discrete frequency, or spectral moment $\int \lvert\omega\rvert^k \, \mathrm{d}F(\omega)$
$\boldsymbol{\omega}_{\boldsymbol{\alpha}} = \Pi_k \omega_k^{\alpha_k}$	multiple spectral moment
$\omega_x(h)$	continuity modulus of function x
$\omega_{x,d}(h)$	continuity modulus of function x in d-distance

Special symbols

$\stackrel{a.s.}{\to}$	almost sure convergence, with probability one
$\stackrel{q.m.}{\to}$	quadratic mean convergence
$\stackrel{P}{\to}$	convergence in probability
$\stackrel{\mathscr{L}}{\to}$	convergence in distribution
$x \stackrel{\mathscr{L}}{=} y$	x and y have the same distribution

Chapter 1

Some probability and process background

This introductory chapter gives examples of some of the main concepts in stationary processes theory that will be dealt with in detail in later chapters. The intention here is not to dwell on the precise mathematical properties, but to point to the statistical content and interpretation of the mathematical constructions, and on the driving ideas. The mathematical details are instead presented in Appendix A, and refreshed when needed.

First of all, we deal with the finite-dimensional distribution functions (d.f.), which uniquely define probabilities for a sufficiently rich family of events, namely events that can be identified through process values at discrete sets of times. In particular, they allow us to find conditions for sample function continuity and differentiability and other properties that are essential in statistical model building. This is the topic of Chapter 2.

Another central topic in the book is the covariance properties of stationary processes and their spectral representations, dealt with in Chapter 3. These also form the tools in the chapters on filter manipulations of stationary processes, Chapters 4 and 5. Prediction and limiting behavior are the topics of Chapter 6. Extensions to multivariate models and stochastic fields are treated in Chapter 7, and the final Chapter 8 deals with techniques and important applications of extremes and crossings.

This chapter ends with four examples of path breaking innovative use of stochastic models in physics and engineering, innovations which helped to shape the presented tools and techniques. The examples are the Wiener process as a model for Brownian motion, stationary processes as models for noise in electronic communication systems, stochastic wave models used in ocean engineering, and finally the link between statistical inference and stochastic processes as a basis for signal processing.

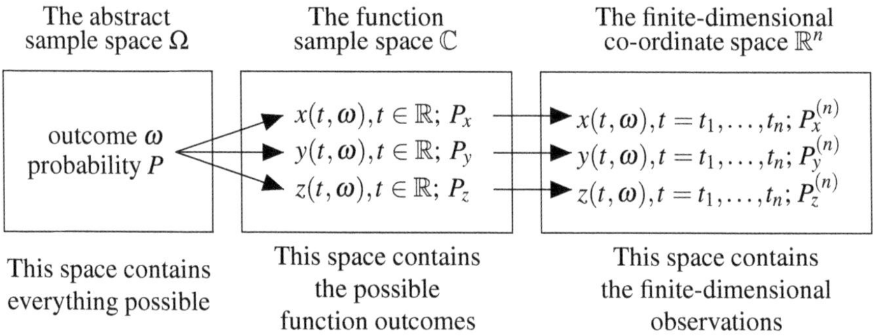

Figure 1.1: *Overview of the three types of worlds in which our processes live.*

1.1 Sample space, sample function, and observables

Stochastic processes are often called random functions; these two notions put emphasis on two different aspects of the theory, namely,

- stochastic processes as families of infinitely many random variables on the same sample space, usually equipped with a fixed probability measure,
- stochastic processes as a means to assign probabilities to sets of functions, for example some specified sets of continuous functions, or sets of piecewise constant functions with unit jumps.

These two aspects of stochastic processes can be illustrated as in Figure 1.1, corresponding to an experiment where the outcomes are continuous functions.

The figure illustrates the three levels of abstraction and observability for a random experiment. To be concrete, think of an experiment controlling the steering of a ship. The sample space Ω can be envisioned to describe the conditions that the ship will meet on the ocean, the wind and wave climate over the lifetime of the ship, the actual weather at a certain trip, etc. The function sample space $\mathbb{C}$ contains the different characteristics that physically characterize the movements of the ship and the steering process, while the co-ordinate space $\mathbb{R}^n$ describes the finite number of measurements that can be taken on these signals.

The general sample space Ω is an abstract set that contains all the possible outcomes of the experiment that can conceivably happen – and it may contain more. A probability measure P is defined on Ω that assigns probabilities to all interesting subsets – we need only one single probability measure to describe our whole world. We may choose to measure many different characteristics, with different statistical distributions, but they are all uniquely defined by the single underlying "world" probability measure P.

During the experiment one can record the time evolution of a number of things, such as rudder angle, which we call $\{x(t), t \in \mathbb{R}\}$, ship head angle, $\{y(t), t \in \mathbb{R}\}$, and roll angle $\{z(t), t \in \mathbb{R}\}$. Each observed function is an observation of a continuous random process. In the figure, the randomness is indicated by the dependence of the experiment outcome ω. The distributions of the different processes are P_x, P_y, P_z – we need one probability measure for each of the phenomena we have chosen to observe.[1]

In practice, the continuous functions are sampled in discrete time steps, $t = t_1, \ldots, t_n$, resulting in an observation vector, $(x_1, \ldots, x_n)$, with an n-dimensional distribution, $P_x^{(n)}$, etc. This familiar type is illustrated in the third box in Figure 1.1.

Since we do not always want to specify a finite value for n, the natural mathematical model for the practical situation is to replace the middle box, the sample space $\mathbb{C}$ of continuous functions, by the sample space $\mathbb{R}^\infty$ of infinite sequences of real numbers $(x_0, x_1, \ldots)$. This is close, as we shall see later, really very close, to the finite-dimensional space $\mathbb{R}^n$, and mathematically not much more complicated.

Taking the set $\mathbb{C}$ of continuous functions as a sample space and assigning probabilities P_x, etc., on it, is not as innocent as it may sound from the description above. Chapter 2 deals with conditions that guarantee that a stochastic process is continuous, i.e., has continuous sample functions. In fact, these conditions are all on the finite-dimensional distributions.

SUMMARY: *The abstract sample space Ω for an experiment contains everything that can happen and is therefore very complex and detailed. Each outcome $\omega \in \Omega$ is unique, and we need only one comprehensive probability measure P to describe every outcome of the experiment.*

An experiment is a way to "observe the world." The function (sequence) sample space $\mathbb{C}$ ($\mathbb{R}^\infty$) is simple. It can be used as sample space for a specified experiment for which the result is a function or a sequence of numbers. We have to define a specific probability measure for each experiment.

[1]The symbol ω is here used to represent the elementary experimental outcome, a practice that is standard in probability theory. In most parts of this book, ω will stand for (angular) frequency; no confusion should arise from this.

1.2 Random variables and stochastic processes

1.2.1 Probability space and random variables

A *probability space* $(\Omega, \mathscr{F}, P)$ is a triple consisting of: a sample space Ω of outcomes, a family $\mathscr{F}$ of events, i.e., subsets of Ω, and a probability measure P, that assigns probabilities to the events in $\mathscr{F}$. The family $\mathscr{F}$ may be chosen in many different ways; it may contain just a few subsets, or it may contain many. It is even possible to let $\mathscr{F}$ contain *all* subsets of Ω.[2] The obligatory properties of the family $\mathscr{F}$ are that it contains Ω and is closed under formation of complements, unions, and intersections of countably many of its elements. A family of subsets with these properties is called a σ*-field*. A less demanding requirement on the family is that it is closed under unions/intersections of finitely many elements. Such a family is called a *field*, denoted $\mathscr{F}_0$.

A probability measure assigns probabilities to the events in $\mathscr{F}$ in such a way that the basic rules for adding probabilities, usually called the *Kolmogorov's axioms of probablities*, are fulfilled. The exact definitions of σ-field and the summation rules are formulated in Definition A.1 in Appendix A.1.

The link between the sample space Ω and the observable world is the random variable, a real valued function $\Omega \ni \omega \mapsto x(\omega) \in \mathbb{R}$. In order that x be called random variable it is essential that the *distribution* of x is well defined, i.e., that the *event* "$x \leq b$" has a probability for each value of $b \in \mathbb{R}$. This means that the set of outcomes $\{\omega \in \Omega; x(\omega) \leq b\}$ must be one of the sets in $\mathscr{F}$, those which have been assigned probability by P. Functions that fulfill this requirement are called *measurable*, or, more precisely, *Borel measurable*.

It is of course possible to define many random variables, $x_1, \ldots, x_n$, on the same probability space $(\Omega, \mathscr{F}, P)$. Since P assigns a probability to each of the individual events

$$\text{“}x_1 \leq b_1\text{”}, \ldots, \text{“}x_n \leq b_n\text{”}$$

it also assigns a probability to the joint occurrence of them all; this is by the property of the σ-field $\mathscr{F}$ that it contains intersections. The result is a multivariate random variable, $\mathbf{x} = \{x_k\}_{k=1}^n$ with an n-dimensional distribution.

In Appendix A.1.1 we explain what more events are given probabilities in this way. These events are the *Borel sets* in $\mathbb{R}^n$ and the family is called the *Borel* σ*-field*, or just *the Borel field*, denoted $\mathscr{B}_n$. Examples of such events are those defined by two-sided inequalities, like the half-open rectangle $a_1 < x_1 \leq b_1, a_2 < x_2 \leq b_2$, unions of rectangles, complements of unions,

[2] If $\mathscr{F}$ contains all subsets of Ω every probability measure has to be discrete.

and many more. The Borel σ-field is a very rich family of events, which all get probabilities defined by the finite-dimensional distribution functions.

1.2.2 Stochastic processes and their finite-dimensional distributions

We are now ready to define stochastic processes in general. There is no difficulty in considering an infinite sequence of random variables $\mathbf{x} = \{x_k\}_{k=1}^{\infty}$, on a single probability space $(\Omega, \mathscr{F}, P)$. One can even consider more than countably many random variables at the same time. Then it is common to index the variables by the letter t instead of k, and let the *parameter* t take values in a general *parameter space* T. Usually one thinks of t as "time" and T as the whole or part of the real line.

Thus, we can consider a family of functions, $\{x(t,\omega) \in \mathbb{R}\}_{t\in T}$, where each $x(t) = x(t,\cdot)$ is a random variable, i.e., a measurable function from Ω to $\mathbb{R}$. Hence, it has a distribution with a distribution function on $\mathbb{R}$, which we denote $F(\cdot;t)$, i.e.,

$$F(b;t) = \text{Prob}(x(t) \le b) = P(\{\omega; x(t,\omega) \le b\}).$$

Taking several variables, observed at times $t_1, \dots, t_n$, we get an n-variate random variable $(x(t_1), \dots, x(t_n))$ with an n-variate distribution in $\mathbb{R}^n$,

$$\begin{aligned} F_{t_1\cdots t_n}(b_1, \dots, b_n) &= \text{Prob}(x(t_1) \le b_1, \dots, x(t_n) \le b_n) \\ &= P\left(\cap_1^n \{\omega; x(t_k,\omega) \le b_k\}\right). \qquad (1.1) \end{aligned}$$

Write $F_{\mathbf{t}^n}$ for the n-dimensional distribution function of $(x(t_1), \dots, x(t_n))$. We summarize the terminology in a formal, but simple, definition.

Definition 1.1. *Let T be a discrete or continuous real parameter set. A stochastic process $\{x(t), t \in T\}$ indexed by the parameter $t \in T$ is a family of random variables $x(t)$ defined on one and the same probability space $(\Omega, \mathscr{F}, P)$. In other words, a stochastic process is a function*

$$T \times \Omega \ni (t,\omega) \mapsto x(t,\omega) \in \mathbb{R},$$

such that, for fixed $t = t_0$, $x(t_0,\cdot)$ is a random variable, i.e., a Borel measurable function, $\Omega \ni \omega \mapsto x(t_0,\omega) \in \mathbb{R}$, and, for fixed $\omega = \omega_0$, $x(\cdot,\omega_0)$ is a function $T \ni t \mapsto x(t,\omega_0) \in \mathbb{R}$, called a sample path or sample function.

We now have the following concepts at our disposal in the three scenes in Figure 1.1:

sample space	events	probability
abstract space: Ω	σ-field $\mathscr{F}$	P
space of continuous functions: $\mathscr{C}$	?	?
real sequences: $\mathbb{R}^\infty$	?	?
real vectors: $\mathbb{R}^n$	Borel sets: $\mathscr{B}_n$	$P_{\mathbf{t}_n}$ from n-dimensional d.f.'s $F_{\mathbf{t}_n}$
real line: $\mathbb{R}$	Borel sets: $\mathscr{B}$	P from a d.f. F

In the table, the ? indicate what we yet have to define – or even show existence of – to reach beyond the elementary probability theory, and into the world of stochastic processes.

1.2.3 The distribution of random sequences and functions

In the previous section we discussed finite-dimensional distributions for one, two, and (finitely) many random variables, how these distributions are defined by the finite-dimensional distribution functions, and what finite-dimensional events get their probabilites defined.

Our aim now is to find the events in $\mathbb{R}^\infty$ (= all real sequences) and $\mathbb{R}^T$ (= all real functions defined on T), and see how one can define a probability measure for these. When this is done, it is legitimate to talk about *the distribution of an infinite random sequence or function*, and of the probability that an experiment results in a sample function that satisfies some rather general requirement, e.g., that its maximum is less than a given constant.

It was this step, from probabilities for one-dimensional or n-dimensional real sets and events, to probabilities in $\mathbb{R}^\infty$ and $\mathbb{R}^T$, that was made axiomatic in A.N. Kolmogorov's celebrated *Grundbegriffe der Wahrscheinlichkeitsrechnung* from 1933 [71].

Generalized rectangles, intervals, and the Borel field

The basic requirement on the events in $\mathbb{R}^\infty$ or $\mathbb{R}^T$ is that they should not be simpler than the events in the finite-dimensional spaces $\mathbb{R}^n$, which means that if an event $B_n \in \mathscr{B}_n$ in $\mathbb{R}^\infty$ is expressed by means of a finite set of random variables $x_1,\dots,x_n$, then it should be an event also in the space $\mathbb{R}^\infty$. Now, such an event can be written

$$\begin{aligned}\{\mathbf{x} = (x_1,x_2,\dots) \in \mathbb{R}^\infty; (x_1,x_2,\dots,x_n) \in B_n\} &= B_n \times \mathbb{R} \times \mathbb{R} \times \dots \\ &= B_n \times \mathbb{R}^\infty.\end{aligned}$$

A set of this form is called a *generalized rectangle in* $\mathbb{R}^\infty$.

Similarly, we define a generalized rectangle in $\mathbb{R}^T$ as a set defined by a condition on any finite set of $x(t)$-values,

$$\{\mathbf{x} \in \mathbb{R}^T; (x(t_1), x(t_2), \dots, x(t_n)) \in B_n\}.$$

Hence, we have to require that the σ-field of events contains at least all generalized rectangles. The natural event field is exactly the smallest σ-field which contains all such sets; cf. Example A.2 in Appendix A.1. This σ-field is denoted $\mathscr{B}_\infty$ and $\mathscr{B}_T$, respectively, and is, naturally, called the *Borel field* in $\mathbb{R}^\infty$ or $\mathbb{R}^T$. Symbolically, we can write, for sequences,

$$\mathscr{B}_\infty = \sigma\left(\cup_{n=1}^{\infty} (\mathscr{B}_n \times \mathbb{R}^\infty)\right).$$

A particularly simple form of rectangles is the *intervals*. These are, in $\mathbb{R}^\infty$, the sets of form

$$I = (a_1, b_1] \times (a_2, b_2] \times \dots \times (a_n, b_n] \times \mathbb{R}^\infty,$$

where each $(a_j, b_j]$ is a real half-open interval. Thus, the half-open intervals in $\mathbb{R}^\infty$ and $\mathbb{R}^T$ are defined by inequalities like

$$a_1 < x_1 \le b_1, a_2 < x_2 \le b_2, \dots, a_n < x_n \le b_n. \tag{1.2}$$

$$a_1 < x(t_1) \le b_1, a_2 < x(t_2) \le b_2, \dots, a_n < x(t_n) \le b_n. \tag{1.3}$$

Sets which are unions of a finite number of intervals are important; they form a field, which we denote $\mathscr{I}$. The σ-field generated by $\mathscr{I}$ is exactly $\mathscr{B}_\infty$, and $\mathscr{B}_T$, respectively, in $\mathbb{R}^\infty$ and $\mathbb{R}^T$. The reader should draw a picture of some sets in the field $\mathscr{I}$ in $\mathbb{R}^2$.

> The sample spaces we are mostly going to work with are $\mathbb{R}^\infty$ and $\mathbb{R}^T$, infinite real sequences, and real functions defined on T, and the Borel σ-fields of events to be given probabilities are $\mathscr{B}_\infty$ and $\mathscr{B}_T$, respectively, generated by the finite-dimensional intervals.

These spaces are important since they provide us with a natural world of objects for stochastic modeling.

Definition 1.2 (The co-ordinate process). *The simplest process on* $(\mathbb{R}^\infty, \mathscr{B}_\infty)$ *is the co-ordinate process, also known as "the canonical process,"* $\{x_n, n \in \mathbb{N}\}$ *defined by* $x_k(\omega) = \omega_k$*, on the outcome* $\omega = (\omega_1, \omega_2, \ldots)$.

Similarly, the simplest process on $(\mathbb{R}^T, \mathscr{B}_T)$ *is the co-ordinate process* $x(t, \omega) = \omega(t)$ *for* $\omega \in \mathbb{R}^T$.

Finding a probability measure

The next big step is to assign probabilities to the Borel sets in $\mathscr{B}_\infty$ and $\mathscr{B}_T$, and this can be done in either of two ways, from the abstract side or from the observable, finite-dimensional side. Going back to Figure 1.1 we shall fill in the two remaining question marks in the following scheme:

$$(\Omega, \mathscr{F}, P) \stackrel{x}{\mapsto} \left\{ \begin{array}{c} (\mathbb{R}^\infty, \mathscr{B}_\infty, ?) \\ (\mathbb{R}^T, \mathscr{B}_T, ?) \end{array} \right\} \leftarrow\!\shortmid (\mathbb{R}^n, \mathscr{B}_n, P_x^{(n)}), n = 1, 2, \ldots.$$

The mapping $\stackrel{x}{\mapsto}$ symbolizes that we have already defined a probability P on some measurable space $(\Omega, \mathscr{F})$ and want to define probabilities on a sample function space for a stochastic process x with discrete or continuous parameter. The mapping $\leftarrow\!\shortmid$ symbolizes that we shall define the probabilities from a family of finite-dimensional distributions on $(\mathbb{R}^n, \mathscr{B}_n), n = 1, 2, \ldots$ so that the co-ordinate process gets exactly these finite-dimensional distributions.

Definition from a random process

Suppose the probability space $(\Omega, \mathscr{F}, P)$ is given a priori, and a process x is defined on $(\Omega, \mathscr{F})$, either as $\{x_n, n \in \mathbb{N}\}$ or as $\{x(t), t \in T\}$. We shall use P on $(\Omega, \mathscr{F})$ to define the distribution of x on the sample function spaces $(\mathbb{R}^\infty, \mathscr{B}_\infty)$ or $(\mathbb{R}^T, \mathscr{B}_T)$. The following lemma may look complicated, but its content is simple. It says that there is a direct connection between the events in $\mathscr{B}_\infty$ or $\mathscr{B}_T$ and certain events in $\mathscr{F}$ on Ω.

Lemma 1.1. *For every set* $A \in \mathscr{B}_\infty$ *or* $A \in \mathscr{B}_T$*, the inverse set* $x^{-1}(A) \in \mathscr{F}$*:*

$$x^{-1}(A) = \{\omega \in \Omega; \mathbf{x}(\cdot, \omega) \in A\} \in \mathscr{F}.$$

Proof. Let $\mathscr{C}$ be the class of sets C in $\mathscr{B}_\infty$ or $\mathscr{B}_T$ such that $x^{-1} \in \mathscr{F}$. By the definition of a stochastic process, $\mathscr{C}$ contains all rectangles, and it is easy to show (do that as Exercise 1:3) that it is a σ-field. Since the Borel field is the smallest σ-field that contains the rectangles, the claim is shown. □

It is now easy to define probabilities on the sample function spaces. If x is a stochastic process, i.e., a function from Ω to $\mathbb{R}^\infty$ or $\mathbb{R}^T$, then a probability measure P_x is defined on $(\mathbb{R}^\infty, \mathscr{B}_\infty)$ and $(\mathbb{R}^T, \mathscr{B}_T)$, respectively, by

$$P_x(B) = P(x^{-1}(B)), \quad \text{for } B \in \mathscr{B}_\infty \text{ or } \mathscr{B}_T. \tag{1.4}$$

Thus, each stochastic process produces a probability measure P_x on $(\mathbb{R}^\infty, \mathscr{B}_\infty)$ or $(\mathbb{R}^T, \mathscr{B}_T)$.

Example 1.1 ("Random amplitude and phase"). *Here is a simple example of a stochastic process that is often used as a building block in more complicated models. Take two random variables A, ϕ, with $A \geq 0$ and ϕ uniformly distributed between 0 and 2π, and define a stochastic process $\{x(t), t \in \mathbb{R}\}$ as $x(t) = A\cos(t+\phi)$. This procedure defines the process x as a function from the sample space of (A, ϕ)-outcomes, $\Omega = [0, \infty) \times [0, 2\pi] = \{(A, \phi)\}$, to the function space $\mathbb{R}^T$, with $T = \mathbb{R}$. Of course, only cosine functions with different amplitudes and phases can appear as sample functions, and the probability measure on $\mathbb{R}^T$ gives probability one to the subset of such functions. The finite-dimensional distributions of $x(t)$ can be calculated from the joint distribution of A and ϕ.*

Example 1.2 ("Poisson process"). *The sample functions of a common Poisson process are piecewise constant and increase in unit steps. It can be explicitly constructed from a sequence of independent exponential variables defining the time that elapses between two jumps, and the finite-dimensional distributions of the Poisson process can be calculated from the distribution of these intervals in $(R^\infty, \mathscr{B}_\infty)$.*

It is essential, in the two examples, that we already have a probability space that allows us to construct the amplitude and phase variables, and the inter-event times, respectively. It is also worth to notice, that the more complex a process we wanted to generate, the more complex did we choose the basic sample space. This causes no problem, since one can always expand the sample space to include the desired variability. There is no need to look for the *simplest* probability space for a particular application.

If we start with a huge sample space $(\Omega, \mathscr{F})$ with a single but complex probability measure P we can construct a rich variety of processes on the single probability space $(\Omega, \mathscr{F}, P)$, and, by (1.4), obtain the probabilities of sets B in $\mathscr{B}_\infty$ or $\mathscr{B}_T$. We can also derive the finite-dimensional distributions. The negative side of this approach is that we don't know how to construct the measure P from experiments or from simple process models like Poisson or Gaussian models. We therefore usually rely on the other alternative to construct process distributions.

Definition from a family of finite-dimensional distributions:

The family $\{F_{\mathbf{t}^n}\}_{n=1}^\infty$ of finite-dimensional distributions is the family of functions

$$F_{t_1\cdots t_n}(b_1,\ldots,b_n) = \mathrm{Prob}(x(t_1) \le b_1,\ldots,x(t_n) \le b_n),$$

for $t_j \in T, t_1 < t_2 < \ldots < t_n, n = 1,2,\ldots$. A family of distribution functions for a stochastic process has to be *consistent* to satisfy certain, trivial requirements.

Definition 1.3. *A family $\{F_{\mathbf{t}^n}\}_{n=1}^\infty$ of finite-dimensional distributions is called consistent if, for every n, and $t_1,b_1,\ldots,t_n,b_n$,*

$$\lim_{b_k\uparrow\infty} F_{t_1,\ldots,t_n}(b_1,\ldots,b_n) = F_{t_1,\ldots,\widetilde{t_k},\ldots,t_n}(b_1,\ldots,\widetilde{b_k},\ldots,b_n),$$

$$F_{t_1,t_2}(b_1,b_2) = F_{t_2,t_1}(b_2,b_1), \text{ etc.}$$

where the notation $\widetilde{\cdot}$ indicates that the variable is missing.

Suppose a family of finite-dimensional distribution functions,

$$\mathbf{F} = \{F_{\mathbf{t}^n}\}_{n=1}^\infty, t_k \in T,$$

$$F_{t_1\cdots t_n}(b_1,\ldots,b_n) = \mathrm{Prob}(x(t_1) \le b_1,\ldots,x(t_n) \le b_n),$$

is given a priori, with one distribution for each n and for each set $(t_1,\ldots,t_n)$ of times. If the family is consistent (Definition 1.3), one can use it to define probabilities for all half-open n-dimensional intervals in $\mathbb{R}^\infty$ or $\mathbb{R}^T$, for example by, for $n = 1,2,\ldots$, taking (cf. (1.1))

$$P_{\mathbf{x}}((a_1,b_1] \times (a_2,b_2] \times \ldots \times (a_n,b_n] \times \mathbb{R}^\infty) = P_n((a_1,b_1] \times (a_2,b_2] \times \ldots \times (a_n,b_n]).$$

This will give us a unique probability measure for each n and each n-dimensional selection of time points $(t_1,\ldots,t_n)$.

Theorem 1.1 (Existence of stochastic process; Kolmogorov, 1933). *To every consistent family of finite-dimensional distribution functions,* $\mathbf{F} = \{F_{\mathbf{t}^n}\}_{n=1}^{\infty}$, $t_k \in T$, *there exists one and only one probability measure* P_x *on* $(\mathbb{R}^T, \mathscr{B}_T)$ *with*

$$P_x(x(t_1) \le b_1, \ldots, x(t_n) \le b_n) = F_{t_1 \cdots t_n}(b_1, \ldots, b_n), \text{ for } n = 1, 2, \ldots.$$

In Appendix A.2 we present an outline of the proof of the existence theorem. The main steps are to show that the finite dimensional distributions give a countably additive probability measure on the field $\mathscr{I}$ of finite unions of intervals, defined for different times $(t_1, \ldots, t_n)$. By the extension property of probability measures on fields, Theorem A.1, one can then conclude that this defines a unique probability measure P_x on the σ-field generated by the intervals, i.e., on the Borel field $\mathscr{B}_\infty$ or $\mathscr{B}_T$, respectively. The finite-dimensional distributions are the distributions in $\mathbf{F}$. The proof of the countable additivity is a significant part of Kolmogorov's existence theorem for stochastic processes; see Theorem A.3.

By this, we have defined events in $\mathbb{R}^\infty$ and $\mathbb{R}^T$ and know how to define probability measures in $(\mathbb{R}^\infty, \mathscr{B}_\infty)$ and $(\mathbb{R}^T, \mathscr{B}_T)$.

Remark 1.1. *It is worth noting that all the well-known processes in elementary probability theory can be introduced by finite dimensional distributions. Going back to Example 1.2, we saw there an explicit way to construct a Poisson process. On the other hand, following the route via the finite-dimensional distribution functions the Poisson process on the real line is characterized by the fact that the increments over a finite number of disjoint intervals are independent Poisson variables with expectation proportional to the interval length. This property can easily be formulated as a property of the finite-dimensional distributions at the interval end points, even if one never does so explicitly.*

Both approaches lead to a probability measure P_x*, defined for all* $B \in \mathscr{B}_T$*, either by* (1.4) *or by the Kolmogorov existence theorem. There is an important difference between the two approaches. With the explicit construction* all *sample functions are piecewise constant with unit jumps, by construction, and the probability measure* P_x *has the potential to take that into account and it can easily be extended to more events (subsets) than just those in* $\mathscr{B}_T$*. The construction by the Kolmogorov theorem does not allow any strong statement about sample function properties. In the next section and in Chapter 2 we discuss some consequences of this fact in more detail.*

Finally, also the explicit construction relies on the Kolmogorov existence theorem, since it requires the existence of an infinite sequence of independent exponential random variables.

1.2.4 An important comment on probabilities on $\mathbb{R}^T$

Processes with continuous time need some extra comments. Even if the existence of such processes can be deduced from Kolmogorov's theorem, the practical applications need some care. The sample space Ω is an abstract space and a mathematical construction, and the link to reality is provided by the random variables. In an experiment, one can observe the values of one or more random variables, x_1, x_2, etc. and also find their distribution, by some statistical procedure. There is no serious difficulty to allow the outcome to be any real number, and to define probability distributions on $\mathbb{R}$ or $\mathbb{R}^n$.

When the result of an experiment is a whole function with continuous parameter, the situation is more complicated. In principle, *all functions* of $t \in T$ are potential outcomes, and the sample space of all functions on T is simply so big that one needs to be restrictive on which sets of outcomes that should be given probability. There are too many realizations that ask for probability.

Here, practice comes to our assistance. In an experiment one can only observe the values of $x(t)$ at a finite number of times, $t_1, t_2, \ldots, t_n$; with $n = \infty$ we allow an unlimited series of observations. The construction of processes with continuous time was built on exactly this fact: the observable events are those that can be defined by countably many $x(t_j)$, $j \in \mathbb{N}$, and the probability measure we constructed assigns probabilities to only such events.

A set of functions that is defined by restrictions on countably many $x(t_j)$ is said to have a *countable basis*; see more on this in Appendix A.1. There are many important function sets that do not have a countable basis and are not Borel sets in $\mathbb{R}^T$. These sets are not given a unique probability by the finite-dimensional distribution functions.

Example 1.3. *Here are some examples of function sets with and without countable basis, when* $T = [0,1]$*:*

- $\{x \in \mathbb{R}^T; \lim_{n\to\infty} x(1/n) \text{ exists}\} \in \mathcal{B}_T$,
- $\{x \in \mathbb{R}^T; \lim_{t\to 0} x(t) \text{ exists}\} \notin \mathcal{B}_T$,
- $\{x \in \mathbb{R}^T; x \text{ is a continuous function}\} \notin \mathcal{B}_T$,
- $\{x \in \mathbb{R}^T; x(t) \leq 2 \text{ for all rational } t\} \in \mathcal{B}_T$.

Approximation by finite-dimensional events

The events in the σ-field $\mathscr{B}_\infty$ in $\mathbb{R}^\infty$ can be approximated in probability by finite-dimensional sets. If $(\mathbb{R}^\infty, \mathscr{B}_\infty, P)$ is a probability space, and $B \in \mathscr{B}_\infty$, then, for every $\varepsilon > 0$, there is a finite n and an event $B_n \in \mathscr{B}_n$ such that

$$P(B \Delta \widetilde{B}_n) \leq \varepsilon,$$

where $\widetilde{B}_n = \{x \in \mathbb{R}^\infty; (x_1, \ldots, x_n) \in B_n\}$ and $A \Delta B = (A - B) \cup (B - A)$.

Similarly, events in $\mathscr{B}_T$ in $\mathbb{R}^T$ can be approximated arbitrarily close by events defined by the values of $x(t)$ for a finite number of t-values: $P(B \Delta \widetilde{B}_n) \leq \varepsilon$, with

$$\widetilde{B}_n = \{x \in \mathbb{R}^T; (x(t_1), \ldots, x(t_n)) \in B_n\}.$$

Remember that every probability measure on $(\mathbb{R}^\infty, \mathscr{B}_\infty)$ is uniquely determined by its finite-dimensional distributions, which implies that also every probability measure P on $(\mathbb{R}^T, \mathscr{B}_T)$ is determined by the finite-dimensional distributions, $\{F_{\mathbf{t}^n}\}_{n=1}^\infty$. In particular, the probability

$$P(\lim_{n \to \infty} x(t_0 + 1/n) \text{ exists and is equal to } x(t_0))$$

is determined by the finite-dimensional distributions. But $x(t_0 + 1/n) \to x(t_0)$ as $n \to \infty$ is almost, but not quite, the same as $x(t) \to x(t_0)$ as $t \to t_0$. To deal with sample function continuity we need a more refined construction of the probability measure from the finite-dimensional distributions. We deal with these matters in Chapter 2.

1.3 Stationary processes and fields

This section summarizes some elementary notation for stationary processes. More details, properties, and proofs of the most important facts will be given in Chapter 3; see also [82]. To avoid too cumbersome notation we sometimes allow ourselves to write "the process $x(t)$," when we should have used the notation $\{x(t), t \in \mathbb{R}\}$ for the process (or any other relevant parameter set T instead of $\mathbb{R}$). The simple notation x will also be frequently used, when the parameter space is clear from the context. If we mean "the random variable $x(t)$," we will say so explicitly. At this stage, we deal with real valued processes, only. Later, we will need to consider also complex valued processes.

We write $m(t) = E(x(t))$ for the *mean value function* and $r(s,t) = C(x(s), x(t))$ for the *covariance function* of the process $\{x(t), t \in \mathbb{R}\}$. If $\{x(t), t \in \mathbb{R}\}$ and $\{y(t), t \in \mathbb{R}\}$ are two processes, $r_{x,y}(s,t) = C(x(s), y(t))$ is the *cross-covariance function*.

1.3.1 Stationary processes, covariance, and spectrum

Definition 1.4 (Stationarity). *A stochastic process* $\{x(t), t \in \mathbb{R}\}$ *is strictly stationary if all n-dimensional distributions of* $x(t_1 + \tau), \ldots, x(t_n + \tau)$ *are independent of* τ. *It is called weakly stationary (or second order stationary) if its mean is constant,* $E(x(t)) = m$, *and if its covariance function,*

$$r(t) = C(x(s+t), x(s)),$$

is a function only of the time lag t.

Here are some simple properties of a covariance function of a stationary process.

Theorem 1.2. *If* $r(t)$ *is the covariance function of a (real) stationary process, then,*

a) $V(x(t+h) \pm x(t)) = 2(r(0) \pm r(h))$, *so* $|r(h)| \leq r(0)$,

b) if $|r(\tau)| = r(0)$ *for some* $\tau \neq 0$, *then* $r(t)$ *is periodic,*

c) if $r(t)$ *is continuous at* $t = 0$, *then it is continuous everywhere.*

Proof. Property (a) is clear, and it implies $|r(t)| \leq r(0)$. From (a) it also follows that if $r(\tau) = \pm r(0)$, then $x(t+\tau) = \pm x(t)$ for all t, so $x(t)$ is periodic with period τ and 2τ, respectively, which is (b). For (c), use the inequality, $C(x,y)^2 \leq V(x)V(y)$, on

$$\begin{aligned}(r(t+h) - r(t))^2 &= (C(x(t+h), x(0)) - C(x(t), x(0)))^2 \\ &= (C(x(t+h) - x(t), x(0)))^2 \\ &\leq V(x(t+h) - x(t))\,V(x(0)) = 2(r(0) - r(h))\,r(0),\end{aligned}$$

which goes to 0 with h if $r(t)$ is continuous for $t = 0$. □

The covariance function is said to define the *time domain* properties of the process. Equally important for the theory is the *frequency domain* properties, expressed by the *spectral distribution*, studied in detail in Chapter 3.

Every continuous covariance function is a Fourier integral,

$$r(t) = \int_{-\infty}^{\infty} e^{i\omega t}\, \mathrm{d}F(\omega), \tag{1.5}$$

where the function $F(\omega)$ is the spectral distribution function.

The spectral distribution is characterized by the properties:[3]

- symmetry: $\mathrm{d}F(-\omega) = \mathrm{d}F(\omega)$,
- monotonicity: $\omega \le \omega'$ implies $F(\omega) \le F(\omega')$,
- boundedness: $F(+\infty) - F(-\infty) = r(0) = V(x(t)) < \infty$.

As indicated by the way we write the three properties, $F(\omega)$ is defined only up to an additive constant, and we usually take $F(-\infty) = 0$. The spectral distribution function is then equal to a cumulative distribution function multiplied by a positive constant, equal to the variance of the process. As we shall see in later chapters, ω is in a natural way interpreted as an *angular frequency*, not to be confused with the elementary event ω in basic probability theory.

In this work we will mainly use the symmetric form (1.5) of the spectral representation when we deal with theoretical aspects of the covariance function and of the spectrum. However, in many examples we will use a one-sided real form with $\omega \ge 0$; for details on this, see Section 3.2.3. In Section 3.3.5 we will meet examples where negative frequencies play an important role in their own right.

If $F(\omega)$ is absolutely continuous with $F(\omega) = \int_{s=-\infty}^{\omega} f(s)\,\mathrm{d}s$, then the spectrum is said to be (absolutely) continuous, and

$$f(\omega) = \frac{\mathrm{d}}{\mathrm{d}\omega} F(\omega)$$

is the *spectral density function*.

The *spectral moments* are defined as

$$\omega_k = \int_{-\infty}^{\infty} |\omega|^k \, \mathrm{d}F(\omega).$$

Note that the odd spectral moments are defined as absolute moments. Since F is symmetric around 0 the signed odd moments are always 0, when they exist.

[3] The differential notation in $\mathrm{d}F(-\omega) = \mathrm{d}F(\omega)$ will be used throughout the book as a convenient way to simplify increment notation.

Spectral moments may be finite or infinite. As we shall see in the next chapter, the finiteness of the spectral moments are coupled to the smoothness properties of the process $x(t)$. For example, the process is differentiable (in quadratic mean), see Section 2.1, if $\omega_2 = -r''(0) < \infty$, and the spectral density of the derivative is

$$f_{x'}(\omega) = \omega^2 f_x(\omega).$$

To ensure that sample functions are differentiable functions one needs just a little bit more than $\omega_2 = \int_{-\infty}^{\infty} \omega^2 f_x(\omega)\, d\omega < \infty$, namely $-r''(t) = \omega_2 - C|t|^\alpha + o(|t|^\alpha)$ as $t \to 0$, for some $\alpha \in (1,2]$; see Theorem 2.10, Section 2.3.2.

The *mean frequency* of a stationary process,

$$\mu^+ = \frac{1}{2\pi}\sqrt{\omega_2/\omega_0}, \tag{1.6}$$

and its inverse, $1/\mu^+$, the *mean period*, characterize the main feature of the spectrum. The width of the spectrum around the mean frequency is measured by the *spectral width* parameter $1/\alpha = (\omega_0\omega_4)^{1/2}/\omega_2$.

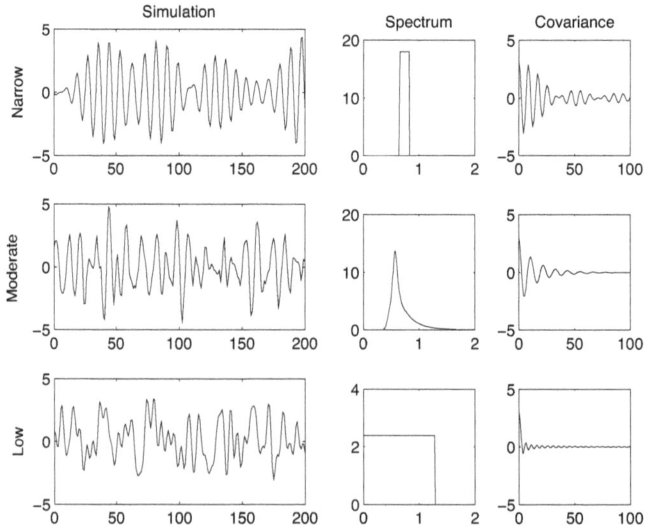

Figure 1.2 *Top: Processes with very narrow band spectrum, oscillating covariance function, and "fading" sample function; Middle: Realistic ocean waves with moderate band width wave spectrum; Bottom: low frequency white noise spectrum. The spectra are illustrated by their one-sided spectral densities.*

Example 1.4. *Here is a first example on the visual characteristics of spectrum, covariance function, and simulated sample function. Figure 1.2 illustrates one very narrow spectrum with oscillating covariance function and* *regular "fading" sample path, one realistic water wave spectrum with less regular sample paths, and one irregular process with "low frequency white noise" spectrum.*

1.3.2 Stationary streams of events

The Poisson process is a *counting process* that counts the number of events that happen at discrete time points. Its characteristic property is that the number of events in any interval of length h has a Poisson distribution with mean λh, with parameter λ called the *intensity*. Further, the number of events in disjoint intervals are statistically independent. The events counted by the Poisson process is an example of a *stationary point process*. This is a natural generalization of the notion "stationary process," since the distribution of the number of events in a set of intervals is the same, regardless of their absolute location.

If one relaxes the independence condition, one obtains a more general class of point processes, where there may be causal or statistical dependence between events. For example, the intensity in a Poisson process may be allowed to vary like a stationary process. It is then possible to describe this dependence by means of a type of covariance function that describes the conditional intensity of events a distance t apart. We shall deal with this covariance density in Section 3.5, where we also introduce the corresponding spectral distribution. For general accounts of points processes, see [37, 39].

1.3.3 Random fields

A random field is a stochastic process $\{x(\mathbf{t})\}$ with multi-dimensional parameter $\mathbf{t} = (t_1, \dots, t_p) \in \mathbf{T}$, which can be discrete or continuous. For example, if $\mathbf{t} = (t_1, t_2)$ is two-dimensional we can think of $(t_1, t_2, x(\mathbf{t}))$, $(t_1, t_2) \in \mathbb{R}^2$, as a random surface. The mean value and covariance functions are defined in the natural way, $m(\mathbf{t}) = E(x(\mathbf{t}))$ and $r(\mathbf{t}, \mathbf{u}) = C(x(\mathbf{t}), x(\mathbf{u}))$.

A random field is called *homogeneous* if it has constant mean value $m(\mathbf{t}) = m$ and the covariance function $r(\mathbf{t}, \mathbf{u})$ depends only on the vector $\mathbf{t} - \mathbf{u}$ between the two observation points, i.e., assuming $m = 0$,

$$r(\mathbf{t}) = r(\mathbf{u} + \mathbf{t}, \mathbf{u}) = E(x(\mathbf{u} + \mathbf{t}) \cdot x(\mathbf{u})).$$

The covariance of the process values at two parameter points depends on *distance* as well as on *direction* of the vector between the two points. This is the equivalent of a stationary process on the real line.

If the covariance between $x(\mathbf{u})$ and $x(\mathbf{v})$ depends only on the distance $\tau = \|\mathbf{u} - \mathbf{v}\|$ between the observation points and not on the direction, the field is called *isotropic*. This requirement poses strong restrictions on the covariance function, as we shall see in Chapter 7, where random fields are treated in more detail.

A digital image can be thought of as a random field with discrete parameter $\mathbf{t} \in \mathbf{T} \subset \mathbb{N}^2$. A time-dependent random surface is a field $(s_1, s_2, x(t, s_1, s_2))$ with $\mathbf{t} = (t, s_1, s_2)$, where t represents time and $(s_1, s_2) \in \mathbb{R}^2$ is location.

1.4 Gaussian processes

1.4.1 Normal distributions and Gaussian processes

The Gaussian, or normal, univariate distribution with mean m and variance σ^2 has probability density function

$$f(x) = \frac{1}{\sigma\sqrt{2\pi}} e^{-(x-m)^2/(2\sigma^2)} = \frac{1}{\sigma}\phi\left(\frac{x-m}{\sigma}\right),$$

where $\phi(x) = \frac{1}{\sqrt{2\pi}} e^{-x^2/2}$ is the standardized normal density function. A distribution concentrated at one point m is regarded as Gaussian with $\sigma^2 = 0$.

Definition 1.5. *A vector $\boldsymbol{\xi} = (\xi_1, \ldots, \xi_p)$ of p random variables is said to have a p-variate Gaussian (normal) distribution if every linear combination of its components $\mathbf{a}\boldsymbol{\xi}' = \sum_k a_k \xi_k$ has a normal distribution. The variables $\xi_1, \ldots, \xi_p$ are said to be "jointly normal."*

With mean vector $\mathbf{m} = E(\boldsymbol{\xi})$ and covariance matrix

$$\boldsymbol{\Sigma} = C(\boldsymbol{\xi}; \boldsymbol{\xi}) = E((\boldsymbol{\xi} - \mathbf{m})'(\boldsymbol{\xi} - \mathbf{m})),$$

the variance of $\mathbf{a}\boldsymbol{\xi}'$ is

$$V(\mathbf{a}\boldsymbol{\xi}') = \mathbf{a}\boldsymbol{\Sigma}\mathbf{a}' \geq 0.$$

The *characteristic function* of a normal variable ξ is (see Appendix A.5)

$$\varphi(s) = E(e^{is\xi}) = e^{ims - \sigma^2 s^2/2},$$

and that of a multi-variate normal $\boldsymbol{\xi}$ is

$$\varphi(\mathbf{s}) = E(e^{i\mathbf{s}\boldsymbol{\xi}'}) = e^{i\mathbf{m}\mathbf{s}' - \mathbf{s}\boldsymbol{\Sigma}\mathbf{s}'/2}. \tag{1.7}$$

Jointly normal uncorrelated variables are independent! It follows from the form of the characteristic function that if two jointly normal variables have zero covariance, then they are also independent. To see this, one can first note that if $\mathbf{\Sigma}$ is diagonal, then the characteristic function factorizes into a product of characteristic functions for independent normals. The uniqueness of characteristic functions does the rest.

If the determinant of $\mathbf{\Sigma}$ is positive, the distribution of $\boldsymbol{\xi}$ is non-singular and has a density

$$f_{\boldsymbol{\xi}}(\mathbf{x}) = \frac{1}{(2\pi)^{p/2}\sqrt{\det \mathbf{\Sigma}}}\, e^{-\frac{1}{2}(\mathbf{x}-\mathbf{m})\mathbf{\Sigma}^{-1}(\mathbf{x}-\mathbf{m})'}.$$

If the determinant is zero, the distribution of $\boldsymbol{\xi}$ is concentrated to a linear subspace of $\mathbb{R}^n$ and there exists at least one linear relationship between the components, i.e., there is at least one $\mathbf{a}$ for which $\mathbf{a}\boldsymbol{\xi}'$ is a constant.

Definition 1.6. *A stochastic process $\{x(t), t \in \mathbb{R}\}$ is a Gaussian process if every linear combination $S = \sum_k a_k x(t_k)$ for real a_k and $t_k \in \mathbb{R}$ has a Gaussian distribution.*

Example 1.5 (Existence of a Gaussian process). *What knowledge is needed before we can ascertain the existence of a specific Gaussian process? In view of Kolmogorov's existence Theorem 1.1, we must present a consistent family of finite-dimensional distribution functions, such that every linear combination $\sum_k a_k x(t_k)$ of variables is Gaussian, i.e., that for every selection, the variables $x(t_1), \ldots, x(t_n)$ are jointly normal.*

Since the multi-variate normal distribution is defined by its characteristic function, it is clear from (1.7) *that it suffices to specify a function $m(t)$ and a function $r(s,t)$ such that*

$$\sum_{j,k} a_j a_k r(t_j - t_k) \geq 0, \tag{1.8}$$

for all real a_k and t_k. Then there exists a Gaussian process with $m(t)$ and $r(s,t)$ as mean value and covariance function, respectively. We will meet condition (1.8) *in Chapter 8.*

It is an easy consequence of the definition that the derivative of a Gaussian process is also Gaussian (when it exists), since it is the limit of the Gaussian variable $z_h = (x(t+h) - x(t))/h$ as $h \to 0$. For a stationary Gaussian process $\{x(t), t \in \mathbb{R}\}$ the mean of z_h is 0 and it has variance $V(z_h) = 2(r(0) - r(h))/h^2$. As we shall prove in Section 2.1.2 this converges to $\omega_2 = \int \omega^2 \, \mathrm{d}F(\omega \leq \infty$. The derivative exists only if this limit is finite.

Also the integral of a Gaussian process is a Gaussian variable; conditions for the existence will be given in Section 2.4.

1.4.2 Conditional normal distributions

The multivariate normal distribution has the very useful property that, conditional on observations of a subset of variables, the unobserved variables are also normal.[4] Further, the conditional expectation is a linear combination of the observations while variances and covariances are independent of the observed values.

Let $\boldsymbol{\xi} = (\xi_1, \dots, \xi_n)$ and $\boldsymbol{\eta} = (\eta_1, \dots, \eta_m)$ be two jointly Gaussian vectors with mean values

$$E(\boldsymbol{\xi}) = m_{\boldsymbol{\xi}}, \quad E(\boldsymbol{\eta}) = m_{\boldsymbol{\eta}},$$

and with covariance matrix (with $\boldsymbol{\Sigma}_{\xi\eta} = \boldsymbol{\Sigma}'_{\eta\xi}$),

$$\boldsymbol{\Sigma} = C((\boldsymbol{\xi}, \boldsymbol{\eta}); (\boldsymbol{\xi}, \boldsymbol{\eta})) = \begin{pmatrix} \boldsymbol{\Sigma}_{\xi\xi} & \boldsymbol{\Sigma}_{\xi\eta} \\ \boldsymbol{\Sigma}_{\eta\xi} & \boldsymbol{\Sigma}_{\eta\eta} \end{pmatrix}.$$

If the determinant of the covariance matrix $\boldsymbol{\Sigma}$ is positive, then the distribution of $(\boldsymbol{\xi}, \boldsymbol{\eta})$ has a non-singular density

$$f_{\xi\eta}(\mathbf{x}, \mathbf{y}) = \frac{1}{(2\pi)^{(m+n)/2}\sqrt{\det \boldsymbol{\Sigma}}} e^{-\frac{1}{2}(\mathbf{x}-m_{\xi}, \mathbf{y}-m_{\eta})\boldsymbol{\Sigma}^{-1}(\mathbf{x}-m_{\xi}, \mathbf{y}-m_{\eta})'}.$$

The density of $\boldsymbol{\eta}$ is

$$f_{\eta}(\mathbf{y}) = \frac{1}{(2\pi)^{m/2}\sqrt{\det \boldsymbol{\Sigma}_{\eta\eta}}} e^{-\frac{1}{2}(\mathbf{y}-m_{\eta})\boldsymbol{\Sigma}_{\eta\eta}^{-1}(\mathbf{y}-m_{\eta})'},$$

and the conditional density $f_{\xi|\eta}(\mathbf{x}|\mathbf{y})$, defined as

$$f_{\xi|\eta}(\mathbf{x} \mid \mathbf{y}) = \frac{f_{\eta\xi}(\mathbf{y}, \mathbf{x})}{f_{\eta}(\mathbf{y})}.$$

[4] A summary of conditional distributions is found in Section A.3.3 in Appendix A.

This is again a Gaussian density and its mean and covariance properties can easily be found by some matrix algebra. We summarize the important facts as a theorem.

Theorem 1.3. *The conditional distribution of* $\boldsymbol{\xi}$ *given* $\boldsymbol{\eta} = \mathbf{y}$ *is Gaussian with conditional mean matrix*

$$E(\boldsymbol{\xi} \mid \boldsymbol{\eta} = \mathbf{y}) = \widehat{\boldsymbol{\xi}}(\mathbf{y}) = E(\boldsymbol{\xi}) + C(\boldsymbol{\xi};\boldsymbol{\eta})\boldsymbol{\Sigma}_{\eta\eta}^{-1}(\mathbf{y} - E(\boldsymbol{\eta}))'$$
$$= m_{\xi} + \boldsymbol{\Sigma}_{\xi\eta}\boldsymbol{\Sigma}_{\eta\eta}^{-1}(\mathbf{y} - m_{\eta})'. \tag{1.9}$$

The conditional covariance is

$$\boldsymbol{\Sigma}_{\xi\xi|\eta} = E((\boldsymbol{\xi} - \widehat{\boldsymbol{\xi}}(\boldsymbol{\eta}))'(\boldsymbol{\xi} - \widehat{\boldsymbol{\xi}}(\boldsymbol{\eta}))) = \boldsymbol{\Sigma}_{\xi\xi} - \boldsymbol{\Sigma}_{\xi\eta}\boldsymbol{\Sigma}_{\eta\eta}^{-1}\boldsymbol{\Sigma}_{\eta\xi}. \tag{1.10}$$

In two dimensions the formulas read

$$m_{x|y} = m_x + \sigma_x \rho_{xy} \cdot \frac{y - m_y}{\sigma_y}, \qquad \sigma_{x|y}^2 = \sigma_x^2(1 - \rho_{xy}^2),$$

with $\rho_{xy} = C(x,y)/\sqrt{V(x)V(y)}$; thus the squared correlation ρ_{xy}^2 gives the relative reduction of the variability (uncertainty) in the random variable x gained by observation of y.

Observe the mnemotechnical friendliness of the formulas for conditional mean and covariance. For example, the covariance matrix $\boldsymbol{\Sigma}_{\xi\xi|\eta}$ has dimension $n \times n$ and the configuration on the right hand side of (1.10) is the only way to combine the matrices involved that matches their dimensions – of course, you have to remember the general structure.

1.4.3 Linear prediction and reconstruction

Prediction and reconstruction are two of the most important applications of stationary process theory. Even though these problems are not main topics in this work, we present one of the basic concepts here; in Section 6.6 we will deal with the more philosophical sides of the prediction problem.

Suppose we have observed the outcomes of a set of random variables, $\boldsymbol{\eta} = (\eta_1, \ldots, \eta_m)$, and that we want to give a statement $\widehat{\xi}$ about the outcome of some other variable ξ, either to be observed sometime in the future, or perhaps a missing observation in a time series. These two cases constitute the framework of *prediction* and *reconstruction*, respectively. Also suppose that we want to make the statement in the best possible way in the mean square sense, i.e., we want $E((\xi - \widehat{\xi})^2)$ to be as small as possible.

It is easy to see (Theorem A.4 in Appendix A) that the best solution in mean square sense is given by the conditional expectation, $\widehat{\xi} = E(\xi \mid \boldsymbol{\eta}) = \varphi(\boldsymbol{\eta})$. On the other hand, if the variables are jointly Gaussian, then we know from formula (1.9) that the conditional expectation of ξ given $\boldsymbol{\eta}$ is linear in $\boldsymbol{\eta}$, so that for Gaussian variables the optimal solution is

$$\widehat{\xi} = E(\xi \mid \boldsymbol{\eta}) = m_\xi + \boldsymbol{\Sigma}_{\xi\boldsymbol{\eta}} \boldsymbol{\Sigma}_{\boldsymbol{\eta}\boldsymbol{\eta}}^{-1} (\boldsymbol{\eta} - m_{\boldsymbol{\eta}})', \tag{1.11}$$

an expression that depends only on the mean values and the second order moments, i.e., variances and covariances.

Looking at the general case, without assuming normality, we restrict ourselves to solutions that are linear functions of the observed variables. It is clear that the solution that is optimal in the mean square sense can only depend on the mean values and variances/covariances of the variables. It therefore has the same form for all variables with the same first and second order moments. Thus, (1.11) gives the best linear predictor in mean square sense, regardless of the distribution.

1.5 Four historical landmarks

1.5.1 Brownian motion and the Wiener process

There is no other Gaussian process with as wide applicability as the Wiener process. Even if it is a non-stationary process it appears repeatedly in the theory of stationary processes, and we spend this section to describe some of its properties and applications.

Definition 1.7 (Wiener process). *The Wiener process $\{w(t); t \geq 0\}$ is a Gaussian process with $w(0) = 0$ such that $E(w(t)) = 0$, and the variance of the increment $w(t+h) - w(t)$ over any interval $[t, t+h]$, $h > 0$, is proportional to the interval length,*

$$V(w(t+h) - w(t)) = h\sigma^2.$$

A Wiener process $\{w(t), t \in \mathbb{R}\}$ over the whole real line is a combination of two independent Wiener processes w_1 and w_2, so that $w(t) = w_1(t)$ for $t \geq 0$, and $w(t) = w_2(-t)$ for $t < 0$.

It is an easy consequence of the definition that the increment $w(t) - w(s)$ is uncorrelated with $w(s)$ for $s < t$; just expand the variance of $x(t)$,

$$\begin{aligned} t\sigma^2 = V(w(t)) &= V(w(s)) + V(w(t) - w(s)) + 2C(w(s), w(t) - w(s)) \\ &= s\sigma^2 + (t-s)\sigma^2 + 2C(w(s), w(t) - w(s)). \end{aligned}$$

It also follows that $C(w(s), w(t)) = \sigma^2 \min(s,t)$. Since the increments over disjoint intervals are jointly normal, by the definition of a normal process, they are also independent.

A characteristic feature of the Wiener process is that its future changes are statistically independent of its actual and previous values. It is intuitively clear that a process with this property cannot be differentiable. The increment over a small time interval from t to $t+h$ is of the order $\sqrt{h}$, which is small enough to make the process continuous, but it is too large to give a differentiable process. In Chapter 2 we will give conditions for continuity and differentiability of sample functions of Gaussian and general processes.

The sample functions of a Wiener process are in fact objects that have *fractal dimension*, and the process is *self similar* in the sense that when magnified with proper scales it retains its statistical geometrical properties. More precisely, for each $a > 0$, the process $\sqrt{a}\,w(t/a)$ has the same distributions as the original process $w(t)$.

The Wiener process is commonly used to model phenomena where the local changes are virtually independent. Symbolically, one can write $\mathrm{d}w(t)$ for the infinitesimal independent increments, or simply use "the derivative" $w'(t)$. In this form, the Wiener process is used as a driving force in many stochastic systems, appearing in Chapter 4. The Brownian motion is a good example of how one can use the Wiener process to get models with more or less physical realism.

The Brownian motion, observed 1827 by the Scottish biologist Robert Brown, is an erratic movement by small particles immersed in a fluid, for example pollen particles in water as in Brown's original experiment. Albert Einstein presented in 1905 a quantitative model for the Brownian movement in his paper *On the movements of small particles in a stationary liquid demanded by the molecular-kinetic theory of heat*, reprinted in [46], based on the assumption that the movements are caused by independent impacts on the particle by the molecules of the surrounding fluid medium.

In Einstein's model the changes in location due to collisions over separate time intervals are supposed to be independent. This requires however that the particles have no mass, which is physically wrong, but sufficiently accurate for microscopic purposes on a human time scale. According to Einstein, the change in location in any of the three directions (x,y,z) over a time interval of length t is random and normal with mean zero, which is not surprising, since the changes are the results of a very large number of independent collisions. What made Einstein's contribution conclusive was that he derived an expression for the variance in terms of other physical parameters, namely

$$V(x(t)) = V(y(t)) = V(z(t)) = t\,\frac{4RT}{Nf} = t\,\sigma^2, \tag{1.12}$$

where T is the absolute temperature, R is the Boltzmann constant, and N is Avogadro's number, i.e., the number of molecules per mole, and the friction coefficient f depends on the shape and size of the particle and on the viscosity of the fluid. Here, the coordinates are independent Wiener processes.

Observations of the Brownian movement and estimation of its variance makes it possible to calculate any of the factors in σ^2, for example N, from the other ones.

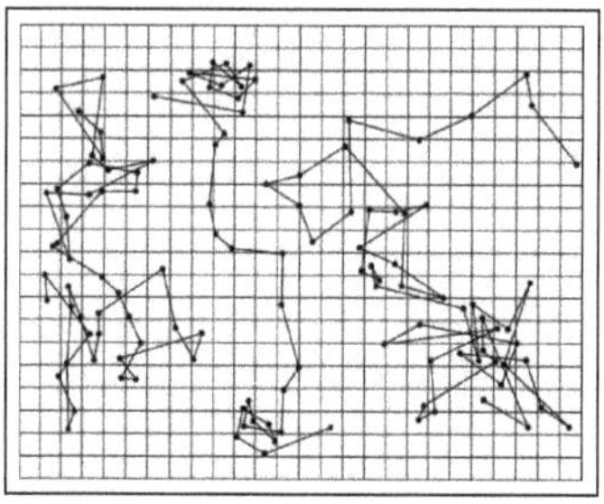

Figure 1.3 *Three particle paths in Perrin's experiment.*

The French physicist J.B. Perrin estimated in a series of experiments 1908-1911 Avogadro's number in this way by observing suspended rubber particles and found an estimate that is correct within about 10%. The figure, reproduced from the book *Les Atomes* by Perrin [97], shows the paths of the motions of three colloidal particles as seen under the microscope. Dots show the positions every 30 seconds. For his experiments and conclusions, Perrin was awarded the Nobel prize in physics, 1926, for his work on the "discontinuous structure of matter."

In a more realistic model, one takes also particle mass and velocity into account. If $v(t)$ denotes the velocity at time t, the fluid offers a resistance from the friction force, which is equal to $fv(t)$, with the friction coefficient as in (1.12). Further, the particle offers a resistance to changes in velocity proportional to its mass m. Finally, one needs to model the independent collisions from the surrounding fluid molecules, and here the Wiener process can again be used, more precisely its increments $\mathrm{d}w(t)$. This gives the *Langevin equation*, named after the French physicist Paul Langevin, for the particle velocity,

$$\mathrm{d}v(t) + \alpha\, v(t)\,\mathrm{d}t = \frac{1}{m}\,\mathrm{d}w(t), \tag{1.13}$$

where $\alpha = f/m$, and $w(t)$ is a standardized Wiener process. It is usually written

$$\frac{\mathrm{d}v(t)}{\mathrm{d}t} + \alpha v(t) = \frac{1}{m} w'(t). \tag{1.14}$$

We will meet the Langevin equation in Example 3.5 on the Ornstein-Uhlenbeck process in Section 3.4.2.

1.5.2 S.O. Rice and electronic noise

The two papers *Mathematical analysis of random noise*, by Steve O. Rice, appeared in Bell System Technical Journal, 1944–1945 [101]. They represent a landmark in the history of stochastic processes in that they bring together and exhibit the wide applicability in engineering of the spectral formulation of a stationary process as a sum, or asymptotically an integral, of harmonic cosine functions with random amplitudes and phases. Correlation functions and their Fourier transforms had been studied at least since the early 1900s, and Rice's work brought together these results in a systematic way. But it also contained many new results, in particular pertaining to crossing related properties, and on the statistical properties of stationary processes "obtained by passing random noise through physical devices."

Rice used the *spectral representation* of a stationary Gaussian process as a sum over discrete positive frequencies $\omega_k > 0$,

$$x(t) = \sum_k a_k \cos \omega_k t + b_k \sin \omega_k t = \sum_k c_k \cos(\omega_k t + \phi_k), \tag{1.15}$$

where the amplitudes a_k, b_k are normal and independent random variables with mean 0 and $E(a_k^2) = E(b_k^2) = \sigma_k^2$, and ϕ_k uniformly distributed over $(0, 2\pi)$, independent of the amplitudes. As we shall see in Chapter 3, such a process has covariance

$$r(t) = \sum_k \sigma_k^2 \cos \omega_k t.$$

The spectral distributions function is a discrete distribution with point mass $\sigma_k^2/2$ at the symmetrically located frequencies $\pm\omega_k$.

The continuous, integral form of the spectral representation is presented as limiting cases in Rice's work. At about the same time, Harald Cramér and Michel Loève gave a probabilistic formulation of the continuous spectral representation, in a mathematically impeccable way [31, 33]. In Chapter 3 we shall deal in more depth with spectral representation, in different forms.

Besides the previously known "Rice's formula" for the expected number of level crossings, Rice's 1945 paper also analyzed crossing and excursion distributions and investigated the joint occurrence of crossings of a fixed level at two distinct points, necessary for calculation of the variance of the number of crossings. The simple Rice's formula can be generalized to vector processes and to random fields, and it has developed into one of the most powerful tools in the statistical analysis of random systems; see [6]. Crossing problems are the topic of Chapter 8.

The flexibility and generality of Rice's methods and examples made correlation and spectral theory fundamental ingredients in communication theory and signal processing for decades to come. An example by Rice himself is the ingenious explanation of the intriguing click noise in analogue FM radio [102].

1.5.3 Gaussian random wave models

Steve Rice's analysis of time dependent stationary processes had, as mentioned, great influence on signal processing in the information sciences. Less well known in the statistical world is the effect his work had in oceanography and naval architecture.

To give the reader a feeling for the dramatic effect Rice's work on random noise had in a quite different field, we cite in extenso (references deleted) the first two paragraphs in Manley St Denis and Willard J. Pierson's paper, *On the motion of ships in confused seas*, which came out 1954 [116]:

> HISTORY
>
> Three years ago the first co-author of the present work collaborated with Weinblum in the writing of a paper entitled "On the motion of ships at sea." In that paper Lord Rayleigh was quoted saying: "The basic law of the seaway is the apparent lack of any law." Having made this quotation, however, the authors then proceed to consider the seaway as being composed of "a regular train of waves defined by simple equations." This artificial substitution of pattern for chaos was dictated by the necessity of reducing the utterly confused reality to a simple form amenable to mathematical treatment.
>
> Yet at the same time and in other fields the challenging study of confusion was being actively pursued. Thus in 1945 Rice was writing on the mathematical analysis of random noise and in 1949 Tukey and Hamming were writing on the properties of stationary time series and their power spectra in connection with colored noise. In the same year Wiener published his now famous book on time series. These works

> were written as contributions to the theory of communication. Nevertheless the fundamental mathematical discipline expounded therein can readily be extended to other fields of scientific endeavor. Thus in 1952 the second co-author, inspired by a contribution of Tukey, was able to apply the foregoing theories to the study of actual ocean waves. As the result of analyses of actual wave records, he succeeded in giving not only a logical explanation as to why waves are irregular, but a statement as well of the laws underlying the behavior of a seaway. There is indeed a basic law of the seaway. Contrary to the obvious inference from the quotation of Lord Rayleigh, the seaway can be described mathematically and precisely, albeit in a statistical way.

If Rice's work had been in the vein of generally accepted ideas in communication theory, the St Denis and Pierson paper represented a complete revolution in common naval practice. Nevertheless, its treatment of irregular water waves as, what is now called, a random field was almost immediately accepted, and set a standard for much of naval architecture. A parallel theory was developed in England where M.S. Longuet-Higgins studied the geometric properties of random wave fields. His long paper [86] from 1957 is still much cited in oceanography, optics, and statistics.

One possible reason for the success of the St Denis-Pierson approach can be that the authors succeeded to formulate and analyze the motions of a ship that moved with constant speed through the random field. The random sea could directly be used as input to a linear (later also non-linear) filter representing the ship. Linear filters are the topic of Chapters 4 and 5.

St Denis and Pierson extended the one-dimensional description of a time dependent process $\{x(t), t \in \mathbb{R}\}$, useful for example to model the waves measured at a single point, to a random field $x(t,(s_1,s_2))$ with time parameter t and location parameter (s_1,s_2). They generalized the sum (1.15) to be a sum of directed waves, with $\boldsymbol{\omega} = (\omega, \kappa_1, \kappa_2)$,

$$\sum_{\boldsymbol{\omega}} A_{\boldsymbol{\omega}} \cos(\omega t - \kappa_1 s_1 - \kappa_2 s_2 + \phi_{\boldsymbol{\omega}}), \tag{1.16}$$

with random amplitudes and phases.

For fixed t each element in (1.16) is a cosine-function in the plane, which is zero along lines $\omega t - \kappa_1 s_1 - \kappa_2 s_2 + \phi_{\boldsymbol{\omega}} = \pi/2 + k\pi$, k integer. The parameters κ_1 and κ_2, called the *wave numbers*, determine the direction and wave length of the waves and ω is the wave *(angular) frequency*, when the sea is observed at a fixed location (s_1,s_2). More on this will follow in Chapter 7.

1.5.4 Detection theory and statistical inference

The first three landmarks illustrated the relation between stochastic model building and physical knowledge, in particular how the concepts of statistical independence and dependence between signal and functions relate to the physical world. About the same time as Rice and St Denis & Pierson advanced physically based stochastic modeling, the statistical inference methodology was placed firmly into a theoretical mathematical framework, as it was by then documented by the classical book by Harald Cramér, *Mathematical Methods of Statistics*, 1945 [32].

A few years later, the connection between the theoretical basis for statistical inference and important engineering questions related to signal detection was elegantly exploited by Ulf Grenander in his PhD thesis from Stockholm, *Stochastic processes and statistical inference* [56]. The classical problem in signal processing of deciding whether a deterministic signal of known shape $s(t)$ is present in an environment of Gaussian dependent, colored as opposed to white, random noise $x(t)$ can be treated as an infinite dimensional decision problem, testing an infinite dimensional statistical hypothesis; see, e.g., the standard work [118] on detection theory.

Suppose one observes a Gaussian stochastic process $x(t)$, $a \leq t \leq b$, with known correlation structure, but with unknown mean value function $m(t)$. If no signal is present, the mean value is 0, but with signal, the mean is equal to a function $s(t)$ of known shape. In statistical terms one has to test the following two hypotheses against each other:

$$H_0: \; m(t) = 0,$$

$$H_1: \; m(t) = s(t).$$

Grenander introduced a series of independent Gaussian *observables*,

$$y_k = \int h_k(t) x(t)\,\mathrm{d}t,$$

by choosing the filter functions h_k as solutions to the integral equation

$$\int r(s,t) h_k(t)\,\mathrm{d}t = c_k h_k(s),$$

with $c_1 \geq c_2 \geq \ldots$, $c_k \to 0$ as $k \to \infty$. Under H_1 the observables will have mean $a_k = \int h_k(t) s(t)\,\mathrm{d}t$ and variance c_k, while under H_0 they will have mean 0, and the same variances. So instead of a continuous problem, we have gotten a denumerable problem, in which one can make a Likelihood-Ratio test of the two alternatives. We will return to this problem in Section 5.3.4.

Exercises

1:1. Consider the sample space $\Omega = [0,1]$ with uniform probability P, $P([a,b]) = b-a$, $0 \le a \le b \le 1$. Construct a stochastic process $\mathbf{y} = (x_1, x_2, \dots)$ on Ω such that the components are independent zero-one variables, with $P(x_k = 0) = P(x_k = 1)$.

1:2. Show that $\mathscr{B}(\mathbb{R}) =$ the class of Borel sets in $\mathbb{R}$ is generated by

a) the open intervals,

b) the closed intervals.

1:3. Prove the claim in Lemma 1.1 that the inverse $x^{-1}(\mathscr{B}) \in \mathscr{F}$.

1:4. A set $A \subset \{1,2,\dots\}$ is said to have asymptotic density θ if

$$\lim_{n\to\infty} n^{-1}|A \cap \{1,2,\dots,n\}| = \theta.$$

(Note, $|B|$ denotes the number of elements in B.) Let $\mathscr{A}$ be the family of sets for which the asymptotic density exists. Is $\mathscr{A}$ a field?

1:5. Take $\mathbb{R}^n$ and show that the family $\mathscr{F}_0$ whose elements are unions of finitely many rectangles $(a_i, b_j]$ (with possibly infinite end points) is a field. Then, let T be an interval and convince yourself that the finite dimensional rectangles in $\mathbb{R}^T$ and unions of finitely many such rectangles, is a field.

1:6. Show that the co-ordinate process $x_k(\omega) = \omega_k$ is a stochastic process on $(\mathbb{R}^\infty, \mathscr{B}^\infty)$.

1:7. Let $\mathbb{Z}$ be the integers, and $\mathscr{A}$ the family of subsets A, such that either A or its complement A^c is finite. Let $P(A) = 0$ in the first case and $P(A) = 1$ in the second case. Show that P can not be extended to a probability to $\sigma(\mathscr{A})$, the smallest σ-field that contains $\mathscr{A}$.

1:8. Take $T = [0,1]$, and consider the set of functions which are continuous on the rational numbers, i.e.,

$$C_Q = \{x \in \mathbb{R}^T; x(q) \to x(q_0) \text{ for all rational numbers } q_0\},$$

where the limit is taken as q tends to q_0 through the rational numbers. Show that $C_Q \in \mathscr{B}_T$.

1:9. Prove that the increments of a Wiener process, as defined as in Definition 1.7, are independent and normal.

1:10. Prove the following useful inequality valid for any non-negative, integer-valued random variable N,

$$E(N) - \frac{1}{2}E(N(N-1)) \le P(N > 0) \le E(N).$$

Generalize it to the following inequalities, where

$$\alpha_i = E(N(N-1)\cdot\ldots\cdot(N-i+1))$$

is the ith factorial moment,

$$\frac{1}{k!}\sum_{i=0}^{2n-1}(-1)^i\frac{1}{i!}\alpha_{(k+i)} \leq P(N=k) \leq \frac{1}{k!}\sum_{i=0}^{2n}(-1)^i\frac{1}{i!}\alpha_{(k+i)}.$$

Chapter 2

Sample function properties

This chapter is the stochastic equivalent of real analysis and integration. As in its deterministic counterpart, limiting concepts and conditions for existence of limits are fundamental. A summary of basic concepts and results on stochastic convergence is given in Appendix A.4.1.

Definition 2.1. *Let $\{x_n\}_{n=1}^{\infty}$ be a random sequence, with the random variables $x_1(\omega)$, $x_2(\omega), \ldots,$ defined on the same probability space as a random variable $x = x(\omega)$. The convergence $x_n \to x$ as $n \to \infty$ can be defined in three ways:*

- **almost surely, with probability one** ($x_n \overset{a.s.}{\to} x$): $P(\{\omega; x_n \to x\}) = 1$;
- **in quadratic mean** ($x_n \overset{q.m.}{\to} x$): $E(|x|^2) < \infty$ *and* $E(|x_n - x|^2) \to 0$;
- **in probability** ($x_n \overset{P}{\to} x$): *for every* $\varepsilon > 0$, $P(|x_n - x| > \varepsilon) \to 0$.

Furthermore, x_n **tends in distribution** *to* x, ($x_n \overset{\mathscr{L}}{\to} x$), *if*

$$P(x_n \le a) \to P(x \le a)$$

for all a *such that* $P(x \le u)$ *is a continuous function of* u *at* $u = a$.

The different modes of convergence lead to different types of statements about the analytic properties of a stochastic process, for example continuity or differentiability. The most striking example is the Poisson process, which is "continuous in quadratic mean," but has sample functions with jump discontinuities. We shall examine different conditions for sample function continuity, differentiability, and integrability, and also give conditions that guarantee that only simple discontinuities occur. In particular, we shall formulate conditions in terms of bivariate distributions, which are easily checked for most standard processes, such as the normal and the Poisson process.

For (weakly) stationary processes, conditions will be formulated in terms of the covariance functions. The conditions for continuity, differentiability, etc., will put restrictions on the smoothness of the covariance function near the origin. For a normal process, the conditions are slightly less restrictive than in general, but they still rely on the smoothness of the covariance function. Conditions valid for non-stationary processes will also be given, and they are local variations of the conditions for stationary processes.

Complex random variables

Many of the results are presented for *complex-valued* processes and variables, i.e., for $x = u + iv$, where the two real variables u and v have a joint distribution. The expectation is $E(x) = m_x = E(u) + iE(v)$.

The covariance between two complex variables is defined as

$$C(x,y) = E((x - m_x)\overline{(y - m_y)}).$$

Note the complex conjugate on the second variable.

This definition leads to the natural property $V(x) = E(|x - m_x|^2) \geq 0$.

2.1 Quadratic mean properties

Continuous or differentiable sample paths are what we expect to encounter in practice when we *observe* a stochastic process. To *prove* that a mathematical model for a random phenomenon has continuous or differentiable sample paths is a quite different matter. Much more simple is to base the stochastic analysis on correlation properties, which could be checked against data, at least in principle. We assume throughout this section, as in most of the chapter, that the process $x(t)$ has mean zero.

First a reminder: in q.m. convergence, $x_n \overset{q.m.}{\to} x$, it is necessary that $E(|x|^2)$ is finite, which implies that $E(|x_n|^2) \to E(|x|^2) < \infty$, and that $E(x_n) \to E(x)$, see Appendix A.4.1. To prove q.m. convergence, we shall often use the Loève criterion (A.16): the sequence x_n converges in quadratic mean if and only if

$$E(x_m \overline{x_n}) \text{ has a finite limit } c, \tag{2.1}$$

when m and n tend to infinity independently of each other.

2.1.1 Quadratic mean continuity

Definition 2.2. *A stochastic process $\{x(t), t \in \mathbb{R}\}$ is said to be continuous in quadratic mean (or $\mathcal{L}^2$-continuous) at time t if*

$$x(t+h) \overset{q.m.}{\to} x(t)$$

as $h \to 0$, i.e., if $E(|x(t+h)-x(t)|^2) \to 0$.

Example 2.1 (Wiener process). *For a Wiener process, the incremental variance is $E((w(t+h)-w(t))^2) = \sigma h \to 0$ as $h \to 0$, so the process is continuous in quadratic mean.*

Example 2.2 (Poisson process). *The Poisson process is a counting process that counts the number of events that occur independently of each other with constant average rate, the intensity, λ, per time unit. If $N(t)$ denotes the number of events that occur in the time interval $[0,t]$, then the increment $N(t+h)-N(t)$ has a Poisson distribution with mean and variance both equal to λh and hence $E((N(t+h)-N(t))^2) \to 0$ as $h \to 0$. Thus, the Poisson process is continuous in quadratic mean even if its sample functions have jumps of size 1 at all times of occurrence.*

Stationary case: The following statement is a direct consequence of the definition of q.m. continuity.

A stationary process $\{x(t), t \in \mathbb{R}\}$ is continuous in quadratic mean if and only if its covariance function $r(t)$ is continuous at $t = 0$:

$$E(|x(t+h)-x(t)|^2) = 2(r(0)-r(h)) \to 0.$$

Example 2.3 (Random telegraph signal). *The random telegraph signal is a stationary process $x(t)$ that starts at time $t = 0$ with equal probability in -1 and in 1, and toggles between -1 and 1 at the time of occurrences in a Poisson process $N(t)$ that has intensity λ. It is easy to show that the random telegraph signal is a stationary process. Its covariance function $r_x(\tau) = e^{-2\lambda|\tau|}$, (Exercise 2:1), is continuous, but the sample functions have jumps of size 2.*

It is to be noted that when a stationary process with continuous covariance function $r(t)$ is observed with measurement error, consisting of independent, additive noise with variance $\sigma^2 > 0$, then the covariance function of the observed process is discontinuous at $t = 0$. The addition of the measurement variance at the origin is called the *nugget effect*; more on that in Chapter 7.

Non-stationary case: For non-stationary processes q.m. continuity depends on the covariance function along the diagonal. We formulate the condition for a complex valued process with mean $E(x(s)) = m(s)$ and covariance function

$$r(s,t) = C(x(s),x(t)) = E((x(s)-m(s))\overline{(x(t)-m(t))}).$$

Theorem 2.1. *A stochastic process $\{x(t), t \in \mathbb{R}\}$ with mean zero is continuous in quadratic mean at t_0 if and only if the covariance function $r(s,t)$ is continuous at the diagonal point $s = t = t_0$.*

Proof. If $r(s,t)$ is continuous at $s = t = t_0$, then

$$\begin{aligned} E(|x(t_0+h)-x(t_0)|^2) &= E(|x(t_0+h)|^2) + E(|x(t_0)|^2) \\ &\quad - E(x(t_0+h)\cdot\overline{x(t_0)}) - E(x(t_0)\cdot\overline{x(t_0+h)}) \\ &= r(t_0+h,t_0+h) - r(t_0+h,t_0) - r(t_0,t_0+h) + r(t_0,t_0) \to 0, \end{aligned}$$

as $h \to 0$, which shows the "if" part.

For the "only if" part, expand

$$\begin{aligned} & r(t_0+h,t_0+k) - r(t_0,t_0) \\ &\quad = E((x(t_0+h)-x(t_0))\cdot\overline{(x(t_0+k)-x(t_0))}) \\ &\qquad + E((x(t_0+h)-x(t_0))\cdot\overline{x(t_0)}) + E(x(t_0)\cdot\overline{(x(t_0+k)-x(t_0))}) \\ &\quad = e_1 + e_2 + e_3, \text{ say.} \end{aligned}$$

Here

$$e_1 \le \sqrt{E(|x(t_0+h)-x(t_0)|^2)\cdot E(|x(t_0+k)-x(t_0)|^2)} \to 0,$$

$$e_2 \le \sqrt{E(|x(t_0+h)-x(t_0)|^2)\cdot E(|x(t_0)|^2)} \to 0,$$

$$e_3 \le \sqrt{E(|x(t_0)|^2)\cdot E(|x(t_0+k)-x(t_0)|^2)} \to 0,$$

so $r(t_0+h,t_0+k) \to r(t_0,t_0)$ as $h,k \to 0$. □

Example 2.4 (Poisson process). *The Poisson process with intensity* λ *has covariance function* $r(s,t) = \lambda \min(s,t)$*, which is continuous on the diagonal, and we can again conclude that it is continuous in quadratic mean.*

2.1.2 *Quadratic mean differentiability*

Definition 2.3. *A stochastic process* $\{x(t), t \in \mathbb{R}\}$ *is called differentiable in quadratic mean at t if there is a random variable* $x'(t)$ *such that*

$$\frac{x(t+h) - x(t)}{h} \overset{q.m.}{\to} x'(t), \text{ as } h \to 0.$$

Of course, $x'(t)$ *is called the (quadratic mean) derivative of* $x(t)$*.*

Example 2.5. *Note that q.m. convergence requires that the variance has a finite limit. The Wiener process is continuous, but not differentiable, in quadratic mean. This follows from* $E\left(\left|\frac{w(t+h)-w(t)}{h}\right|^2\right) = \frac{\sigma^2}{h} \to \infty$*, as* $h \to 0$*; see also Exercise 2.7.*

Theorem 2.2. *a) A stationary process* $\{x(t), t \in \mathbb{R}\}$ *is quadratic mean differentiable if and only if its covariance function* $r(t)$ *is twice continuously differentiable in a neighborhood of* $t = 0$*. The derivative process* $x'(t)$ *is stationary with mean zero and it has covariance function*

$$r_{x'}(t) = C(x'(s+t), x'(s)) = -r''(t) = \int_{-\infty}^{\infty} \omega^2 e^{i\omega t} \, \mathrm{d}F(\omega).$$

If the spectrum is continuous, the derivative has spectral density function

$$f_{x'}(\omega) = \omega^2 f_x(\omega). \tag{2.2}$$

b) The condition for q.m. differentiability is equivalent to $\int_{-\infty}^{\infty} \omega^2 \, \mathrm{d}F(\omega) < \infty$*.*

Proof. For the "if" part we use the Loève criterion, and show that if $h,k \to 0$ independently of each other, then

$$E\left(\frac{x(t+h)-x(t)}{h} \cdot \overline{\frac{x(t+k)-x(t)}{k}}\right) = \frac{r(h-k)-r(h)-r(-k)+r(0)}{hk} \tag{2.3}$$

has a finite limit c. Define

$$\begin{aligned} f(h,k) &= r(h)-r(h-k), \\ f_1'(h,k) &= \frac{\partial}{\partial h} f(h,k) = r'(h)-r'(h-k), \\ f_{12}''(h,k) &= \frac{\partial^2}{\partial h \partial k} f(h,k) = r''(h-k). \end{aligned}$$

By applying the mean value theorem we see that there exist $\theta_1, \theta_2 \in (0,1)$, such that (2.3) is equal to

$$\begin{aligned} -\frac{f(h,k)-f(0,k)}{hk} &= -\frac{f_1'(\theta_1 h,k)}{k} = -\frac{f_1'(\theta_1 h,0)+kf_{12}''(\theta_1 h,\theta_2 k)}{k} \\ &= -f_{12}''(\theta_1 h,\theta_2 k) = -r''(\theta_1 h-\theta_2 k). \end{aligned} \tag{2.4}$$

Since $r''(t)$ by assumption is continuous, this tends to $-r''(0)$ as $h,k \to 0$, which is the required limit in the Loève criterion.

To prove the "only if" part we need a fact about Fourier integrals and the spectral representation $r(t) = \int_{-\infty}^{\infty} e^{i\omega t}\, \mathrm{d}F(\omega) = \int_{-\infty}^{\infty} \cos \omega t \, \mathrm{d}F(\omega)$.

Lemma 2.1. *a)* $\lim_{t\to 0} \frac{2(r(0)-r(t))}{t^2} = \omega_2 = \int_{-\infty}^{\infty} \omega^2 \, \mathrm{d}F(\omega) \le \infty$.

b) If $\omega_2 < \infty$ then $r''(0) = -\omega_2$ and $r''(t)$ exists for all t.

c) If $r''(0)$ exists, finite, then $\omega_2 < \infty$ and then, by (b), $r''(t)$ exists for all t.

Proof of lemma: If $\omega_2 < \infty$, (a) follows from

$$\frac{2(r(0)-r(t))}{t^2} = \int_{-\infty}^{\infty} \omega^2 \frac{1-\cos \omega t}{\omega^2 t^2/2} \, \mathrm{d}F(\omega)$$

by dominated convergence, since $0 \le \frac{1-\cos \omega t}{\omega^2 t^2/2} \le 1$. If $\omega_2 = \infty$, the result follows from Fatou's lemma, since $\lim_{t\to 0} \frac{1-\cos \omega t}{\omega^2 t^2/2} = 1$.

To prove (b), suppose $\omega_2 < \infty$. Then it is possible to differentiate twice under the integral sign in $r(t) = \int_{-\infty}^{\infty} \cos \omega t \, dF(\omega)$ to obtain that $r''(t)$ exists with

$$-r''(t) = \int_{-\infty}^{\infty} \omega^2 \cos(\omega t) \, dF(\omega),$$

and in particular $-r''(0) = \omega_2$.

For (c), suppose $r(t)$ has a finite second derivative at the origin, and implicitly is differentiable near the origin. By the same argument as for (2.4), with $h = k = t$, we see that

$$\frac{2(r(0) - r(t))}{t^2} = -r''((\theta_1 - \theta_2)t) \to -r''(0)$$

as $t \to 0$. Then part (a) shows that $\omega_2 < \infty$, and the lemma is proved. □

Proof of the "only if" part of Theorem 2.2: If $x(t)$ is q.m. differentiable, $(x(t+h) - x(t))/h \overset{q.m.}{\to} x'(t)$ with $E(|x'(t)|^2)$ finite, and (see (A.15) in Appendix A.4.4),

$$E(|x'(t)|^2) = \lim_{h \to 0} E\left\{ \left| \frac{x(t+h) - x(t)}{h} \right|^2 \right\} = \lim_{h \to 0} \frac{2(r(0) - r(h))}{h^2} = \omega_2 < \infty.$$

Part (b) of the lemma shows that $r''(t)$ exists.

To see that the covariance function of $x'(t)$ is equal to $-r''(t)$, just take the limit of

$$\begin{aligned} E & \left(\frac{x(s+t+h) - x(s+t)}{h} \cdot \overline{\frac{x(s+k) - x(s)}{k}} \right) \\ & = \frac{1}{hk} \left(r(t+h-k) - r(t+h) - r(t-k) + r(t) \right) \\ & = -f''_{12}(t + \theta_1 h, \theta_2 k) = -r''(t + \theta_1 h - \theta_2 k) \to -r''(t), \end{aligned}$$

for some θ_1, θ_2, as $h, k \to 0$. □

2.1.3 *Higher order derivatives and their correlations*

The finiteness of the second spectral moment ω_2 restricts the amount of rapid oscillations in a random process to the extent that its becomes differentiable (in quadratic mean). The meaning of this statement will become clear from the spectral representation in Chapter 3.

To obtain higher order derivatives one just has to examine the covariance function $r_{x'}(t) = -r_x''(t)$ for the conditions in the existence theorem, etc. Alternatively, one may examine the spectral moments, $\omega_4 = \int \omega^4 \, \mathrm{d}F(\omega)$, and higher.

Before we go into details, a word of warning is appropriate. Many of the spectra that are used in practice have a functional form that only have finite spectral moment up to second order. Of course, that should not be taken as proof that the process they intend to model is not sufficiently differentiable, only that one has to be careful to use that spectrum for statements about derivatives.

The derivative of a (mean square) differentiable stationary process is also stationary and has covariance function $r_{x'}(t) = -r_x''(t)$, from Theorem 2.2. Working recursively, one can easily derive the cross-covariance relations between the derivative and the process (strictly, we again need (A.15) from Appendix A.4.4).

Theorem 2.3. *a) The kth (q.m.) derivative of a stationary process $\{x(t), t \in \mathbb{R}\}$ has covariance function*

$$r_{x^{(k)}}(t) = (-1)^k r_x^{(2k)}(t) = \int_{-\infty}^{\infty} \omega^{2k} e^{i\omega t} \, \mathrm{dF}(\omega).$$

If spectrum is continuous, the spectral density is

$$f_{x^{(k)}}(\omega) = \omega^{2k} f_x(\omega). \tag{2.5}$$

b) The cross-covariance between derivatives $x^{(j)}(s)$ and $x^{(k)}(t)$ is

$$\mathrm{C}(x^{(j)}(s), x^{(k)}(t)) = (-1)^k r_x^{(j+k)}(s-t). \tag{2.6}$$

In particular,

$$\mathrm{C}(x(s), x'(t)) = -r_x'(s-t), \tag{2.7}$$

and, hence, $x(t)$ and $x'(t)$ are uncorrelated random variables.

In Chapter 8 we will need the covariances between the process and its first two derivatives. The covariance matrix for $(x(t), x'(t), x''(t))$ is given in terms of the spectral moments $\omega_{2k} = (-1)^k r_x^{2k}(0) = \int \omega^{2k} \mathrm{d}F(\omega)$, as

$$\begin{pmatrix} \omega_0 & 0 & -\omega_2 \\ 0 & \omega_2 & 0 \\ -\omega_2 & 0 & \omega_4 \end{pmatrix}.$$

Thus, the slope (= derivative) at a specified point is uncorrelated both with the process value at that particular point (but not, of course, with the process at neighboring points!), and with the curvature, while process value and curvature have negative correlation. For a Gaussian process Theorem1.3 gives

$$\begin{aligned} E(x''(0) \mid x(0) = u, x'(0) = z) &= -u\omega_2/\omega_0, \\ V(x''(0) \mid x(0), x'(0)) &= \omega_4 - \omega_2^2/\omega_0, \\ E(x''(t) \mid x(0) = u, x'(0) = z) &= ur''(t)/\omega_0 - zr'''(t)/\omega_2, \\ V(x''(t) \mid x(0), x'(0)) &= \omega_4 - r''(t)^2/\omega_0 - r'''(t)^2/\omega_2. \end{aligned}$$

2.2 Sample function continuity

2.2.1 *Countable and uncountable events*

The examples in the previous section show that we have a need for a stronger continuity and differentiability concept, valid for the sample functions, at least with probability one.

The first problem that we encounter with sample function continuity is that the event of interest, namely the set of continuous functions,

$$\mathbb{C} = \{x \in \mathbb{R}^T; x(\cdot) \text{ is a continuous function}\},$$

does not have a countable basis, and is not a Borel set, i.e., $\mathbb{C} \notin \mathscr{B}_T$. If $\{x(t), t \in T\}$ is a stochastic process on a probability space $(\Omega, \mathscr{F}, P)$, then the probability $P(\mathbb{C})$ need not be defined – it depends on the structure of $(\Omega, \mathscr{F})$ and on how complicated the process x is in itself, as a function on Ω. In particular, even if $P(\mathbb{C})$ is defined, it is not uniquely determined by the finite-dimensional distributions. To see this, take a process on a sufficiently rich sample space, i.e., one that contains enough sample points ω, and suppose we have defined a stochastic process $\{x(t), t \in \mathbb{R}\}$ that has, with probability one, continuous sample paths. Then x has a certain family of finite-dimensional distributions. Now, take a random time, τ, independent of x, and with continuous distribution, for example an exponential distribution,[1] and define a new

[1] Here it is necessary that Ω is rich enough so one can define an independent τ. For example, if x is defined on Ω_0 and we define τ as a random variable on a space Ω_1, then it is enough to take $\Omega = \Omega_0 \times \Omega_1$ and define a new probability measure on Ω by a product measure.

process $\{y(t), t \in \mathbb{R}\}$, such that

$$\begin{aligned} y(t) &= x(t), \qquad \text{for } t \neq \tau, \\ y(\tau) &= x(\tau) + 1. \end{aligned}$$

Then y has the same finite-dimensional distributions as x but its sample functions are always discontinuous at the random time point τ.

Equivalence

In the just constructed example, the two processes x and y differ only at a single point τ, and as we constructed τ to be random with continuous distribution, we have

$$P(x(t) = y(t)) = P(\tau \neq t) = 1, \quad \text{for all } t. \tag{2.8}$$

Two processes x and y which satisfy (2.8) are called *equivalent.* The sample paths of two equivalent process always coincide, with probability one, when observed at a fixed, pre-determined time point. (In the example above, the time τ where they differ is random.) Two processes that are equivalent have the same finite-dimensional distributions (Exercise 2:2).

Separability

The annoying fact that a stochastic process can fail to fulfill some natural regularity condition, such as continuity, even if it by all natural standards should be regular, can be partly neutralized by the concept of *separability*, introduced by J. Doob; see [41]. It uses the approximation by sets with countable basis mentioned in Section 1.2.4. Loosely speaking, a process $\{x(t), t \in \mathbb{R}\}$ is separable in an interval I if there exists a countable set of t-values, $T = \{t_k\} \subset I$, such that the process, with probability one, does not behave more irregularly on I than it does already on T. An important consequence is that for all t in the interior of I, there are sequences $\tau_1 < \tau_2 < \ldots \tau_n \uparrow t$ and $\tau_1' > \tau_2' > \ldots \tau_n' \downarrow t$ such that, with probability one,

$$\liminf_{n\to\infty} x(\tau_n) = \liminf_{\tau \uparrow t} x(\tau) \leq \limsup_{\tau \uparrow t} x(\tau) = \limsup_{n\to\infty} x(\tau_n),$$

with a similar set of relations for the sequence τ_n'. Hence, if the process is continuous on any discrete set of points in an interval then it is continuous in the entire interval. Every process has an equivalent separable version; see [41].

2.2.2 *Conditions for sample function continuity*

The finite-dimensional distributions of any two equivalent processes $\{x(t)\}$ and $\{y(t)\}$ are always the same. We shall now seek conditions on the finite-dimensional distribution functions, under which we can assume a stochastic process to have continuous sample paths. Conditions will be given both in terms of the bivariate distributions directly, and in terms of probabilistic bounds on the process increments. As we have seen, one has to be satisfied if, among all equivalent processes, one can find one which has continuous sample paths. The conditions for sample function continuity go back to Kolmogorov (proof published in [115]); we use the formulation and proof from [35].

We first formulate and prove a sufficient condition for sample function continuity that puts a probabilistic restriction on the size of the small oscillations of the process. Later in the chapter we will give examples for special processes, like Gaussian processes.

Theorem 2.4. *Let $\{x(t), 0 \le t \le 1\}$ be a real stochastic process. If there exist two non-decreasing functions, $g(h)$ and $q(h)$, $0 \le h \le 1$, such that*

$$\sum_{n=1}^{\infty} g(2^{-n}) < \infty, \quad \sum_{n=1}^{\infty} 2^n q(2^{-n}) < \infty,$$

and, for all $t < t+h$ in $[0,1]$,

$$P(|x(t+h) - x(t)| \ge g(h)) \le q(h), \tag{2.9}$$

then there exists an equivalent stochastic process $\{y(t), 0 \le t \le 1\}$, whose sample paths are continuous on $[0,1]$.

Proof. Start with the process x with given finite-dimensional distributions. Such a process exists, and what is questioned is whether it has continuous sample functions if its bivariate distributions satisfy the condition in the theorem. We shall now explicitly construct a process y, equivalent to x, and with continuous sample paths. Then y will automatically have the same finite-dimensional distributions as x. The process y shall be constructed as the continuous limit of a sequence of piecewise linear functions x_n, which have the

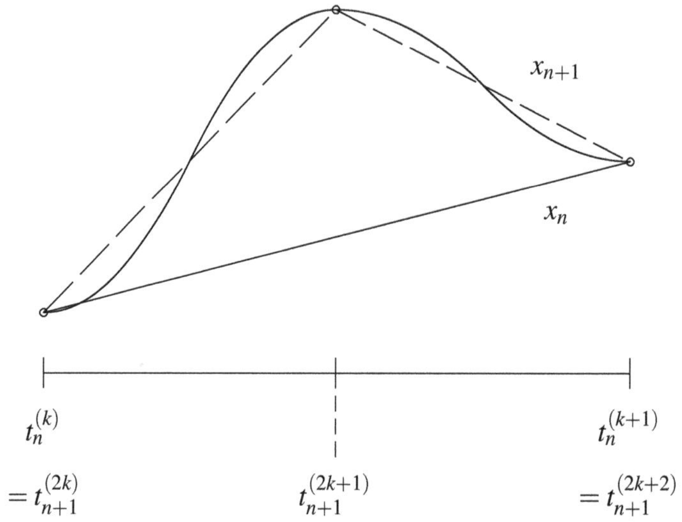

Figure 2.1: *Successive approximations with piecewise linear functions.*

correct distribution at the dyadic time points of order n,

$$t_n^{(k)} = k/2^n, \quad k = 0, 1, \ldots, 2^n; \quad n = 1, 2, \ldots.$$

Define the process x_n equal to x at the dyadic points,

$$x_n(t) = x(t), \text{ for } t = t_n^{(k)}, \, k = 0, 1, \ldots, 2^n,$$

and let it be linear between these points; see Figure 2.1.

Now, we can estimate the maximal distance between two successive approximations. As is obvious from the figure, the maximal difference between two successive approximations for t between $t_n^{(k)}$ and $t_n^{(k+1)}$ occurs in the middle of the interval, and hence

$$|x_{n+1}(t) - x_n(t)| \leq \left| x(t_{n+1}^{(2k+1)}) - \frac{1}{2}\left(x(t_n^{(k)}) + x(t_n^{(k+1)}) \right) \right|$$
$$\leq \frac{1}{2}\left| x(t_{n+1}^{(2k+1)}) - x(t_{n+1}^{(2k)}) \right| + \frac{1}{2}\left| x(t_{n+1}^{(2k+2)}) - x(t_{n+1}^{(2k+1)}) \right| = \frac{1}{2}A + \frac{1}{2}B,$$

say.

The tail distribution of the maximal difference between two successive approximations,

$$M_n^{(k)} = \max_{t_n^{(k)} \leq t \leq t_n^{(k+1)}} |x_{n+1}(t) - x_n(t)| \leq \frac{1}{2}A + \frac{1}{2}B,$$

can therefore be estimated by

$$P(M_n^{(k)} \geq c) \leq P(A \geq c) + P(B \geq c),$$

since if $M_n^{(n)} \geq c$, then either $A \geq c$ or $B \geq c$, or both.

Now, take $c = g(2^{-n-1})$ and use the bound (2.9) on the events A and B, to get

$$P(M_n^{(k)} \geq g(2^{-n-1})) \leq 2q(2^{-n-1}),$$

for each $k = 0, 1, \ldots, 2^n - 1$. By Boole's inequality[2] we get, since there are 2^n intervals,

$$P\left(\max_{0\leq t\leq 1} |x_{n+1}(t) - x_n(t)| \geq g(2^{-n-1})\right) = P\left(\bigcup_{k=0}^{2^n-1} M_n^{(k)} \geq g(2^{-n-1})\right)$$
$$\leq 2^{n+1} q(2^{-n-1}).$$

Now $\sum_n 2^{n+1} q(2^{-n-1}) < \infty$ by assumption, and then the Borel-Cantelli lemma (Section A.4.2 in Appendix A) gives that, with probability one, only finitely many of the events

$$\max_{0\leq t\leq 1} |x_{n+1}(t) - x_n(t)| \geq g(2^{-n})$$

occur for $n = 1, 2, \ldots$, (remember that $g(2^{-n}) \geq g(2^{-n-1})$). This means that there is a set Ω_0 with $P(\Omega_0) = 1$, such that for every outcome $\omega \in \Omega_0$, from some random integer N (depending of the outcome, $N = N(\omega)$) and onwards, $(n \geq N)$,

$$\max_{0\leq t\leq 1} |x_{n+1}(t) - x_n(t)| < g(2^{-n}).$$

First of all, this shows that there exists a limiting function $y(t)$ for all $\omega \in \Omega_0$; the condition (A.12) for almost sure convergence, given in Appendix A.4.3, says that $\lim_{n\to\infty} x_n(t)$ exists, with probability one.

It also shows that the convergence is uniform:[3] for $\omega \in \Omega_0$ and $n \geq N$, $m > 0$,

$$|x_{n+m}(t) - x_n(t)| \leq |x_{n+1}(t) - x_n(t)| + |x_{n+2}(t) - x_{n+1}(t)| + \ldots$$
$$+ |x_{n+m}(t) - x_{n+m-1}(t)|$$
$$\leq \sum_{j=0}^{m-1} g(2^{-n-j}) \leq \sum_{j=0}^{\infty} g(2^{-n-j}).$$

---

[2] I.e., the sub-additivity of P: $P(\cup A_k) \leq \sum P(A_k)$.

[3] I.e., $\sup_{[0,1]} |x_n(t) - y(t)| \to 0$, as $n \to \infty$. The uniform limit of a sequence of continuous functions is itself continuous; see page 294 and Definition A.6.

Letting $m \to \infty$, so that $x_{n+m}(t) \to y(t)$, and observing that the inequalities hold for all $t \in [0,1]$, we get that

$$\max_{0 \le t \le 1} |y(t) - x_n(t)| \le \sum_{j=0}^{\infty} g(2^{-n-j}) = \sum_{j=n}^{\infty} g(2^{-j}).$$

Since this bound tends to 0 as $n \to \infty$, we have the uniform convergence, and since all x_n are continuous functions, we also have that y is continuous for all $\omega \in \Omega_0$. For $\omega \notin \Omega_0$, define $y(t) \equiv 0$, making y a continuous function for all $\omega \in \Omega$.

It remains to prove that x and y are equivalent processes, i.e., $P(x(t) = y(t)) = 1$, for all $t \in [0,1]$. For that sake, take any $t \in [0,1]$ and find a sequence of dyadic numbers $t_n^{(k_n)} \to t$ such that

$$t_n^{(k_n)} \le t < t_n^{(k_n)} + 2^{-n}.$$

Since both $g(h)$ and $q(h)$ are non-decreasing, we have from (2.9),

$$\begin{aligned} P\left(\left|x(t_n^{(k_n)}) - x(t)\right| \ge g(2^{-n})\right) &\le P\left(\left|x(t_n^{(k_n)}) - x(t)\right| \ge g(t - t_n^{(k_n)})\right) \\ &\le q(t - t_n^{(k_n)}) \le q(2^{-n}). \end{aligned}$$

Adding over n gives

$$\sum_{n=1}^{\infty} P\left(\left|x(t_n^{(k_n)}) - x(t)\right| \ge g(2^{-n})\right) \le \sum_{n=1}^{\infty} q(2^{-n}) < \infty,$$

and it follows from the Borel-Cantelli lemma that it can happen only finitely many times that $|x(t_n^{(k_n)}) - x(t)| \ge g(2^{-n})$. Since $g(2^{-n}) \to 0$ as $n \to \infty$, we have proved that $x(t_n^{(k_n)}) \to x(t)$ with probability one. Further, since $y(t)$ is continuous, $y(t_n^{(k_n)}) \to y(t)$. But $x(t_n^{(k_n)}) = y(t_n^{(k_n)})$, and therefore the two limits are equal, with probability one, as was to be proved. □

The theorem says that for each process $\{x(t)\}$ that satisfies the conditions there exists at least one other equivalent process $\{y(t)\}$ with continuous sample paths, and with exactly the same finite-dimensional distributions. Of course it seems unnecessary to start with $x(t)$ and immediately change to an equivalent continuous process $y(t)$. In the future we assume that we only have the continuous version, whenever the sufficient conditions for sample function continuity are satisfied.

2.2.3 Moment conditions for continuity

Theorem 2.4 is simple to use, since it depends only on the distribution of the increments of the process, and involves only bivariate distributions. For special processes, conditions that put bounds on the moments of the increments are even simpler to use. One such condition is given in the following corollary.

Corollary 2.1. *If there exist constants C, and $r > p > 0$, such that, for all small enough $h > 0$,*

$$E\left(|x(t+h)-x(t)|^p\right) \le C\frac{|h|}{|\log|h||^{1+r}}, \tag{2.10}$$

then the condition in Theorem 2.4 is satisfied and the process has continuous sample paths.

Note, that many processes satisfy a stronger inequality than (2.10), namely

$$E\left(|x(t+h)-x(t)|^p\right) \le C|h|^{1+c}, \tag{2.11}$$

for some constants C, and $c > 0, p > 0$. Then (2.10) is automatically satisfied with any $r > p$, and the process has continuous sample paths.

Proof. Markov's inequality (a generalization of Chebyshev's inequality, see Exercise A:1 in Appendix A) states that for all random variables U, $P(|U| \ge \lambda) \le E(|U|^p)/\lambda^p$. Applying the theorem with $g(h) = |\log|h||^{-b}$, $1 < b < r/p$, one gets,

$$P\left(|x(t+h)-x(t)| > g(h)\right) \le \frac{C|h|}{|\log|h||^{1+r-bp}}.$$

Since $b > 1$, one has $\sum g(2^{-n}) = \sum \frac{1}{(n\log 2)^b} < \infty$, and, with

$$q(h) = C|h|/|\log|h||^{1+r-bp},$$

and $1+r-bp > 1$, one can bound the sum,

$$\sum 2^n q(2^{-n}) = \sum \frac{C}{(n\log 2)^{1+r-bp}} < \infty,$$

which proves the assertion. □

Example 2.6. *We show that the Wiener process $W(t)$ has, with probability one, continuous sample paths. In the standard Wiener process, the increment $w(t+h)-w(t), h>0$, is Gaussian with mean 0 and variance h. Thus,*

$$E(|w(t+h)-w(t)|^p) = C|h|^{p/2},$$

with $C = E(|U|^p) < \infty$, for a standard normal variable U, giving the moment bound

$$E(|w(t+h)-w(t)|^4) = C|h|^2 < \frac{|h|}{|\log|h||^6},$$

for small h. We see that condition (2.10) in the corollary is satisfied with $r = 5 > 4 = p$. Condition (2.11) is satisfied with $p = 3$, $c = 1/2$.

Continuity of stationary processes

In a stationary process with covariance function $r(t) = C(x(s+t), x(s))$, the increments have variance

$$E((x(t+h)-x(t))^2) = 2(r(0)-r(h)), \tag{2.12}$$

so it is clear that continuity conditions can be formulated in terms of the covariance function. Equivalent conditions can be formulated by means of the spectral distribution function $F(\omega)$, which is such that $r(t) = \int_{-\infty}^{\infty} e^{i\omega t}\, \mathrm{d}F(\omega)$.

A first consequence of (2.12) is, as we saw in Section 2.1, that $x(t+h) \overset{q.m.}{\to} x(t)$ as $h \to 0$ if and only if the covariance function $r(t)$ is continuous for $t = 0$. For sample function continuity, a sufficient condition in terms of the covariance function follows directly from Corollary 2.1.

Theorem 2.5. *If the covariance function $r(t)$ of a stationary stochastic process $\{x(t), t \in \mathbb{R}\}$ is such that, as $t \to 0$,*

$$r(t) = r(0) - O\left(\frac{|t|}{|\log|t||^a}\right), \tag{2.13}$$

for some $a > 3$, then $x(t)$ has continuous sample functions.[4]

[4]The "big ordo" notation $f(x) = g(x) + O(h(x))$ as $x \to 0$ means that $|(f(x)-g(x))/h(x)|$ is bounded by some finite constant C as $x \to 0$.

2.2.4 *Continuity of stationary Gaussian processes*

For stationary Gaussian processes the conditions for sample function continuity can be considerably weakened, to require slightly more that just continuity of the covariance function. We state the sufficient condition both in terms of the covariance function and in terms of the spectrum, and to this end, we formulate an analytic lemma by Belyaev [8, 9], the proof of which can be found in [35, Sect. 9.3]. Note, in particular, the weak restriction (2.14) on the amount of high frequency content in the process.

Lemma 2.2. *a) If, for some* $a > 0$,

$$\int_0^\infty (\log(1+\omega))^a \, \mathrm{d}F(\omega) < \infty, \tag{2.14}$$

then, for any $b \leq a$,

$$r(t) = r(0) - O(|\log|t||^{-b}), \quad \text{as } t \to 0. \tag{2.15}$$

b) If (2.15) *holds for some* $b > 0$, *then* (2.14) *holds for any* $a < b$.

Theorem 2.6. *A stationary Gaussian process* $\{x(t), t \in \mathbb{R}\}$ *has continuous sample paths if, for some* $a > 3$, *any of the following conditions is satisfied:*

$$r(t) = r(0) - O(|\log|t||^{-a}), \quad \text{as } t \to 0, \tag{2.16}$$

$$\int_0^\infty (\log(1+\omega))^a \, \mathrm{d}F(\omega) < \infty. \tag{2.17}$$

Proof. In a stationary Gaussian process, $x(t+h) - x(t)$ is normal with mean zero and variance $\sigma_h^2 = 2(r(0) - r(h))$, and, when (2.16) holds,

$$\sigma_h \leq \frac{C}{|\log|h||^{a/2}},$$

for some constant $C > 0$. Writing $\Phi(x) = \int_{-\infty}^x \phi(y)\,\mathrm{d}y = \frac{1}{\sqrt{2\pi}} \int_{-\infty}^x e^{-y^2/2}\,\mathrm{d}y$ for the standard normal distribution function, we have

$$P(|x(t+h)-x(t)|>g(h)) = 2\left\{1-\Phi\left(\frac{g(h)}{\sigma_h}\right)\right\}.$$

Now, take $g(h) = |\log|h|/\log 2|^{-b}$, where b is chosen so that $1 < b < (a-1)/2$, which is possible since $a > 3$ by assumption. From the simple bound $1-\Phi(x) \leq \phi(x)/x$, of the normal distribution tail, we then get

$$P(|x(t+h)-x(t)|>g(h)) \leq 2\left\{1-\Phi\left(\frac{g(h)|\log|h||^{a/2}}{C}\right)\right\}$$

$$\leq \frac{2C}{g(h)|\log|h||^{a/2}}\,\phi\left(\frac{g(h)|\log|h||^{a/2}}{C}\right) = q(h), \text{ say.}$$

The reader should complete the proof and show that

$$\sum g(2^{-n}) < \infty \quad \text{and} \quad \sum 2^n q(2^{-n}) < \infty,$$

and then apply Theorem 2.4 to see that (2.16) is sufficient. Lemma 2.2 shows that also (2.17) is sufficient for sample function continuity for a Gaussian process. □

Example 2.7. *Any stationary process with*

$$r(t) = r(0) - C|t|^\alpha + o(|t|^\alpha),$$

for some $C > 0$, as $t \to 0$, has continuous sample paths if $1 < \alpha \leq 2$.[5] *If it furthermore is a Gaussian process it is continuous if $0 < \alpha \leq 2$. The random telegraph signal from Example 2.3 is of this type with $\alpha = 1$ and is not continuous.*

Remark 2.1. *The sufficient conditions for sample function continuity given in the theorems are satisfied for almost all covariance functions that are encountered in applied probability. But for Gaussian processes, even the weak condition (2.16) for $a > 3$ can be relaxed to require only that $a > 1$, which is very close to being necessary; see [35, Sect. 9.5] for a series of weak conditions.*

Gaussian stationary processes that are not continuous behave with necessity very badly, and their sample functions are, with probability one, unbounded in any interval. This was shown by Yu.K. Belyaev [10], but is also a consequence of a theorem by Dobrushin [40]; see also [35, Ch. 9.5].

---

[5] At this stage you should convince yourself that $\alpha > 2$ is impossible! Hint: Consider the analogy between characteristic function and covariance function, and between a spectral density and a probability density, and then take a look at Theorem A.8.

2.2.5 *Probability measures on* $\mathbb{C}[0,1]$

We can now complete the table at the end of Section 1.2.2 and define probabilities on $\mathbb{C}[0,1]$, the space of continuous functions on the interval $[0,1]$. A stochastic process $\{x(t), 0 \le t \le 1\}$ on a probability space $(\Omega, \mathscr{F})$ has its realizations in $\mathbb{R}^{[0,1]}$. If the finite-dimensional distributions of x satisfy any of the sufficient conditions for sample function continuity, either x or an equivalent process y has continuous sample functions, and hence has its realizations in $\mathbb{C}[0,1] \subset \mathbb{R}^{[0,1]}$.

In the same way as the finite-dimensional distributions define a probability measure on $(\mathbb{R}^{[0,1]}, \mathscr{B}_{[0,1]})$, assigning probabilities to all Borel sets, we can now define a probability measure on $\mathbb{C}[0,1]$. The question is, what is the σ-field of events which get probability? In fact, we can take the simplest choice,

$$\mathscr{B} = \mathscr{B}_{[0,1]} \cap \mathbb{C}[0,1] = \{B \cap \mathbb{C}[0,1]; B \in \mathscr{B}_{[0,1]}\},$$

i.e., take those parts of the Borel sets which intersect $\mathbb{C}[0,1]$. Thus, we have the following existence theorem.

Theorem 2.7. *If* $\{F_{\mathbf{t}_n}\}_{n=1}^{\infty}$ *is a family of finite-dimensional distributions that satisfy any of the sufficient conditions for sample functions continuity, then there exists a probability measure on* $(\mathbb{C}[0,1], \mathscr{B})$ *such that the co-ordinate process* $\{x(t), 0 \le t \le 1\}$ *has the given finite-dimensional distributions.*

The family $\mathscr{B}$ can be described alternatively in terms of open sets in $\mathbb{C}[0,1]$. Take a continuous function $x(t) \in \mathbb{C}[0,1]$. By an ε-surrounding of x we mean the set of functions which are in an ε-band around x,

$$\mathscr{S}(x,\varepsilon) = \{y \in \mathbb{C}[0,1]; \max_{0 \le t \le 1} |y(t) - x(t)| < \varepsilon\}. \tag{2.18}$$

A set of functions $A \subseteq \mathbb{C}[0,1]$ is called *open* if for every $x \in A$, there is an ε-surrounding of x that is completely in A. The open sets in $\mathbb{C}[0,1]$ generate a σ-field, the smallest σ-field that contains all open sets, and that σ-field is exactly $\mathscr{B}$.

Remark 2.2 (Convergence of probability measures on $\mathbb{C}$). *The introduction of open sets from the maximum norm in* (2.18) *makes it possible to study continuous functionals on* $\mathbb{C}$. *For example* $\max_{[0,1]} x(t)$ *is a continuous functional, since if two functions* $x(t)$ *and* $y(t)$ *are uniformly close, then their maxima* *are also close. An important question is then, if $x_n(t)$ is a sequence of processes with finite-dimensional distributions converging to those of a limiting process $x(t)$, does that imply the convergence in distribution of, for example, $\max_{[0,1]} x(t)$? The reader is encouraged to construct a simple example to the contrary; Exercise 2.19.*

To determine if finite-dimensional convergence implies distributional convergence of continuous functionals one has to impose the constraint on the processes $x_n(t)$ that their family of distributions is tight, *in essence, that the continuity modulii for the processes are uniformly small; [14, Thm. 8.2]. For more on this problem the reader is referred to the book by P. Billingsley [14].*

2.3 Derivatives, tangents, and other characteristics

2.3.1 *Differentiability – general conditions*

When does a continuous stochastic process have continuously differentiable sample functions? The answer can be given as conditions similar to those for sample function continuity, but now with bounds on the second order differences. By pasting together piecewise linear approximations by means of smooth arcs, one can prove the following theorem; see [35, Sect. 4.3].

Theorem 2.8. *a) Suppose the stochastic process $\{x(t), 0 \le t \le 1\}$ satisfies the conditions for sample function continuity in Theorem 2.4. If, furthermore, for all $t-h < t+h$ in $[0,1]$,*

$$P(|x(t+h) - 2x(t) + x(t-h)| \ge g_1(h)) \le q_1(h), \tag{2.19}$$

where g_1 and q_1 are two non-decreasing functions, such that

$$\sum_{n=1}^{\infty} 2^n g_1(2^{-n}) < \infty \quad \text{and} \quad \sum_{n=1}^{\infty} 2^n q_1(2^{-n}) < \infty,$$

then there exists an equivalent process $\{y(t), 0 \le t \le 1\}$ with continuously differentiable sample paths.

b) The sufficient condition in (a) is satisfied if

$$E(|x(t+h) - 2x(t) + x(t-h)|^p) \le \frac{K|h|^{1+p}}{|\log|h||^{1+r}}, \tag{2.20}$$

for some constants $p < r$ and K.

Remark 2.3. *Many processes satisfy a stronger inequality than (2.20), namely*

$$E\left(|x(t+h)-2x(t)+x(t-h)|^p\right) \le C|h|^{1+p+c}, \tag{2.21}$$

for some constants C, and $c > 0$, $p > 0$. Then (2.20) is satisfied, and the process has continuously differentiable sample paths.

Q.m. and sample function derivative

In Section 2.1 we considered conditions for quadratic mean (q.m.) differentiability of a stationary process. One may ask: What is the relation between the q.m.-derivative and the sample function derivative? They are both limits of the differential quotient $(x(t+h)-x(t))/h$ as $h \to 0$. Now, it is easy to prove that if the limit exists in both quadratic mean and as a sample function limit with probability one, then the two limits are equal (also with probability one); from the q.m. convergent sequence, select, by Theorem A.6, a subsequence that converges, with probability one, to the same limit. One concludes that the two derivative processes are equivalent, and have the same finite-dimensional distributions and the same covariance function.

In fact, if the covariance function $r_x(t)$ is twice differentiable, which is necessary for q.m.-differentiability, then condition (2.11) for sample function continuity is satisfied. This can, of course, be extended to higher derivatives. If the derivative of order $k+1$ exists as a q.m.-derivative, then the sample functions of the kth order derivatives are continuous.

2.3.2 *Differentiable Gaussian processes*

In order that a stationary process has continuously differentiable sample functions it is necessary that its covariance function has a smooth higher order Taylor expansion. We give here the conditions for a Gaussian process. Without the Gaussian assumption a slightly stricter condition is needed, just as is the case for continuity, cf (2.13) and (2.16).

As we shall see in the following theorem, demanding just slightly more than a finite second spectral moment guarantees that a Gaussian process has continuously differentiable sample paths. The proof relies on some useful relations between the covariance function and the spectrum, proved by Belyaev [8, 9], similar to those in Lemma 2.2. For a proof, and for further extensions, the reader is referred to [35, Sect. 9.3]. For easy reference, we give the extended version here.

Lemma 2.3. *a) If, for some $a > 0$,*

$$\int_0^\infty \omega^2 (\log(1+\omega))^a \, dF(\omega) < \infty, \tag{2.22}$$

then $r(t)$ has a second derivative $r''(t)$ satisfying

$$r''(t) = -\omega_2 + O\left(\frac{1}{|\log|t||^b}\right), \quad \text{as } t \to 0, \tag{2.23}$$

for any $b \le a$.

b) If $r(t)$ has a second derivative $r''(t)$ satisfying (2.23) for some $b > 0$, then (2.22) is satisfied for any $a < b$.

c) If (2.22) holds for some $a > 0$, then

$$r(t) = r(0) - \frac{\omega_2 t^2}{2} + O\left(\frac{t^2}{|\log|t||^b}\right) \tag{2.24}$$

holds for any $b \le a$. Conversely, if (2.24) holds for some $b > 0$ and some constant ω_2, then ω_2 is the second spectral moment and (2.22) holds for any $a < b$.

Theorem 2.9. *A stationary Gaussian process is continuously differentiable if, for some $a > 3$, any of the following conditions are satisfied:*

$$r(t) = r(0) - \frac{\omega_2 t^2}{2} + O\left\{\frac{t^2}{|\log|t||^a}\right\}, \tag{2.25}$$

$$\int_0^\infty \omega^2 (\log(1+\omega))^a \, dF(\omega) < \infty, \tag{2.26}$$

where $\omega_2 = -r''(0) = \int_{-\infty}^{\infty} \omega^2 \, dF(\omega) < \infty$.

As for sample function continuity, it suffices that the constant a is greater than 1 in order to have sample function differentiability for Gaussian processes.

Example 2.8. *Here is an example that shows that correlation conditions* *alone are not sufficient to determine continuity or differentiability, since, as we will show, $r(t) = e^{-\alpha|t|}$ can be the covariance function for discontinuous, continuous but non-differentiable, and differentiable stationary processes – the sample function properties will depend on the complete distribution, not only on the correlation function.*

The common exponential covariance function $r(t) = e^{-\alpha|t|}$ admits an expansion $r(t) = 1 - \alpha|t| + o(t^2)$ near the origin. It is not differentiable at the origin, and hence not twice differentiable either.

The random telegraph signal with this covariance function is not Gaussian and it has discontinuous sample functions. The Ornstein-Uhlenbeck process is Gaussian and continuous but its sample functions are nowhere differentiable. Finally, consider the following process,

$$x(t) = Y\cos(t\Gamma + \phi), \tag{2.27}$$

where Y, Γ, ϕ are independent random variables, $E(Y) = 0, E(Y^2) = 2$, and ϕ uniformly distributed in $[0, 2\pi]$. This process has sample functions that are pure cosine functions, and hence are infinitely differentiable. The task in Exercise 3.14 is to show that there is a distribution for Γ, namely the Cauchy distribution, that makes the covariance function of $x(t)$ equal to $e^{-\alpha|t|}$.

We encourage the reader to prove the following result that gives sufficient condition in terms of $-r''(t)$, the covariance function of the derivative.

Theorem 2.10. *A stationary process $\{x(t), t \in \mathbb{R}\}$ is continuously differentiable if any of the following two conditions[6] hold as $t \to 0$,*

a) $-r''(t) = -r''(0) - C|t|^\alpha + o(|t|^\alpha)$ *with* $1 < \alpha \le 2$,

b) it is Gaussian and $-r''(t) = -r''(0) - C|t|^\alpha + o(|t|^\alpha)$ *with* $0 < \alpha \le 2$.

2.3.3 *Jump discontinuities, Hölder conditions, and tangents*

What type of discontinuities are possible for processes which do not have continuous sample functions? For example, the Poisson process can be defined as a piecewise constant process with unit jumps after independent, exponential waiting times. Obviously, such a process has sample functions that increase

[6]The "little ordo" notation $f(x) = g(x) + o(h(x))$ as $x \to 0$ means that $|f(x) - g(x)|/h(x) \to 0$ as $x \to 0$.

only with jumps of size one! But suppose we don't know the constructive definition, but are only presented the finite-dimensional distributions. Could we still conclude that it has only jump discontinuities?

We would like to have a condition that guarantees that only simple discontinuities are possible, and such a condition exists, with restriction on the increment over two adjacent intervals. Similarly, for a process that is not continuously differentiable, how far from differentiable are the sample functions? We will answer that question by means of a probabilistic statement about the continuity modulus. As a byproduct we can prove a general lemma about the existence of certain tangents, useful in Chapter 8, on level crossings.

Jump discontinuities

The proof of the following theorem is indicated in [35, Sect. 4.4].

Theorem 2.11. *If there are positive constants C, p, r, such that for all s, t with $0 \le t < s < t+h \le 1$,*

$$E\left\{|x(t+h) - x(s)|^p \cdot |x(s) - x(t)|^p\right\} \le C|h|^{1+r}, \tag{2.28}$$

then the process $\{x(t), 0 \le t \le 1\}$ has, with probability one, sample functions with at most jump discontinuities, i.e., $\lim_{t \downarrow t_0} x(t)$ and $\lim_{t \uparrow t_0} x(t)$ exist for every $t_0 \in [0,1]$.

Example 2.9. *The Poisson process $\{N(t), 0 \le t < \infty\}$ with intensity λ has independent increments and, hence,*

$$\begin{aligned} &E\left(|N(t+h) - N(s)|^2 \cdot |N(s) - N(t)|^2\right) \\ &= E\left(|N(t+h) - N(s)|^2\right) \cdot E\left(|N(s) - N(t)|^2\right) \\ &= (\lambda(t+h-s) + (\lambda(t+h-s))^2) \cdot (\lambda(s-t) + (\lambda(s-t))^2) \le C\lambda^2 h^2. \end{aligned}$$

The conditions of the theorem are obviously satisfied.

Remark 2.4. *In Section 2.2.5 we discussed how to define a probability measure on the space $\mathscr{C}[0,1]$ of continuous functions on the interval $[0,1]$. By the same means it is possible to define probability measures on the space $\mathscr{D}[0,1]$ of function with right continuous functions with left limits; see [14].*

Hölder continuity and the continuity modulus

How large are the increments in a non-differentiable process? In fact, moment bounds on the increments give precise estimates for the distribution of the *continuity modulus* $\omega_x(h)$ of the sample functions, defined as

$$\omega_x(h) = \sup_{|s-t| \leq h} |x(s) - x(t)|, \tag{2.29}$$

where the supremum is taken over $0 \leq s,t \leq 1$.

Definition 2.4 (Hölder continuity). *Functions for which there exist constants A, a, such that for all $t, t+h \in [0,1]$,*

$$|x(t+h) - x(t)| \leq A|h|^a,$$

are said to be Hölder continuous of order a.[7]

Next, we shall present stochastic estimates of the continuity modulus for a stochastic process, and also give a sufficient condition for a stochastic process to be Hölder continuous.

Theorem 2.12. *If there are constants C, $p \geq r > 0$, such that*

$$E\{|x(t+h) - x(t)|^p\} \leq C|h|^{1+r}, \tag{2.30}$$

then, with probability one, $\sup_{h>0} \omega_x(h)/h^a$ is finite for all $a < r/p$.

Proof. First examine the increments over the dyadic numbers

$$t_n^{(k)} = k/2^n, \quad k = 0, 1, \ldots, 2^n; \quad n = 1, 2, \ldots.$$

Take an $a < r/p$ and write $\delta = 2^{-a}$. Then, by Markov's inequality (Exercise A:1),

$$P\left(\left|x(t_n^{(k+1)}) - x(t_n^{(k)})\right| > \delta^n\right) \leq \frac{E\left(\left|x(t_n^{(k+1)}) - x(t_n^{(k)})\right|^p\right)}{\delta^{(np)}}$$

$$\leq \frac{C(2^{-n})^{1+r}}{2^{-anp}} = \frac{C}{2^{n(1+r-ap)}}.$$

[7]The term "Lipschitz continuous" is also used.

From Boole's inequality, summing over n, we obtain in succession,

$$P(\max_{0\le k\le 2^n-1}|x(t_n^{(k+1)})-x(t_n^{(k)})|>\delta^n)\le \frac{C2^n}{2^{n(1+r-ap)}}=\frac{C}{2^{n(r-ap)}},$$

$$\sum_{n=0}^{\infty}P(\max_{0\le k\le 2^n-1}|x(t_n^{(k+1)})-x(t_n^{(k)})|>\delta^n)\le \sum_{n=0}^{\infty}\frac{C}{2^{n(r-ap)}}<\infty,$$

since $r-ap>0$. The Borel-Cantelli lemma gives that only finitely many events

$$A_n=\{\max_{0\le k\le 2^n-1}|x(t_n^{(k+1)})-x(t_n^{(k)})|>\delta^n\},\quad n=1,2,\ldots,$$

occur, which means that there exists a random index ν such that for all $n>\nu$,

$$|x(t_n^{(k+1)})-x(t_n^{(k)})|\le\delta^n,\quad \text{for all } k=0,1,\ldots,2^n-1. \tag{2.31}$$

Next, we estimate the increment from a dyadic point $t_n^{(k)}$ to an arbitrary point t. To that end, take $t\in[t_n^{(k)},t_n^{(k+1)})$, and consider its dyadic expansion, ($\alpha_m=0$ or 1),

$$t=t_n^{(k)}+\sum_{m=1}^{\infty}\frac{\alpha_m}{2^{n+m}}.$$

Summing all the inequalities (2.31), we obtain that the increment from $t_n^{(k)}$ to t is bounded (for $n>\nu$),

$$|x(t)-x(t_n^{(k)})|\le\sum_{m=1}^{\infty}\delta^{n+m}=\frac{\delta^{n+1}}{1-\delta}. \tag{2.32}$$

The final estimate relates $t+h$ to the dyadic points. Let $\nu<\infty$ be the random index just found to exist. Then, suppose $h<2^{-\nu}$ and find n,k such that $2^{-n}\le h<2^{-n+1}$ and $k/2^n\le t<(k+1)/2^n$. We see that $n>\nu$ and that

$$t_n^{(k)}\le t<t_n^{(k+1)}<t+h\le t_n^{(k+\ell)},$$

where ℓ is either 2 or 3. As above, we obtain

$$|x(t+h)-x(t_n^{(k+1)})|\le\frac{\delta^{n+1}}{1-\delta}+\delta^n. \tag{2.33}$$

Summing the three estimates (2.31-2.33), we see that

$$|x(t+h)-x(t)|\le\delta^n+\frac{\delta^{n+1}}{1-\delta}+\frac{\delta^{n+1}}{1-\delta}+\delta^n=\frac{2}{1-\delta}(2^{-n})^a\le\frac{2}{1-\delta}h^a,$$

for $2^{-n} \le h < 2^{-\nu}$. For $h \ge 2^{-\nu}$ it is always true that

$$|x(t+h) - x(t)| \le M \le \frac{M}{2^{-\nu}} h^a,$$

for some random M. Take $A = \max(M2^{\nu}, 2/(1-\delta))$, and then complete the proof by combining the last two inequalities, to get $|x(t+h) - x(t)| \le Ah^a$, with the random constant $A < \infty$. □

Example 2.10. *For the Wiener process,*

$$E(|x(t+h) - x(t)|^p) = C_p|h|^{p/2} = C_p|h|^{1+(p/2-1)},$$

so the Wiener process is Hölder continuous of order $a < (p/2-1)/p = 1/2 - 1/p$ for every $p > 0$. This means that it is Hölder continuous of all orders $a < 1/2$. It can be shown that the sample functions of the Wiener process are everywhere non-differentiable.

In the next section we shall investigate the non-existence of tangents of a predetermined level. We shall then need a small lemma on the size of the continuity modulus of a continuous stochastic process.

Theorem 2.13. *Let $\{x(t), t \in \mathbb{R}\}$ be a stochastic process with continuous sample functions on $0 \le t \le 1$, and let $\omega_x(h)$ be its (random) continuity modulus, defined by (2.29). Then, to every $\varepsilon > 0$, there is a (deterministic) function $\omega_\varepsilon(h)$, such that $\omega_\varepsilon(h) \downarrow 0$ as $h \downarrow 0$, and*

$$P(\omega_x(h) < \omega_\varepsilon(h) \text{ for } 0 < h \le 1) > 1 - \varepsilon.$$

Proof. Every sample function $x(t)$ is continuous on $[0,1]$ and is therefore also uniformly continuous on that interval, i.e., its continuity modulus tends to 0 for $h \to 0$,

$$\lim_{h \to 0} P(\omega_x(h) < c) = 1$$

for every fixed $c > 0$. Take a sequence $c_1 > c_2 \ldots > c_n \downarrow 0$. For a given $\varepsilon > 0$, we can find a decreasing sequence $h_n \downarrow 0$ such that $P(\omega_x(h_n) < c_n) > 1 - \varepsilon/2^{n+1}$. Since $\omega_x(h)$ is non-increasing as h decreases, then also

$$P(\omega_x(h) < c_n \text{ for } 0 < h \le h_n) > 1 - \varepsilon/2^{n+1},$$

for $n = 1, 2, \ldots$. Summing the exceptions, we get that

$$P(\omega_x(h) < c_n \text{ for } 0 < h \leq h_n, n = 1, 2, \ldots) > 1 - \sum_{n=1}^{\infty} \varepsilon/2^{n+1} = 1 - \varepsilon/2. \tag{2.34}$$

Now we can define the deterministic function $\omega_\varepsilon(h)$ from the sequences c_n and h_n. Let $h_0 = 1$ and take $\omega_\varepsilon(n) = c_n$ for $h_{n+1} < h \leq h_n$, $n = 0, 1, 2, \ldots$. If we take c_0 large enough to make $P(\omega_x(h) < c_0 \text{ for } h_1 < h \leq 1) > 1 - \varepsilon/2$, and combine with (2.34) we get the desired estimate. □

A lemma about tangents

We start with a theorem due to E.V. Bulinskaya, 1961 [23], on the non-existence of tangents of a pre-specified level.

Theorem 2.14 (Bulinskaya's lemma). *Suppose the density $f_t(x)$ of $x(t)$ is bounded in the interval $0 \leq t \leq 1$, $f_t(x) \leq c_0 < \infty$, and that $x(t)$ has, with probability one, continuously differentiable sample paths. Then, for any level u,*

a) the probability is zero that there exists a $t \in [0,1]$ such that simultaneously $x(t) = u$ and $x'(t) = 0$, i.e., there exists no points where $x(t)$ has a horizontal tangent line on the level u in $[0,1]$,

b) there are only finitely many $t \in [0,1]$ for which $x(t) = u$.

Proof. a) By assumption, $x(t)$ has continuously differentiable sample paths. We identify the location of those t-values for which $x'(t) = 0$ and $x(t)$ is close to u. For that sake, take an integer n and a constant $h > 0$, let H_τ be the event

$$H_\tau = \{x'(\tau) = 0\} \cap \{|x(\tau) - u| \leq h\},$$

and define, for $k = 1, 2, \ldots, n$,

$$\begin{aligned} A_h(k,n) &= \{H_\tau \text{ occurs for at least one } \tau \in \left[\tfrac{k-1}{n}, \tfrac{k}{n}\right]\}, \\ A_h &= \{H_\tau \text{ occurs for at least one } \tau \in [0,1]\} = \cup_{k=1}^{n} A_h(k,n). \end{aligned}$$

Now, take a sample function that satisfies the conditions for $A_h(k,n)$ and let $\omega_{x'}$ be the continuity modulus of its derivative. For such a sample function, the mean value theorem implies that

$$x(k/n) = x(\tau) + (k/n - \tau)x'(\tau + \theta(k/n - \tau)),$$

for some $\theta, 0 \le \theta \le 1$, and hence, on $A_h(k,n)$,

$$|x(k/n) - u| \le h + n^{-1}\omega_{x'}(n^{-1}). \tag{2.35}$$

We now use Theorem 2.13 to bound $\omega_{x'}$. For any function $\omega(t) \downarrow 0$ as $t \downarrow 0$, define B_ω to be the set of sample functions for which $\omega_{x'}(t) \le \omega(t)$ for all t in $[0,1]$. By the theorem, given $\varepsilon > 0$, there exists at least one function $\omega_\varepsilon(t) \downarrow 0$ such that $P(B_{\omega_\varepsilon}) > 1 - \varepsilon/2$. For outcomes satisfying (2.35) we use the bound ω_ε, and obtain

$$\begin{aligned} P(A_h) &\le \sum_{k=1}^{n} P(A_h(k,n) \cap B_{\omega_\varepsilon}) + (1 - P(B_{\omega_\varepsilon})) \\ &\le \sum_{k=1}^{n} P(|x(k/n) - u| \le h + n^{-1}\omega_\varepsilon(n^{-1})) + \varepsilon/2 \\ &\le 2nc_0(h + n^{-1}\omega_\varepsilon(n^{-1})) + \varepsilon/2, \end{aligned}$$

where c_0 is the bounding constant for the density $f_t(x)$.

Since $\omega_\varepsilon(t) \to 0$ as $t \to 0$, we can select first an n_0 and then an h_0 to make $P(A_{h_0}) \le \varepsilon$. But if there exists a time point t for which $x'(t) = 0$ and $x(t) = u$, simultaneously, then certainly A_h has occurred for any $h > 0$, and the event of interest has probability less than ε, which was arbitrary. The probability of simultaneous occurrence is therefore 0 as stated.

b) To prove that there are only a finite number of u-values in $[0,1]$, assume, on the contrary, that there is an infinite sequence of points $t_i \in [0,1]$ with $x(t_i) = u$. There is then at least one limit point $t_0 \in [0,1]$ of $\{t_i\}$ for which, by continuity, $x(t_0) = u$. Since the derivative of $x(t)$ is assumed continuous, we also have $x'(t_0) = 0$, and we have found a point where $x(t_0) = u$, $x'(t_0) = 0$. By (a), that event has probability 0. □

2.3.4 *Continuity conditions based on entropy*

In our treatment of sample function continuity, Theorem 2.4 and Theorems 2.12–2.13, we exploited the linear structure of the real line, to restrict the oscillation $|x(t+h) - x(t)|$ of the process over a small interval, typically of length $h = 2^{-n}$.

It is possible to formulate these results in a form that can easily be generalized to random fields with parameter space $\mathbf{T} = [0,1]^p$ and to an even more general space. Instead of the Euclidian distance $e(\mathbf{s},\mathbf{t}) = |\mathbf{s}-\mathbf{t}| = \sqrt{\sum(s_k - t_k)^2}$, the restriction is formulated directly in terms of the expected size of the oscillations,

$$d(\mathbf{s},\mathbf{t}) = \sqrt{E(|x(\mathbf{s}) - x(\mathbf{t})|^2)},$$

which is used as a new distance measure.

Let $\{\mathbf{s} \in \mathbf{T}; d(\mathbf{s},\mathbf{t}) < \varepsilon\}$ be an open ball in the d-metric, centered at $\mathbf{t}$ and with d-radius ε. For $\varepsilon > 0$, let $N(\mathbf{T},\varepsilon)$ be the *entropy* of $\mathbf{T}$, defined as the smallest number of balls with radius ε that is needed to cover $\mathbf{T}$. Define the log-entropy function

$$H(\varepsilon) = \log N(\mathbf{T},\varepsilon).$$

The following theorem restricts the random d-continuity modulus

$$\omega_{x,d}(\delta) = \sup_{\mathbf{s},\mathbf{t}\in\mathbf{T}, d(\mathbf{s},\mathbf{t})<\delta} |x(\mathbf{s}) - x(\mathbf{t})|,$$

by means of the log-entropy $H(\varepsilon)$. It is a variant of [3, Thm. 1.3.5]; the proof contains the same elements as the proof of our Theorem 2.12.

Theorem 2.15. *Let $\{x(\mathbf{t}); \mathbf{t} \in \mathbf{T}\}$ be a Gaussian random field with $E(x(\mathbf{t})) = 0$, and suppose $\mathbf{T}$, for each $\varepsilon > 0$, can be covered by a finite number of balls with d-radius ε. Then, there exists a random variable $\eta < \infty$, and a constant K, such that*

$$\omega_{x,d}(\delta) \le K \int_0^\delta \sqrt{H(\varepsilon)}\, d\varepsilon,$$

for all $\delta < \eta$.

While the general Theorem 2.4 has its roots in Kolmogorov's work in the 1930's (see Slutsky [115]), the entropy formulations for Gaussian processes are based on the work by Fernique in the 1960's; see [49].

Summary of smoothness conditions The following table summarizes some crude and simple sufficient conditions for different types of quadratic mean and sample function smoothness.

Condition on $r(t)$ as $t \to 0$	further condition	property
$r(t) = r(0) - o(1)$		q.m. cont.
$r(t) = r(0) - C\|t\|^\alpha + o(\|t\|^\alpha)$	$1 < \alpha \le 2$	a.s. cont.
$r(t) = r(0) - C\|t\|^\alpha + o(\|t\|^\alpha)$	$0 < \alpha \le 2$ Gaussian	a.s. cont.
$r(t) = r(0) - \omega_2 t^2/2 + o(t^2)$		q.m. diff.
$-r''(t) = \omega_2 - C\|t\|^\alpha + o(\|t\|^\alpha)$	$1 < \alpha \le 2$	a.s. diff.
$-r''(t) = \omega_2 - C\|t\|^\alpha + o(\|t\|^\alpha)$	$0 < \alpha \le 2$ Gaussian	a.s. diff.

2.4 Stochastic integration

In this section we shall define the two simplest types of stochastic integrals, of the form

$$J_1 = \int_a^b g(t)x(t)\,\mathrm{d}t, \qquad J_2 = \int_a^b g(t)\,\mathrm{d}x(t),$$

where $g(t)$ is a deterministic function and $x(t)$ a stochastic process with mean 0. The integrals can be defined either as quadratic mean limits of approximating Riemann or Riemann-Stieltjes sums, and depending on the type of convergence we require, the process $x(t)$ has to satisfy suitable regularity conditions.

The two types of integrals are sufficient for our needs in this work, in particular for the simple linear stochastic differential equations we shall study in Chapter 4. A third type of stochastic integrals, needed for more general stochastic differential equations, are those of the form

$$J_3 = \int_a^b g(t, x(t))\,\mathrm{d}x(t),$$

in which the integrand g and the integrator $x(t)$ are both random and dependent. These will not be dealt with here; see, e.g., [27].

The integrals are defined as limits in quadratic mean of the approximating sums

$$J_1 = \lim_{n\to\infty} \sum_{k=1}^{n} g(t_k)x(t_k)(t_k - t_{k-1}),$$

$$J_2 = \lim_{n\to\infty} \sum_{k=1}^{n} g(t_k)(x(t_k) - x(t_{k-1})),$$

when $a = t_0 < t_2 < \ldots < t_n = b$, and $\max|t_k - t_{k-1}| \to 0$ as $n \to \infty$, provided the limits exist, and are independent of the subdivision $\{t_k\}$. To simplify the writing we have suppressed the double index in the sequences of t_k-values; there is one sequence for each n.

Assume $E(x(t)) = 0$. The limits exist as quadratic mean limits if the corresponding integrals of the covariance function $r(s,t) = C(x(s), x(t)) = E(x(s)\overline{x(t)})$ are finite, as formulated in the following theorem. Since we shall mainly use the integrals together with the complex function $g(t) = e^{i\omega t}$, we formulate the theorem for complex processes.

In the theorem we will meet the integral

$$\int_a^b \int_a^b g(s)\overline{g(t)}\, \mathrm{d}_{s,t} r(s,t),$$

which is defined in an analogous way to J_2, when $r(s,t)$ gives weight

$$r(s_2,t_2) - r(s_1,t_2) - r(s_2,t_1) + r(s_1,t_1)$$

to the rectangle $s_1 < s \leq s_2$, $t_1 < t \leq t_2$.

Theorem 2.16. *a) If $r(s,t)$ is continuous in $[a,b] \times [a,b]$, and $g(t)$ is such that the Riemann integral*

$$Q_1 = \int_a^b \int_a^b g(s)\overline{g(t)} r(s,t)\, \mathrm{d}s\, \mathrm{d}t < \infty,$$

then $J_1 = \int_a^b g(t)x(t)\, \mathrm{d}t$ exists as a quadratic mean limit, and $E(J_1) = 0$ and $E(|J_1|^2) = Q_1$.

b) If $r(s,t)$ has bounded variation[8] *in $[a,b] \times [a,b]$ and $g(t)$ is such that the Riemann-Stieltjes integral*

$$Q_2 = \int_a^b \int_a^b g(s)\overline{g(t)}\, \mathrm{d}_{s,t} r(s,t) < \infty,$$

then $J_2 = \int_a^b g(t)\, \mathrm{d}x(t)$ exists as a quadratic mean limit, and $E(J_2) = 0$ and $E(|J_2|^2) = Q_2$.

[8] That $f(t)$ is of bounded variation in $[a,b]$ means that $\sup \sum |f(t_k) - f(t_{k-1})|$ is bounded, with the sup taken over all possible partitions.

Proof. The simple proof uses the Loève criterion (A.16) for quadratic mean convergence: take two sequences of partitions of $[a,b]$ with points $s_1, s_2, \dots, s_m$, and $t_1, t_2, \dots, t_n$, respectively, and consider

$$E(S_m \overline{S_n}) = \sum_{k=1}^{m} \sum_{j=1}^{n} g(s_k)\overline{g(t_j)} r(s_k, t_j)(s_k - s_{k-1})(t_j - t_{j-1}).$$

If Q_1 exists finite, then this expectation converges to Q_1 as the division becomes infinitely fine. This proves (a). The reader should complete the proof for (b). □

Example 2.11 (Wiener process). *Take $x(t) = w(t)$, as the Wiener process. Since $r_w(s,t) = \sigma^2 \min(s,t)$, the integral Q_2 is all concentrated to the diagonal, $s = t$, and we see that $\int_a^b g(t)\,\mathrm{d}w(t)$ exists for all square integrable $g(t)$.*

Example 2.12 (Partial integration). *If $g(t)$ is a continuously differentiable function and $x(t) = w(t)$, then*

$$\int_a^b g(t)\,\mathrm{d}w(t) = g(b)w(b) - g(a)w(a) - \int_a^b g'(t)w(t)\,\mathrm{d}t.$$

To prove this, consider

$$\begin{aligned} S_2 &= \sum_{k=1}^{m} g(t_k)(w(t_k) - w(t_{k-1})) \\ &= g(t_m)w(t_m) - g(t_1)w(t_0) - \sum_{k=2}^{m} (g(t_k) - g(t_{k-1}))w(t_{k-1}). \end{aligned}$$

Since $g(t)$ is continuously differentiable, there is a ρ_k such that

$$g(t_k) - g(t_{k-1}) = (t_k - t_{k-1})(g'(t_k) + \rho_k),$$

and $\rho_k \to 0$ uniformly in $k = 1, \dots, m$, as $m \to \infty$, $\max_k |t_k - t_{k-1}| \to 0$. Thus

$$\begin{aligned} S_2 = {} & g(t_m)w(t_m) - g(t_1)w(t_0) - \sum_{k=2}^{m} g'(t_k)w(t_{k-1})(t_k - t_{k-1}) \\ & + \sum_{k=2}^{m} \rho_k w(t_{k-1})(t_k - t_{k-1}) \to g(b)w(b) - g(a)w(a) - \int_a^b g'(t)w(t)\,\mathrm{d}t. \end{aligned}$$

Remark 2.5. *A natural question is: are quadratic mean integrals and ordinary integrals equal? If a stochastic process has a continuous covariance function, and continuous sample paths, with probability one, and if $g(t)$ is, for* *example, continuous, then* $\int_a^b g(t)x(t)\,\mathrm{d}t$ *exists both as a regular Riemann integral and as a quadratic mean integral. Both integrals are random variables and they are limits of the same approximating Riemann sum, the only difference being the mode of convergence – with probability one, and in quadratic mean, respectively. But then the limits are equivalent, i.e., equal with probability one.*

The proofs of the following two theorems are left to the reader.

Theorem 2.17. *If* $\{x(t), t \in \mathbb{R}\}$ *and* $\{y(t), t \in \mathbb{R}\}$ *are stochastic processes with cross-covariance* $r_{x,y}(s,t) = C(x(s), y(t))$, *and if the conditions of Theorem 2.16 are satisfied, then*

$$E\left(\int_a^b g(s)x(s)\,\mathrm{d}s \cdot \int_c^d \overline{h(t)y(t)}\,\mathrm{d}t\right) = \int_a^b \int_c^d g(s)\overline{h(t)}\,r_{x,y}(s,t)\,\mathrm{d}s\,\mathrm{d}t,$$

$$E\left(\int_a^b g(s)\,\mathrm{d}x(s) \cdot \int_c^d \overline{h(t)\,\mathrm{d}y(t)}\right) = \int_a^b \int_c^d g(s)\overline{h(t)}\,\mathrm{d}_{s,t}r_{x,y}(s,t).$$

Theorem 2.18. *For the Wiener-process with* $r_{w,w}(s,t) = \min(s,t)$ *one has* $\mathrm{d}_{s,t}r_{w,w}(s,t) = \mathrm{d}s$, *for* $s = t$ *and* 0 *otherwise, which gives, for* $a < c < b < d$,

$$E\left(\int_a^b g(s)\,\mathrm{d}w(s) \cdot \int_c^d h(t)\,\mathrm{d}w(t)\right) = \int_c^b g(t)h(t)\,\mathrm{d}t. \tag{2.36}$$

The integrals are well defined if $\int_a^b g^2(s)\,\mathrm{d}s < \infty$, $\int_c^d h^2(t)\,\mathrm{d}t < \infty$.

Gaussian white noise – a preview

Theorem 2.18 can be formulated in a very convenient way that will be much used in the coming chapters. Write $\mathrm{d}w(t) = w'(t)\,\mathrm{d}t$, and write (2.36) as

$$C\left(\int_A g(s)w'(s)\,\mathrm{d}s, \int_B h(t)w'(t)\,\mathrm{d}t\right) = \int_{A\cap B} g(t)h(t)\,\mathrm{d}t. \tag{2.37}$$

This relation is a characteristic property of "Gaussian white noise," denoted $w'(t)$. At this stage, we only use it as shorthand for the covariance formula

(2.36), but in Chapter 4, on linear filters, an integral of the form $\int g(s)w'(s)\,\mathrm{d}s$ will turn out to be a handy way to denote the solution to a differential equation driven by very irregular random disturbances.

Gaussian white noise can be given a precise mathematical meaning as a generalized stochastic process, similar to a "generalized function." In the same way as a generalized function is a real or complex valued linear functional defined on a suitable family of functions, a generalized stochastic process is a linear functional whose values are random variables. It can be "measured" by a physical device, like $\int g(t)w'(t)\mathrm{d}t$. An introduction to the topic, expressed in non-technical language, can be found in [125, Appendix I].

2.5 An ergodic result

An ergodic theorem deals with convergence properties of *time or sample function averages* to *ensemble averages*, i.e., to *statistical expectation*:

$$\frac{1}{T}\int_0^T f(x(t))\,\mathrm{d}t \to E(f(x(0))), \quad \text{as } T \to \infty,$$

for a function of a stationary stochastic process $x(t)$. Such theorems will be the theme of the entire Chapter 6, but we show already here a simple such result based only on covariance properties. The process $x(t)$ need not even be stationary, but we assume $E(x(t)) = 0$ and that the covariance function $r(s,t) = C(x(s),x(t))$ exists.

Theorem 2.19. *a) If $r(s,t)$ is continuous for all s,t and*

$$\frac{1}{T^2}\int_0^T\int_0^T r(s,t)\,\mathrm{d}s\,\mathrm{d}t \to 0, \quad \text{as } T \to \infty, \tag{2.38}$$

then $\frac{1}{T}\int_0^T x(t)\,\mathrm{d}t \overset{q.m.}{\to} 0$.

b) If there exist constants K, α, β, such that $0 \le 2\alpha < \beta < 1$, and

$$|r(s,t)| \le K\frac{s^\alpha + t^\alpha}{1+|s-t|^\beta}, \quad \text{for } s,t \ge 0, \tag{2.39}$$

then $\frac{1}{T}\int_0^T x(t)\,\mathrm{d}t \overset{a.s.}{\to} 0$, as $T \to \infty$.

Proof. a) This is immediate from Theorem 2.17, since, by assumption,

$$\sigma_T^2 = E\left(\left(\frac{1}{T}\int_0^T x(t)\,\mathrm{d}t\right)^2\right) = \frac{1}{T^2}\int_0^T\int_0^T r(s,t)\,\mathrm{d}s\,\mathrm{d}t \to 0.$$

b) Before we prove the almost sure convergence, note that the condition $2\alpha < \beta < 1$ puts a limit on how fast $E(x(t)^2) = r(t,t)$ is allowed to increase as $t \to \infty$, and it limits the amount of dependence between $x(s)$ and $x(t)$ for large $|s-t|$. If the dependence is too strong, it may happen that $\frac{1}{T}\int_0^T x(t)\,\mathrm{d}t$ converges, but not to 0 but to a (random) constant different from 0.

We show here only part of the theorem, and refer the reader to [35, p. 95] for a completion. What we show is that there exists a subsequence of times, $T_n \to \infty$, such that $\frac{1}{T_n}\int_0^{T_n} x(t)\,\mathrm{d}t \overset{a.s.}{\to} 0$.

First estimate σ_T^2 from the proof of (a):

$$\begin{aligned}
\sigma_T^2 &= E\left(\left(\frac{1}{T}\int_0^T x(t)\,\mathrm{d}t\right)^2\right) = \frac{1}{T^2}\int_0^T\int_0^T r(s,t)\,\mathrm{d}s\,\mathrm{d}t \\
&\le \frac{K}{T^2}\int_0^T\int_0^T \frac{s^\alpha + t^\alpha}{1+|s-t|^\beta}\,\mathrm{d}s\,\mathrm{d}t \\
&= \frac{K}{T^{\beta-\alpha}}\cdot\frac{1}{T^{2-\beta}}\int_0^T\int_0^T \frac{(s/T)^\alpha + (t/T)^\alpha}{1+|s-t|^\beta}\,\mathrm{d}s\,\mathrm{d}t \\
&\le \frac{K}{T^{\beta-\alpha}}\cdot\frac{2}{T^{1-\alpha}}\int_0^T \frac{1}{1+u^\beta}\,\mathrm{d}u.
\end{aligned}$$

Here $(2/T^{1-\alpha})\int_0^T (1+u^\beta)^{-1}\,\mathrm{d}u$ tends to zero as $T \to \infty$, which implies that $\sigma_T^2 \le K'/T^{\beta-\alpha}$ for some constant K'.

Take the constant γ such that $\gamma(\beta-\alpha) > 1$, which is possible from the properties of α and β, and put $T_n = n^\gamma$, with

$$\sum_{n=1}^\infty \sigma_{T_n}^2 \le \sum_{n=1}^\infty \frac{K'}{n^{\gamma(\beta-\alpha)}} < \infty. \tag{2.40}$$

That the sum (2.40) is finite implies, by the Borel-Cantelli lemma and the Chebyshev inequality, that $T_n^{-1}\int_0^{T_n} x(t)\,\mathrm{d}t \overset{a.s.}{\to} 0$.

In order to complete the proof, see [35, p. 95] on how to show that

$$\sup_{T_n \le T \le T_{n+1}} \left| \frac{1}{T}\int_0^T x(t)\,\mathrm{d}t - \frac{1}{T_n}\int_0^{T_n} x(t)\,\mathrm{d}t \right| \overset{a.s.}{\to} 0, \quad \text{as } n \to \infty.$$

□

Corollary 2.2. *a) If $x(t)$ is stationary and $\frac{1}{T}\int_0^T r(t)\,dt \to 0$ as $T \to \infty$, then*

$$\frac{1}{T}\int_0^T x(t)\,dt \overset{q.m.}{\to} 0.$$

b) If, moreover, there is a constant $K > 0$ and a $\beta > 0$, such that $|r(t)| \le \frac{K}{|t|^\beta}$ as $t \to \infty$, then the convergence is with probability one.

Exercises

2:1. Show that the covariance function of the Random Telegraph Signal in Example 2.3 is $r(\tau) = e^{-2\lambda|\tau|}$. Why can you not use Theorem 2.5 to conclude that the process has continuous sample functions?

2:2. Show that equivalent processes have the same finite-dimensional distributions.

2:3. Let $\{x(t), t \in \mathbb{R}\}$ and $\{y(t), t \in \mathbb{R}\}$ be equivalent processes which both have, with probability one, continuous sample paths. Prove that

$$P(x(t) = y(t), \text{ for all } t \in \mathbb{R}) = 1.$$

2:4. Find the values of the constants a and b that make a Gaussian process twice continuously differentiable if its covariance function is $r(t) = e^{-|t|}(1 + a|t| + bt^2)$.

2:5. Give the argument that a function with $r(t) = r(0) - |t|^a + o(|t|^a)$ as $t \to 0$ cannot be a covariance function if $a > 2$.

2:6. Complete the proof of Theorem 2.6 and show that, in the notation of the proof, $\sum_n g(2^{-n}) < \infty$, and $\sum_n 2^n q(2^{-n}) < \infty$.

2:7. Show that the Wiener process has infinite variation, a.s., by showing the stronger statement that if

$$Y_n = \sum_{k=0}^{2^n-1} \left| w\left(\frac{k+1}{2^n}\right) - w\left(\frac{k}{2^n}\right) \right|$$

then $\sum_{n=1}^\infty P(Y_n < n) < \infty$.

2:8. Show that a non-stationary process is continuous in quadratic mean at $t = t_0$ only if its mean value function $m(t) = E(x(t))$ is continuous at t_0 and its covariance function $r(s,t) = C(x(s), x(t))$ is continuous at $s = t = t_0$.

2:9. Find the covariance function of the (non-stationary) Brownian bridge $BB_T(t) = w(t) - tw(T)/T$, where $T > 0$ is a constant time point.

2:10. a) Show that the standard Wiener process conditioned on $w(T) = 0$ is a Brownian bridge.

b) Also show that if $BB_T(t), 0 \le t \le T$, is a Brownian bridge over $[0,T]$, and Z is normal with mean zero and variance 1, independent of BB_T, then $BB_T(t) + \frac{t}{\sqrt{T}} Z$ is a standard Wiener process.

2:11. Convince yourself of the "trivial" fact that if a sequence of normal variables $\{x_n, n \in \mathbb{N}\}$, such that $E(x_n)$ and $V(x_n)$ have finite limits, then the sequence converges in distribution to a normal variable.

2:12. Show how Theorem 2.10 follows from Theorem 2.8.

2:13. Give an example of a stationary process that violates the sufficient conditions in Bulinskaya's lemma, Theorem 2.14, and for which the sample functions can be tangents of the level $u = 1$.

2:14. In Chapter 8 we will study crossing events of a stochastic process $\{x(t), t \in \mathbb{R}\}$, in particular the first crossing time $T = \inf\{t > 0; x(t) = u\}$. When is T a well defined random variable?

2:15. Prove Theorem 2.17. Hint: Consider the approximating sums, J_1 and J_2, and read Section A.4.4.

2:16. Assume that sufficient conditions on $r(s,t) = E(x(s)\overline{x(t)})$ are satisfied so that the integral

$$\int_0^T g(t)x(t)\,dt$$

exists for all T, both as a quadratic mean integral and as a sample function integral. Show that, if

$$\int_0^\infty |g(t)|\sqrt{r(t,t)}\,dt < \infty,$$

then the generalized integral $\int_0^\infty g(t)x(t)\,dt$ exists as a limit as $T \to \infty$, both in quadratic mean and with probability one.

2:17. Find an example of two dependent normal random variables U and V such that $C(U,V) = 0$; obviously you cannot let (U,V) have a bivariate normal distribution.

2:18. Prove that Corollary 2.2 on page 67 follows from Theorem 2.19.

2:19. Take the trivial process $x(t) = 0$ for $0 \le t \le 1$. Give an example of a sequence of processes $x_n(t)$ for which the finite-dimensional distributions tend to those of $x(t)$ but $\max_{[0,1]}(x_n)$ does not tend in distribution to 0.

Chapter 3

Spectral representations

This chapter deals with the spectral representation of weakly stationary processes – stationary in the sense that the mean is constant and the covariance $C(x(s),x(t))$ only depends on the time difference $s-t$. For real-valued Gaussian processes, the mean and covariance function determine all finite-dimensional distributions, and hence the entire process distribution. However, the spectral representation requires complex-valued processes, and then it is necessary that one specifies also the correlation structure between the real and the imaginary part of the process. We therefore start with a summary of the basic properties of complex-valued processes, in general, and in the Gaussian case, before we introduce and prove the spectral representation of the covariance function, and subsequently, of the process itself. We remind the reader of the classical memoirs by S.O. Rice [101], which can be recommended to anyone with the slightest historical interest. That work also contains many old references.

3.1 Complex-valued stochastic processes

3.1.1 Moments in complex-valued processes

In many applications of stationary stochastic processes, a random signal is represented in complex form, $x(t) = A(t)\,e^{i\Theta(t)}$, where $A(t)$ is a time varying random amplitude process and $\Theta(t)$ a random phase function. The work on noise in radio transmission by S.O. Rice, mentioned in Section 1.5.2, is an example of this type of complex process. In the spectral theory presented in this chapter we will have use of a spectral process of similar complex form, but which is a function, not of time t, but of frequency ω. We start this chapter with some general notions of correlation between complex-valued variables and processes.

As defined in Chapter 2, a complex-valued random variable $x = y + iz$ is composed of two real-valued variables, with some joint, two-dimensional distribution. The expectation is defined as $E(x) = E(y) + iE(z) = m$ and the covariance between two complex-valued variables x_1 and x_2 is defined as

$$C(x_1,x_2) = E((x_1 - m_1)\overline{(x_2 - m_2)}) = E(x_1\overline{x_2}) - m_1\overline{m_2}.$$

Two complex variables with $C(x_1,x_2) = 0$ are called *orthogonal*, stochastically uncorrelated. It is a surprising consequence of this terminology that if the real and imaginary part are uncorrelated with the same variance, then x and its complex conjugate $\overline{x}$ are orthogonal, although they are totally dependent: with $E(y) = E(z) = 0$ we have

$$C(x,\overline{x}) = E(x \cdot \overline{\overline{x}}) = E((y+iz)^2) = E(y^2) - E(z^2) + 2iE(yz) = 0.$$

A complex-valued process $x(t) = y(t) + iz(t)$ of a parameter t, not necessarily denoting time, is formed by two real-valued processes $\{y(t), t \in \mathbb{R}\}$ and $\{z(t), t \in \mathbb{R}\}$ with some joint distribution.

Definition 3.1. *A complex-valued process $\{x(t), t \in \mathbb{R}\}$ is strictly stationary if all 2n-dimensional distributions of*

$$y(t_1 + \tau), z(t_1 + \tau), \ldots, y(t_n + \tau), z(t_n + \tau)$$

are independent of τ. It is called weakly stationary or second order stationary if $E(x(t)) = m$ is constant, and

$$E(x(s)\overline{x(t)}) = C(x(s), x(t)) + |m|^2$$

only depends on the time difference $s - t$. The covariance function

$$r(\tau) = E\left((x(s+\tau) - m)\overline{(x(s) - m)}\right)$$

is Hermitian, i.e., $r(-\tau) = \overline{r(\tau)}$.

For a real-valued process $\{x(t), t \in \mathbb{R}\}$, the covariance function $r(\tau)$ determines all covariances between $x(t_1), \ldots, x(t_n)$,

$$\mathbf{\Sigma}(t_1, \ldots, t_n) = \begin{pmatrix} V(x(t_1)) & C(x(t_1), x(t_2)) & \ldots & C(x(t_1), x(t_n)) \\ \vdots & \vdots & \ddots & \vdots \\ C(x(t_n), x(t_1)) & C(x(t_n), x(t_2)) & \ldots & V(x(t_n)) \end{pmatrix}$$

$$= \begin{pmatrix} r(0) & r(t_1 - t_2) & \dots & r(t_1 - t_n) \\ \vdots & \vdots & \ddots & \vdots \\ r(t_n - t_1) & r(t_n - t_2) & \dots & r(0) \end{pmatrix}. \tag{3.1}$$

3.1.2 *Non-negative definite functions*

It is a characteristic property of a covariance function that it is *non-negative definite*.

Definition 3.2 (Non-negative definite function). *A (real or complex) function $r(t)$ is said to be non-negative definite if, for all finite set of time points $t_1, \dots, t_n$, and arbitrary complex numbers $a_1, \dots, a_n$,*

$$\sum_{j,k} a_j \overline{a_k} r(t_j - t_k) \geq 0.$$

Note that the requirement that the sum is non-negative implies that it is real and that $r(t)$ is a Hermitian function.

It is easy to see that the covariance function $r(t)$ of a complex-valued stationary process $\{x(t), t \in \mathbb{R}\}$ is always non-negative definite. Assuming, for simplicity, that $E(x(t)) = 0$, we have

$$\sum_{j,k} a_j \overline{a_k} r(t_j - t_k) = E\left(\sum_{j,k} a_j x(t_j) \overline{a_k x(t_k)} \right) = E\left(\left| \sum_{j=1}^{n} a_j x(t_j) \right|^2 \right),$$

which obviously is non-negative.

Theorem 3.1 (Characterization of covariance function). *Every non-negative definite function $r(\tau)$ is the covariance function for a strictly stationary process. In fact, we will construct a stationary Gaussian process with $r(\tau)$ as its covariance function, but there may be non-Gaussian processes with the same property. Thus, the class of covariance functions is identical to the class of non-negative definite functions.*

Proof. We have to show that if $r(\tau)$ is a non-negative definite function then there are finite-dimensional distributions for a process $x(t)$ with $E(x(t)) = 0$, such that

$$r(\tau) = E(x(s+\tau)\overline{x(s)}).$$

Suppose first that $r(\tau)$ is a real function. By Kolmogorov's existence theorem, see Appendix A.2, we only have to show that for every selection of time points $t_1, \dots, t_n$, there is an n-dimensional distribution with mean 0 and covariances given by $\mathbf{\Sigma}(t_1, \dots, t_n)$ as defined by (3.1), such that the obtained distributions form a consistent family, i.e., for example $F_{x,y}(u,v) = F_{y,x}(v,u)$ and $F_{x,y}(u,\infty) = F_x(u)$.

If $r(t)$ is a real function and $\mathbf{u} = (u_1, \dots, u_n)$ a real vector, consider the non-negative quadratic form

$$Q(\mathbf{u}) = \sum_{j,k} u_j u_k r(t_j - t_k).$$

Then we recognize $\exp(-Q(\mathbf{u})/2)$ as the characteristic function (see Appendix A.5), for an n-variate normal distribution with covariance matrix $\mathbf{\Sigma}(t_1, \dots, t_n)$, and we have found a distribution with the specified properties.

If $r(t)$ is complex, with $r(-t) = \overline{r(t)}$, there are real functions $p(t) = p(-t)$, $q(t) = -q(-t)$ such that

$$r(t) = p(t) + iq(t).$$

We now need to find $2n$-dimensional distributions for the real and imaginary components of the process.

Take $a_j = u_j - iv_j$ and consider the non-negative quadratic form in real variables $(\mathbf{u}, \mathbf{v}) = (u_1, \dots, u_n, v_1, \dots, v_n)$,

$$\begin{aligned} Q(\mathbf{u},\mathbf{v}) &= \sum_{j,k} a_j \overline{a_k}\, r(t_j - t_k) \\ &= \sum_{j,k} (u_j - iv_j)(u_k + iv_k)(p(t_j - t_k) + iq(t_j - t_k)) \\ &= \sum_{j,k} \{p(t_j - t_k)(u_j u_k + v_j v_k) - q(t_j - t_k)(u_j v_k - u_k v_j)\}; \end{aligned}$$

note that the imaginary part vanishes since Q is assumed to be non-negative, and hence real. Similarly, as in the real case,

$$\exp(-Q(\mathbf{u},\mathbf{v})/2) = E\left(\exp(i \sum_j (u_j y_j + v_j z_j)) \right)$$

is recognized as the characteristic function of the $2n$-dimensional normal vector

$$(y_1,\dots,y_n,z_1,\dots,z_n)$$

with the specified properties:

$$E(y_jy_k) = E(z_jz_k) = p(t_j - t_k), \quad E(y_jz_k) = -E(y_kz_j) = -q(t_j - t_k).$$

With $x_j = (y_j + iz_j)/\sqrt{2}$, we have $E(x_j) = 0$ and

$$E(x_j\overline{x_k}) = p(t_j - t_k) + iq(t_j - t_k) = r(t_j - t_k),$$

as required. Since the Gaussian distribution of the y- and z-variables is determined by the covariances, and these depend only on the time difference, the process is strictly stationary. □

3.1.3 *Strict and weak stationarity for Gaussian processes*

Since the first two moments determine a real normal distribution, it is clear that each weakly (covariance) stationary real-valued normal process is strictly stationary. For complex processes matters are not that easy, and one has to impose two stationarity conditions in order to guarantee strict stationarity.

Theorem 3.2. *A complex normal process $x(t) = y(t) + iz(t)$ with mean zero is strictly stationary if and only if the two functions*

$$r(s,t) = E(x(s)\overline{x(t)}), \quad q(s,t) = E(x(s)x(t)),$$

only depend on $s - t$.

Proof. To prove the "if" part, express $r(s,t)$ and $q(s,t)$ in terms of y and z,

$$r(s,t) = E(y(s)y(t) + z(s)z(t)) + iE(z(s)y(t) - y(s)z(t)),$$

$$q(s,t) = E(y(s)y(t) - z(s)z(t)) + iE(z(s)y(t) + y(s)z(t)).$$

Since these only depend on $s - t$, the same is true for the sums and differences of their real and imaginary parts, i.e., for $E(y(s)y(t))$, $E(z(s)z(t))$, $E(z(s)y(t))$, $E(y(s)z(t))$. Therefore, the $2n$-dimensional distribution of

$$y(t_1),\dots,y(t_n),z(t_1),\dots,z(t_n)$$

only depends on time differences, and $x(t)$ is strictly stationary. The converse is trivial. □

Example 3.1. *If* $\{x(t), t \in \mathbb{R}\}$ *is a real and stationary normal process, and* μ *is a constant, then*

$$x^*(t) = e^{i\mu t} x(t)$$

is a weakly, but not strictly, stationary complex normal process, since the second of the conditions in Theorem 3.2 is violated,

$$E(x^*(s)\overline{x^*(t)}) = e^{i\mu(s-t)} E(x(s)x(t)),$$

$$E(x^*(s)x^*(t)) = e^{i\mu(s+t)} E(x(s)x(t)).$$

Complex-valued processes have many interesting properties and are used as standard tools in many electrical engineering applications; the book by Miller [93] is a a good reference.

3.2 Bochner's theorem and the spectral distribution

3.2.1 The spectral distribution

We have seen that covariance functions for stationary processes are characterized by the property of being non-negative definite. However, checking the non-negativeness is not straightforward, and a more practical characterization is therefore needed.

We shall now formulate and prove such a useful characterization, namely that every *continuous* covariance function is the Fourier transform of its *spectral distribution.* The theorem due to Bochner [16] is central not only for stationary processes but also for the mathematics of harmonic functions. The proof presented here is that given by Cramér [30], and it relies on a basic theorem in probability theory on convergence of a sequence of characteristic functions, Theorem A.7, page 297.

Theorem 3.3 (Bochner's theorem). *A continuous function* $r(t)$*, real or complex, is non-negative definite, and hence a covariance function, if and only if there exists a non-decreasing, right continuous, and bounded real function* $F(\omega)$*, such that*

$$r(t) = \int_{-\infty}^{\infty} e^{i\omega t}\, \mathrm{d}F(\omega).$$

The function $F(\omega)$ is the spectral distribution function of the process, and it has all the properties of a statistical distribution function except that $F(+\infty) - F(-\infty) = r(0)$ need not be equal to one. Symbolically, $\mathrm{d}F(\omega)$ is the "spectral mass at ω." The function $F(\omega)$ is defined only up to an additive constant, and one usually takes $F(-\infty) = 0$.

Proof. The "if" part is clear, since if $r(t) = \int_{-\infty}^{\infty} \exp(i\omega t)\,\mathrm{d}F(\omega)$, then

$$\begin{aligned}\sum_{j,k} a_j\overline{a_k}r(t_j - t_k) &= \sum_{j,k} a_j\overline{a_k}\int_{-\infty}^{\infty} e^{i\omega t_j}\cdot e^{-i\omega t_k}\,\mathrm{d}F(\omega)\\ &= \int_{-\infty}^{\infty}\sum_{j,k} a_j e^{i\omega t_j}\cdot\overline{a_k e^{i\omega t_k}}\,\mathrm{d}F(\omega)\\ &= \int_{-\infty}^{\infty}\left|\sum_j a_j e^{i\omega t_j}\right|^2\mathrm{d}F(\omega) \geq 0.\end{aligned}$$

For the "only if" part we shall show that, given a continuous covariance function $r(t)$, there exists a proper distribution function $F_\infty(\omega) = F(\omega)/F(\infty)$ such that

$$F_\infty(\infty) - F_\infty(-\infty) = 1, \quad \int_{-\infty}^{\infty} e^{i\omega t}\,\mathrm{d}F_\infty(\omega) = \frac{r(t)}{r(0)}.$$

To this end, take a real $A > 0$, and define

$$\begin{aligned}g(\omega, A) &= \frac{1}{2\pi A}\int_0^A\int_0^A r(t-u)\,e^{-i\omega(t-u)}\,\mathrm{d}t\,\mathrm{d}u\\ &= \frac{1}{2\pi A}\lim\sum_{j,k} r(t_j - t_k)e^{-i\omega t_j}\overline{e^{-i\omega t_k}}\,\Delta t_j\Delta t_k\\ &= \frac{1}{2\pi A}\lim\sum_{j,k} r(t_j - t_k)\Delta t_j e^{-i\omega t_j}\cdot\overline{\Delta t_k e^{-i\omega t_k}} \geq 0,\end{aligned}$$

since $r(t)$ is non-negative definite. (Here, of course, the t_j define a successively finer subdivision of the interval $[0, A]$.) When we let $A \to \infty$, the limit of $g(\omega, A)$ will give us the desired spectral distribution.

Before we proceed with the proof, we express $g(\omega, A)$ as

$$\begin{aligned}g(\omega, A) &= \frac{1}{2\pi A}\int_0^A\int_0^A r(t-u)e^{-i\omega(t-u)}\,\mathrm{d}t\,\mathrm{d}u\\ &= \frac{1}{2\pi}\int_{-A}^{A}\left(1 - \frac{|t|}{A}\right)r(t)e^{-i\omega t}\,\mathrm{d}t = \frac{1}{2\pi}\int_{-\infty}^{\infty}\mu(t/A)r(t)e^{-i\omega t}\,\mathrm{d}t,\end{aligned}$$

where

$$\mu(t) = \begin{cases} 1-|t|, & \text{for } |t| \leq 1, \\ 0, & \text{otherwise.} \end{cases}$$

The proof is divided into three steps:

Step 1) Prove that $g(\omega,A) \geq 0$ is integrable, and

$$\int_{\omega} g(\omega,A)\,\mathrm{d}\omega = r(0), \tag{3.2}$$

so $f_A(\cdot) = g(\cdot,A)/r(0)$ is a regular statistical density function.

Step 2) Show that

$$\mu(t/A)\frac{r(t)}{r(0)} = \int_{-\infty}^{\infty} \frac{g(\omega,A)}{r(0)} e^{it\omega}\,\mathrm{d}\omega, \tag{3.3}$$

so the function $\mu(t/A)\,r(t)/r(0)$ is the characteristic function for the density $g(\omega,A)/r(0)$.

Step 3) Take limits as $A \to \infty$,

$$\lim_{A\to\infty}\left(1-\frac{|t|}{A}\right) r(t) = r(t), \tag{3.4}$$

and use the fact that the limit of a convergent sequence of characteristic functions is also a characteristic function, provided it is continuous. That concludes the proof that there exists a statistical distribution such that $r(t)/r(0)$ is its characteristic function.

We now show steps (1) and (2). Multiply $g(\omega,A)$ by $\mu(\omega/2M)$, integrate, and change the order of integration (since $\mu(\omega/2M)\mu(t/A)r(t)e^{-i\omega t}$ is bounded and has support in $[-2M,2M]\times[-A,A]$, Fubini's theorem permits this):

$$\begin{aligned}\int_{-\infty}^{\infty}\mu(\omega/2M)g(\omega,A)\,\mathrm{d}\omega &= \frac{1}{2\pi}\int_{-\infty}^{\infty}\mu(\omega/2M)\int_{-\infty}^{\infty}\mu(t/A)r(t)e^{-i\omega t}\,\mathrm{d}t\,\mathrm{d}\omega \\ &= \frac{1}{2\pi}\int_{-\infty}^{\infty}\mu(t/A)r(t)\int_{-\infty}^{\infty}\mu(\omega/2M)e^{-i\omega t}\,\mathrm{d}\omega\,\mathrm{d}t.\end{aligned} \tag{3.5}$$

Here, the inner integral is a Fourier transform of the square of the "sinc" function,[1]

$$\int_{-\infty}^{\infty} \mu(\omega/2M)e^{-i\omega t}\,\mathrm{d}\omega = \int_{-2M}^{2M}\left(1-\frac{|\omega|}{2M}\right)e^{-i\omega t}\,\mathrm{d}\omega$$

$$= \int_{-2M}^{2M}\left(1-\frac{|\omega|}{2M}\right)\cos\omega t\,\mathrm{d}\omega = 2M\left(\frac{\sin Mt}{Mt}\right)^2,$$

so (3.5) is equal to

$$\frac{M}{\pi}\int_{-\infty}^{\infty}\mu(t/A)r(t)\left(\frac{\sin Mt}{Mt}\right)^2\mathrm{d}t = \frac{1}{\pi}\int_{-\infty}^{\infty}\mu(s/MA)r(s/M)\left(\frac{\sin s}{s}\right)^2\mathrm{d}s$$

$$\leq \frac{1}{\pi}r(0)\int_{-\infty}^{\infty}\left(\frac{\sin s}{s}\right)^2\mathrm{d}s = r(0).$$

Now, the integrand in $\mu(\omega/2M)g(\omega,A)$ in the left hand side of (3.5) increases, as $M\uparrow\infty$, to $g(\omega,A)$, so taking limits inside the integral gives

$$\int_{-\infty}^{\infty} g(\omega,A)\,\mathrm{d}\omega = \lim_{M\to\infty}\int_{-\infty}^{\infty}\mu(\omega/2M)g(\omega,A)\,\mathrm{d}\omega \leq r(0).$$

We have now shown that $g(\omega,A)$ and $\mu(t/A)r(t)$ are both absolutely integrable over the whole real line. Since they form a Fourier transform pair, i.e.,

$$g(\omega,A) = \frac{1}{2\pi}\int_{-\infty}^{\infty}\mu(t/A)r(t)e^{-i\omega t}\,\mathrm{d}t,$$

we can use the Fourier inversion theorem, Theorem A.9, and conclude that

$$\mu(t/A)r(t) = \int_{-\infty}^{\infty} g(\omega,A)e^{i\omega t}\,\mathrm{d}\omega,$$

which is step (2) in the proof.[2]

By taking $t=0$ we also get step (1), and $f_A(\omega) = g(\omega,A)/r(0)$ is a probability density function for some probability distribution with characteristic function

$$\varphi_A(t) = \int_{-\infty}^{\infty}\frac{g(\omega,A)}{r(0)}e^{i\omega t}\,\mathrm{d}\omega = \frac{\mu(t/A)}{r(0)}r(t).$$

---

[1] $\mathrm{sinc}(t) = \frac{\sin t}{t}$ or the normalized version $\mathrm{sinc}(t) = \frac{\sin \pi t}{\pi t}$; see the table in Appendix C.

[2] See [26, Sect. 6.5] for a proof of (2) without reference to the Fourier inversion theorem.

For step (3) we use the convergence property of characteristic functions. Theorem A.7 in Appendix A.5 says that if $F_A(x)$ is a family of distribution functions such that the characteristic functions $\varphi_A(t)$ converge to a continuous function $\varphi(t)$, as $A \to \infty$, then there exists a distribution function $F(x)$ with characteristic function $\varphi(t)$ and $F_A(x) \to F(x)$, for all x where $F(x)$ is continuous.

Here, the characteristic functions $\varphi_A(t) = \mu(t/A)r(t)/r(0)$ converge to $\varphi(t) = r(t)/r(0)$, and since we have assumed $r(t)$ to be continuous, we know from the characteristic function convergence that $F_A(x) = \int_{-\infty}^{x} f_A(\omega)\,\mathrm{d}\omega$ converges to a distribution function $F_\infty(x)$ as $A \to \infty$, with characteristic function

$$\varphi(t) = \frac{r(t)}{r(0)} = \int_{-\infty}^{\infty} e^{i\omega t}\,\mathrm{d}F_\infty(\omega).$$

We get the desired spectral representation with $F(\omega) = r(0)F_\infty(\omega)$. □

To remember: The pair *covariance function – spectral distribution* shares (except for the normalization of F) all properties of the pair *characteristic function – statistical distribution function*, and everything you know about one relation can be applied to the other. For example, the product of two covariance functions is also a covariance function with a spectral distribution that is the convolution of the original spectra. Similarly, the product of two squared integrable spectral densities is a spectral density.

The continuity of the covariance function is essential for the existence of a proper (bounded) spectral distribution. This restricts somewhat the use of the spectrum when a process in continuous time is observed with independent measurement noise, which would be impossible in practice anyway.

3.2.2 *The inversion theorem and the spectrum-covariance relation*

The spectral distribution function $F(\omega)$ is non-decreasing, and by our construction always continuous to the right, $F(\omega) = \lim_{h \downarrow 0} F(\omega + h)$. We can distinguish between two main classes, continuous and discrete spectra. Mixtures of the two exist in practice when a clean signal is present together with a background of noise. See also Section 6.6.2 for more theoretical aspects of mixed type spectra.

Continuous spectrum: The spectral distribution function $F(\omega)$ is absolutely continuous with $F(\omega) = \int_{s=-\infty}^{\omega} f(s)\,\mathrm{d}s$, and $f(\omega) \geq 0$ is the spectral density; then

$$r(t) = \int_{-\infty}^{\infty} e^{i\omega t} f(\omega)\,\mathrm{d}\omega. \tag{3.6}$$

For real processes, the spectral density is symmetric, $f(-\omega) = f(\omega)$.

Discrete spectrum: The spectral distribution function $F(\omega)$ is piecewise constant with jumps ΔF_k at $\omega = \omega_k$; then

$$r(t) = \sum_k e^{i\omega_k t}\,\Delta F_k. \tag{3.7}$$

Note that $F(\omega)$ can have only a denumerable number of discontinuity points. For real processes, the jumps are symmetrically located and can be numbered $\omega_0 = 0$ and $\omega_{-k} = -\omega_k$, $k = 1, \ldots$, with $\omega_k > 0$, and $\Delta F_{-k} = \Delta F_k$.

Spectra of mixed continuous and discrete type are possible, in particular in signal processing and audio applications, where the discrete part carries the information and the continuous part corresponds to noise. A third type of "singular" spectra of theoretical interest will be dealt with in Section 6.6.2.

The covariance function $r(t)$ and the spectral density $f(\omega)$ form a Fourier transform pair. In general, the spectral distribution is uniquely determined by the covariance function but the precise relationship is somewhat complicated if the spectrum is not absolutely continuous. To formulate a general *inversion theorem* we need to identify those ω for which the spectral distribution function is not continuous. Write $\Delta F_\omega = F(\omega) - F(\omega - 0) \geq 0$ for the jump (possibly 0) at ω, and define $\widetilde{F}(\omega)$ as the average between the left and right limits of $F(\omega)$,

$$\widetilde{F}(\omega) = \frac{F(\omega) + F(\omega - 0)}{2} = F(\omega) - \frac{1}{2}\Delta F_\omega. \tag{3.8}$$

Theorem 3.4. *a) If the covariance function* $r(t), t \in \mathbb{R}$*, is absolutely integrable, i.e.,* $\int_{-\infty}^{\infty} |r(t)|\,\mathrm{d}t < \infty$*, then the spectrum is absolutely continuous and the Fourier inversion formula holds,*

$$f(\omega) = \frac{1}{2\pi}\int_{-\infty}^{\infty} e^{-i\omega t} r(t)\,\mathrm{d}t. \tag{3.9}$$

b) For all covariance functions, when $\omega_1 < \omega_2$*,*

$$\widetilde{F}(\omega_2) - \widetilde{F}(\omega_1) = \frac{1}{2\pi}\lim_{T\to\infty}\int_{-T}^{T} \frac{e^{-i\omega_2 t} - e^{-i\omega_1 t}}{-it} r(t)\,\mathrm{d}t. \tag{3.10}$$

Proofs of the theorem are given in most probability books [19, 26, 45], and we present one in Section A.5.2 in Appendix A.

Remark 3.1 (Cauchy's principal value). *The inversion formula (3.10) defines the spectral distribution for all continuous covariance functions. One can also use (3.9) to calculate the spectral density in case* $r(t)$ *is absolutely integrable, but if it is not, one may use (3.10) and take* $\widetilde{f}(\omega) = \lim_{h\to 0}(\widetilde{F}(\omega+h) - \widetilde{F}(\omega))/h$*. This is always possible, but one has to be careful in case* $f(\omega)$ *is not continuous. Even when the limit* $\widetilde{f}(\omega)$ *exists it need not be equal to* $f(\omega)$ *as the following example shows. The limit in* (3.10)*, which always exists, is called the Cauchy principal value.*

Example 3.2. *We use (3.10) to find the spectral density of low frequency white noise, with covariance function* $r(t) = \frac{\sin t}{t}$*. This covariance function is not absolutely integrable. We get from the general inversion formula,*

$$\begin{aligned}
\frac{\widetilde{F}(\omega+h) - \widetilde{F}(\omega-h)}{2h} &= \frac{1}{2\pi}\frac{1}{2h}\lim_{T\to\infty}\int_{-T}^{T} \frac{e^{-i(\omega+h)t} - e^{-i(\omega-h)t}}{-it}\,\frac{\sin t}{t}\,\mathrm{d}t \\
&= \frac{1}{2\pi}\int_{-\infty}^{\infty} e^{-i\omega t}\,\frac{\sin ht}{ht}\,\frac{\sin t}{t}\,\mathrm{d}t \\
&= \begin{cases} \frac{1}{2}, & \text{for } |\omega| < 1-h, \\ \frac{1}{4}(1 + (1-|\omega|)/h), & \text{for } 1-h < |\omega| < 1+h, \\ 0, & \text{for } |\omega| > 1+h. \end{cases}
\end{aligned}$$

The limit as $h \to 0$ is $1/2$, $1/4$, and 0, respectively, giving the density

$$f(\omega) = \begin{cases} 1/2, & \textit{for } |\omega| < 1, \\ 1/4, & \textit{for } |\omega| = 1, \\ 0, & \textit{for } |\omega| > 1. \end{cases}$$

Note that the inversion formula (3.9) gives $1/4$ for $\omega = 1$ as the Cauchy principal value,

$$\lim_{T\to\infty} \frac{1}{2\pi} \int_{-T}^{T} e^{-i\omega t} r(t)\,\mathrm{d}t = \lim_{T\to\infty} \frac{1}{2\pi} \int_{-T}^{T} \frac{\sin 2t}{2t}\,\mathrm{d}t = 1/4.$$

Remark 3.2. *It should be noted that $\int |r(t)|\,\mathrm{d}t < \infty$ is a sufficient but not necessary condition for the inversion formula* (3.9). *If the spectrum is absolutely continuous, the spectral density is given by* (3.9) *at each frequency ω where f is continuous with left and right derivatives; see Appendix A.5.2.*

Appendix C contains a list of commonly used spectra and corresponding covariance functions.

3.2.3 The one-sided real form

A real stationary process has a symmetric covariance function and a symmetric spectrum. In practical applications one often uses only the positive side of the spectrum. The one-sided spectral distribution will be denoted by $G(\omega)$,

$$G(\omega) = \begin{cases} 0, & \text{for } \omega < 0, \\ F(\omega) - F(-\omega - 0) = 2F(\omega) - r(0), & \text{for } \omega \geq 0. \end{cases} \tag{3.11}$$

Then $G(0-) = 0$ and $G(\infty) = r(0)$. If $F(\omega)$ is discontinuous at $\omega = 0$ then $G(\omega)$ will have a jump $F(0) - F(-0) = G(0+)$ at $\omega = 0$. For discontinuity points $\omega > 0$ the jump of $G(\omega)$ will be twice that of $F(\omega)$. If $F(\omega)$ has a density $f(\omega)$, then $G(\omega)$ has density $g(\omega) = 2f(\omega)$ for $\omega \geq 0$.

Any real covariance function can be expressed as

$$r(t) = \int_{0-}^{\infty} \cos \omega t \,\mathrm{d}G(\omega) = \int_{-\infty}^{\infty} \cos \omega t \,\mathrm{d}F(\omega), \tag{3.12}$$

and the inversion formula (3.10) gives that for any continuity point of G,

$$G(\omega) = F(\omega) - F(-\omega) = \frac{2}{\pi}\int_0^\infty \frac{\sin \omega t}{t}\, r(t)\,\mathrm{d}t,$$

and, when there is a spectral density, under the conditions in Theorem 3.4, or Remark 3.2,

$$g(\omega) = \frac{1}{\pi}\int_{-\infty}^{\infty} \cos(\omega t)\, r(t)\,\mathrm{d}t, \quad \omega \geq 0.$$

Note that $\int_{-\infty}^{\infty} \sin \omega t \,\mathrm{d}F(\omega) = 0$, since F is symmetric.

One should note that in almost all applications it is the real form that is presented in illustrations and the covariance function is represented as

$$r(t) = \int_0^\infty \cos(\omega t)\, g(\omega)\,\mathrm{d}\omega.$$

3.2.4 *Spectrum for stationary sequences*

For a stationary sequence $\{x_n, n \in \mathbb{Z}\}$, the covariance function $r(t)$ is defined only for $t \in \mathbb{Z}$. Instead of Bochner's theorem we have the following theorem, in the literature called *Herglotz' lemma.* The proof is similar to that of Bochner's theorem, and can be found, for example, in [22, §4.3].

Theorem 3.5 (Herglotz' lemma)**.** *A function $r(t), t \in \mathbb{Z}$, defined on the integers, is non-negative definite, and hence a covariance function for a stationary sequence, if and only if there exists a non-decreasing, right-continuous, and bounded real function $F(\omega)$ on $(-\pi, \pi]$, such that*

$$r(t) = \int_{-\pi+0}^{\pi} e^{i\omega t}\,\mathrm{d}F(\omega). \tag{3.13}$$

Note that we define the spectrum over the half-open interval to keep the right-continuity of $F(\omega)$. It is possible to move half the spectral mass in π to $-\pi$ without changing the representation (3.13).

One important consequence of (3.13) is that for every stationary sequence $\{x_n, n \in \mathbb{Z}\}$, there exists a stationary process $\{x(t), t \in \mathbb{R}\}$, defined for all $t \in \mathbb{R}$, that coincides in distribution with x_n for integer t. As we shall see in the next section, there are many such processes.

Theorem 3.6 (Inversion theorem). *a) If $\sum_{t=-\infty}^{\infty} |r(t)| < \infty$, the spectrum is absolutely continuous with spectral density given by*

$$f(\omega) = \frac{1}{2\pi} \sum_{t=-\infty}^{\infty} e^{-i\omega t}\, r(t).$$

b) In general, for $-\pi < \omega_1 < \omega_2 \le \pi$,

$$\begin{aligned} \widetilde{F}(\omega_2) - \widetilde{F}(\omega_1) &= \frac{1}{2\pi} r(0)(\omega_2 - \omega_1) \\ &\quad + \lim_{T\to\infty} \frac{1}{2\pi} \sum_{\substack{t=-T,\\ t\neq 0}}^{T} r(t)\, \frac{e^{-i\omega_2 t} - e^{-i\omega_1 t}}{-it}, \end{aligned}$$

where, as before, $\widetilde{F}(\omega)$ is defined as the average of left and right hand side limits of $F(\omega)$.

Definition 3.3 (Discrete white noise). *A stationary sequence of uncorrelated variables with mean zero is called a white noise sequence. It has $r(t) = 0$ for $t \neq 0$, and the spectral density is constant $f(\omega) = r(0)/2\pi, -\pi < \omega \le \pi$.*

3.2.5 Sampling and the alias effect

If a stationary process $\{x(t), t \in \mathbb{R}\}$, with spectral distribution $F(\omega)$ and covariance function $r(t) = \int_{-\infty}^{\infty} e^{i\omega t}\, \mathrm{d}F(\omega)$, is observed only at integer time points $t \in \mathbb{Z}$ one obtains a stationary sequence $\{x_n, n \in \mathbb{N}\}$ for which the covariance function is the same as that of $x(t)$, but restricted to integer values of t. In the spectral representation of $r_x(t)$ the factor $e^{i\omega t}$ is equal to $e^{i(\omega+2k\pi)t}$ for all integer t and k, and the spectrum may therefore be restricted to the interval $(-\pi, \pi]$:

$$\begin{aligned} r(t) &= \int_{-\infty}^{\infty} e^{i\omega t}\, \mathrm{d}F(\omega) = \sum_{k=-\infty}^{\infty} \int_{2k\pi-\pi+0}^{2k\pi+\pi} e^{i\omega t}\, \mathrm{d}F(\omega) \\ &= \int_{-\pi+0}^{\pi} e^{i\omega t} \sum_{k=-\infty}^{\infty} \mathrm{d}F(\omega + 2k\pi), \quad \text{for } t = 0, \pm 1, \pm 2, \ldots. \end{aligned} \tag{3.14}$$

This means that the spectral masses at all frequencies $\omega + 2k\pi$ for $k \neq 0$ are lumped together with the spectrum at the frequency ω and cannot be individually distinguished. This is the *aliasing* or *folding* effect of sampling a continuous time process.

If the sampling interval is not unity but observations are taken a time $d > 0$ apart, the spectrum can be folded on to the half open interval $(-\pi/d, \pi/d]$. If $x(t)$ has spectral density $f_x(\omega)$, $\omega \in \mathbb{R}$, the covariance function of the sampled sequence $x(kd), k \in \mathbb{Z}$, will have a spectral representation,

$$r(kd) = \int_{-\pi/d+0}^{\pi/d} e^{i\omega kd} \left\{ \sum_{k=-\infty}^{\infty} f_x(\omega + 2k\pi/d) \right\} \mathrm{d}\omega, \quad \text{for } k = 0, \pm 1, \pm 2, \ldots.$$

If $f_x(\omega)$ is zero (or approximately zero) for $|\omega| > \pi/d$, only the central term contributes to the lumped sum; otherwise the frequency content of the process is virtually shifted from high to low frequencies.

The critical frequency π/d is called the *Nyquist frequency*, and it is the highest frequency that can be uniquely identified when sampling a stationary process at regular intervals. However, in practice, sampling at irregular intervals may be more common than not, and then it is possible to identify also higher frequencies. We return to this problem in Section 3.3.7, Remark 3.4, in connection with spectral density estimation.

Example 3.3 (Undersampling). *Figure 3.1 shows the dramatic effect of undersampling a stationary process. The spectrum has its main mass between* 0.4 *and* 1, *and the two top rows show the full spectrum and a continuous simulation, sampled at distances $d = 0.04$ and $d = 1$, for which the Nyquist frequency is well above the maximum frequencies in the process. The sampling distance $d = 5$ with $\pi/d = 0.6382$ gives a somewhat distorted sequence, while $d = 10$ moves all energy to frequencies near* 0.

3.3 Spectral representation of a stationary process

3.3.1 The spectral process

A process with discrete spectrum

The spectral representation of stationary stochastic process is a subtle combination of a very concrete physical object, as used by S.O. Rice in [101], namely a simple harmonic sum,

$$x(t) = \sum_{k=1}^{N} A_k \cos(\omega_k t + \phi_k), \tag{3.15}$$

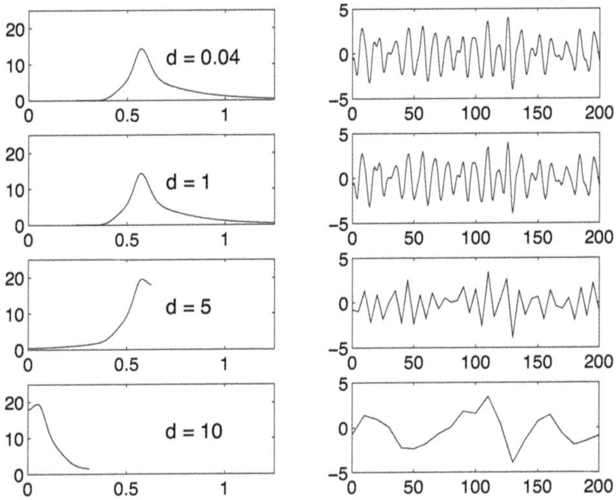

Figure 3.1 *Drastic effect of undersampling a stationary process with sampling interval* d *larger than the Nyquist frequency; left: spectrum of sampled sequence, right: sampled sequence.*

and a mathematically sophisticated generalization as a stochastic integral, introduced by H. Cramér [31] and M. Loève [84].

In the sum of random cosine functions, the $\omega_k > 0$, $k = 1, \ldots, N \leq \infty$, are fixed frequencies, and the amplitudes A_k and phases ϕ_k are independent random variables. At each repetition of an experiment, i.e., for each ω in the sample space Ω, all amplitudes and phases get new random values, and these remain fixed during the whole time span of the experiment. A new repetition gives new amplitudes and phases.

The statistical distribution of the amplitudes is not crucial for the theory, but in order that the process is stationary, it is necessary that the phases are uniformly distributed over the interval $(0, 2\pi)$. Otherwise the terms in (3.15) would have some preferred location relative to the origin. Thus, all ϕ_k are assumed to be uniformly distributed in $(0, 2\pi)$ and independent of the A_k.

To find the covariance function of the simple harmonic sum, it suffices to find the covariance function for each of the terms, since different terms are independent. Intuitively (and easily proved), $E(A_k \cos(\omega_k t + \phi_k)) = 0$ when the cosine function is randomly shifted over its period of 2π. Furthermore,[3]

$$E(A_k \cos(\omega_k(s+t) + \phi_k) \cdot A_k \cos(\omega_k s + \phi_k))$$

[3] $2 \cos\alpha \cos\beta = \cos(\alpha+\beta) + \cos(\alpha-\beta)$.

$$
\begin{aligned}
&= E(A_k^2)\cdot E(\cos(\omega_k(s+t)+\phi_k)\cos(\omega_k s+\phi_k))\\
&= \frac{E(A_k^2)}{2}\cdot(E(\cos\omega_k t)+E(\cos(\omega_k(2s+t)+2\phi_k)))\\
&= \frac{E(A_k^2)}{2}\cos\omega_k t = \frac{E(A_k^2)}{4}\left(e^{i\omega_k t}+e^{-i\omega_k t}\right).
\end{aligned}
$$

Since the terms in (3.15) are independent, the covariance function of the sum is

$$
r(t) = \sum_k E(A_k^2/2)\cos\omega_k t = \sum_k \frac{E(A_k^2)}{4}\left(e^{i\omega_k t}+e^{-i\omega_k t}\right), \tag{3.16}
$$

and the variance is $r(0) = \sum_k E(A_k^2/2)$. Thus, the contributions to the variance and covariances in the process come from random cosine functions with frequencies concentrated at $\{\pm\omega_k\}$, $k = 1,\dots,N$, with equal weights $E(A_k^2)/4$ on each side.

The spectral distribution is found from (3.16) and it has spectral mass $E(A_k^2)/4$ at $\pm\omega_k$, and the one-sided spectral distribution function is

$$
G(\omega) = \sum_{k;\omega_k\le\omega} E(A_k^2/2), \quad \omega \ge 0.
$$

In the two-sided spectral distribution function $F(\omega)$, the jumps have been split between positive and negative frequencies, and have size $E(A_k^2)/4$ at $\pm\omega_k$.

In the process (3.15), the cosine functions have been amplified by a random amplitude A_k and shifted by a random phase ϕ_k. To prepare for the generalization to a stochastic integral, we write the terms in complex form,

$$
x(t) = \sum_{k=1}^N e^{i\omega_k t}\frac{A_k}{2}e^{i\phi_k} + \sum_{k=1}^N e^{-i\omega_k t}\frac{A_k}{2}e^{-i\phi_k}, \tag{3.17}
$$

where the harmonic functions $e^{i\omega_k t}$ and $e^{-i\omega_k t}$ have been modulated by multiplication by the complex uncorrelated random factors $A_k e^{i\phi_k}$ and $A_k e^{-i\phi_k} = \overline{A_k e^{i\phi_k}}$. These modulating factors build the spectral process in the case of a discrete spectrum.

Basic properties of the spectral process

The general form of the spectral representation is as an integral, valid both for discrete and for continuous spectrum,

$$
x(t) = \int_{-\infty}^{\infty} e^{i\omega t}\,\mathrm{d}Z(\omega), \tag{3.18}
$$

where $\{Z(\omega), \omega \in \mathbb{R}\}$ is a complex-valued stochastic spectral process. Its parameter is *frequency*, ω, and it has mean zero and *orthogonal increments*, i.e.,

$$E((Z(\omega_4) - Z(\omega_3)) \cdot \overline{(Z(\omega_2) - Z(\omega_1))}) = 0,$$

for $\omega_1 < \omega_2 < \omega_3 < \omega_4$. In the discrete case (3.17), it is piecewise constant with jumps of size $A_k e^{\pm i\phi_k}$ at $\omega = \pm\omega_k$.

The variance of an increment in the spectral process is equal to the increment of the spectral distribution, i.e., for $\omega_1 < \omega_2$,

$$E(|Z(\omega_2) - Z(\omega_1)|^2) = F(\omega_2) - F(\omega_1). \tag{3.19}$$

One can summarize the relations between the spectral process $Z(\omega)$ and the spectral distribution $F(\omega)$ in an efficient symbolic way as

$$E\left(\mathrm{d}Z(\omega) \cdot \overline{\mathrm{d}Z(\mu)}\right) = \begin{cases} \mathrm{d}F(\omega), & \text{if } \omega = \mu, \\ 0, & \text{if } \omega \neq \mu. \end{cases} \tag{3.20}$$

To construct a stationary process from a complex spectral process

Now, let us start with a complex stochastic (spectral) process $\{Z(\omega), \omega \in \mathbb{R}\}$, with $E(Z(\omega)) = 0$ and with orthogonal increments, and define the function $F(\omega)$ by

$$F(\omega) = \begin{cases} E(|Z(\omega) - Z(0-)|^2), & \text{for } \omega \geq 0, \\ -E(|Z(\omega) - Z(0-)|^2), & \text{for } \omega < 0. \end{cases}$$

Since only the increments of $Z(\omega)$ are used in the theory, we can fix its value at any point, and we take $Z(0+) = 0$. Following the definition of a stochastic integral in Section 2.4, we can define the stochastic process $x(t)$ by letting

$$x(t) = \int_{-\infty}^{\infty} e^{i\omega t}\, \mathrm{d}Z(\omega) = \lim \sum e^{i\omega_k t}(Z(\omega_{k+1}) - Z(\omega_k)),$$

where the limit is in quadratic mean as the ω_k become asymptotically dense. It is then easy to prove that this gives a process with $E(x(t)) = 0$ and that its covariance function is given by the Fourier-Stieltjes transform of $F(\omega)$: use Theorem 2.17, and (3.20), to get

$$\begin{aligned} C(x(s),x(t)) &= E\left(\int_{\omega=-\infty}^{\infty} e^{i\omega s}\,\mathrm{d}Z(\omega)\cdot\overline{\int_{\mu=-\infty}^{\infty} e^{i\mu t}\,\mathrm{d}Z(\mu)}\right) \\ &= \int_{\omega=-\infty}^{\infty}\int_{\mu=-\infty}^{\infty} e^{i(\omega s-\mu t)}E\left(\mathrm{d}Z(\omega)\cdot\overline{\mathrm{d}Z(\mu)}\right) \\ &= \int_{\omega=-\infty}^{\infty} e^{i\omega(s-t)}\,\mathrm{d}F(\omega). \end{aligned}$$

This covariance obviously depends only on the difference $s-t$, and hence the process $x(t)$ is weakly stationary. It may be complex; it will be real if the increments of $Z(\omega)$ are Hermitian, $\mathrm{d}Z(-\omega)=\overline{\mathrm{d}Z(\omega)}$.

3.3.2 *The spectral theorem*

We are now ready to prove the central result in the theory of stationary processes, namely that every mean square continuous, weakly stationary process $\{x(t), t\in\mathbb{R}\}$ with mean zero has a spectral representation, $x(t) = \int_{-\infty}^{\infty} e^{i\omega t}\,\mathrm{d}Z(\omega)$, where $Z(\omega)\in\mathscr{H}(x)=\mathscr{S}(x(t),t\in\mathbb{R})$, i.e., $Z(\omega)$ is an element in the Hilbert space which consists of limits of linear combinations of $x(t)$-values.[4]

In fact, one can define $Z(\omega)$ explicitly for a continuity point ω of the spectral distribution function F,

$$Z(\omega)=\lim_{T\to\infty}\frac{1}{2\pi}\int_{-T}^{T}\frac{e^{-i\omega t}-1}{-it}x(t)\,\mathrm{d}t, \tag{3.21}$$

and prove that it has all the required properties. This is the technique used in Yaglom's classical book [125]; see Exercise 3:16.

We shall present a functional analytic proof, as in [35], and find a relation between the Hilbert space $\mathscr{H}(x)$ and the function Hilbert space $\mathscr{H}(F)=\mathscr{L}_2(F)$, the set of all functions $g(\omega)$ with $\int|g(\omega)|^2\,\mathrm{d}F(\omega)<\infty$. The inner products in the two spaces are

$$(u,v)_{\mathscr{H}(x)}=E(u\overline{v}),\quad (g,h)_{\mathscr{H}(F)}=\int_{-\infty}^{\infty} g(\omega)\overline{h(\omega)}\,\mathrm{d}F(\omega). \tag{3.22}$$

[4]See Appendix A.6 for basic facts about Hilbert spaces. Note in particular the interpretation of zero mean random variables as elements in a Hilbert space where scalar product between elements is defined as their covariance, and convergence is equivalent to convergence in quadratic mean of random variables.

We start by the definition of an *isometry*.

Definition 3.4. *A linear one-to-one mapping f between two Hilbert spaces X and Y is called an isometry if it conserves the inner product, i.e., $(u,v)_X = (f(u), f(v))_Y$. In particular $\|u-v\|_X = \|f(u) - f(v)\|_Y$, so distances are also preserved.*

Theorem 3.7 (The spectral theorem). *If $\{x(t), t \in \mathbb{R}\}$ is a continuous stationary process with mean zero and spectral distribution $F(\omega)$, there exists a complex-valued spectral process $\{Z(\omega), \omega \in \mathbb{R}\}$, $Z(\omega) \in \mathscr{H}(x)$, with orthogonal increments, such that*

$$E\left(|Z(\omega_2) - Z(\omega_1)|^2\right) = F(\omega_2) - F(\omega_1),$$

for $\omega_1 < \omega_2$ and

$$x(t) = \int_{-\infty}^{\infty} e^{i\omega t}\, \mathrm{d}Z(\omega).$$

Proof. We shall build an isometry between the Hilbert space of random variables $\mathscr{H}(x) = \mathscr{S}(x(t), t \in \mathbb{R})$, and the Hilbert space $\mathscr{H}(F) = \mathscr{L}_2(F)$, of complex functions with scalar products defined as (3.22). The corresponding norms in the two spaces are

$$\|y\|^2_{\mathscr{H}(x)} = E(|y|^2), \quad \|g\|^2_{\mathscr{H}(F)} = \int_{-\infty}^{\infty} |g(\omega)|^2\, \mathrm{d}F(\omega).$$

Now take a *fixed* time t and consider the random variable $x(t)$ and the function $e^{i \cdot t}$, regarded as a "function of the dot." Then, note that

$$\|x(t)\|^2_{\mathscr{H}(x)} = E(|x(t)|^2) = r(0),$$

$$\|e^{i \cdot t}\|^2_{\mathscr{H}(F)} = \int_{-\infty}^{\infty} |e^{i\omega t}|^2\, \mathrm{d}F(\omega) = r(0),$$

which means that $x(t)$ has the same length, regarded as an element of $\mathscr{H}(x)$, as $e^{i \cdot t}$ has, regarded as an element of $\mathscr{H}(F)$, i.e., $\|x(t)\|_{\mathscr{H}(x)} = \|e^{i \cdot t}\|_{\mathscr{H}(F)}$.

Furthermore, scalar products are preserved,

$$(x(s),x(t))_{\mathscr{H}(x)} = E(x(s)\overline{x(t)}) = \int_{-\infty}^{\infty} e^{i\omega s}\overline{e^{i\omega t}}\, \mathrm{d}F(\omega) = (e^{i\cdot s}, e^{i\cdot t})_{\mathscr{H}(F)}.$$

This is the start of our isometry: $x(t)$ and $e^{i\cdot t}$ are corresponding elements in the two spaces. Instead of looking for random variables $Z(\omega_0)$ in $\mathscr{H}(x)$ we shall look for functions $g_{\omega_0}(\cdot)$ in $\mathscr{H}(F)$ with the corresponding properties.

Step 1: Start with the random variable $x(t)$, for a fixed t, and the ω-function $e^{i\omega t}$, and extend the correspondence to finite linear combinations of $x(t)$ and $e^{i\omega t}$ by letting y and $g(\cdot)$, with

$$y = \alpha_1 x(t_1) + \ldots + \alpha_n x(t_n), \tag{3.23}$$

$$g(\omega) = \alpha_1 e^{i\omega t_1} + \ldots + \alpha_n e^{i\omega t_n}, \tag{3.24}$$

be corresponding elements. Check by yourself that scalar product is preserved, i.e.,

$$(y_1, y_2)_{\mathscr{H}(x)} = (g_1, g_2)_{\mathscr{H}(F)}.$$

Step 2: Distances are preserved, i.e., $\|y_1 - y_2\|_{\mathscr{H}(x)} = \|g_1 - g_2\|_{\mathscr{H}(F)}$, so $y_1 = y_2$ if and only if $g_1 = g_2$, where equality means equal with probability one, and almost everywhere, respectively.

Step 3: Convergence in the two spaces means the same. If $y_1, y_2, \ldots$ converges towards y in $\mathscr{H}(x)$, and $g_1, g_2, \ldots$ are the corresponding elements in $\mathscr{H}(F)$, then

$$\|y_n - y_m\|_{\mathscr{H}(x)} \to 0 \quad \text{implies} \quad \|g_n - g_m\|_{\mathscr{H}(F)} \to 0,$$

and since $\mathscr{H}(F)$ is complete in the sense of Definition A.8(5), there exists a limit element $g \in \mathscr{H}(F)$ such that $\|y\|_{\mathscr{H}(x)} = \|g\|_{\mathscr{H}(F)}$. The reverse implication also holds. Thus, we have extended the correspondence between the two spaces to all limits of sums of $x(t)$-variables and $e^{i\omega t}$-functions.[5]

Step 4: The correspondence can be extended to all of $\mathscr{H}(F)$ and $\mathscr{H}(x)$. The set $\mathscr{H}(x)$ consists by definition of limits of linear combinations of $x(t_k)$, and every function in $\mathscr{L}_2(F)$ can be approximated by a polynomial in $e^{i\omega t_k}$ for

[5]Remember that in $\mathscr{H}(F)$ all $e^{i\omega t}$ are functions of frequency ω, and that we have one function for every t. Similarly, in $\mathscr{H}(x)$, $x(t) = x(t,\omega)$ is a function of outcome ω, not to be confused with the frequency ω.

different t_k-s. This is the famous *Stone-Weierstrass theorem*. We have then found the isometry between $\mathscr{H}(x)$ and $\mathscr{H}(F)$: if u and v are elements in $\mathscr{H}(x)$ and $f(u)$ and $f(v)$ the corresponding elements in $\mathscr{H}(F)$, then $\|u - v\|_{\mathscr{H}(x)} = \|f(u) - f(v)\|_{\mathscr{H}(F)}$.

Step 5: We now turn our interest to a special function g_{ω_0} in $\mathscr{H}(F)$, which will correspond to $Z(\omega_0)$,

$$g_{\omega_0}(\omega) = \begin{cases} 1 \text{ for } \omega \leq \omega_0, \\ 0 \text{ for } \omega > \omega_0. \end{cases}$$

Obviously, $\|g_{\omega_0}\|^2_{\mathscr{H}(F)} = \int_{-\infty}^{\infty} |g_{\omega_0}(\omega)|^2 \, dF(\omega) = \int_{-\infty}^{\omega_0} \mathrm{d}F(\omega)$, and, with $\omega_1 < \omega_2$,

$$\|g_{\omega_2} - g_{\omega_1}\|^2_{\mathscr{H}(F)} = F(\omega_2) - F(\omega_1).$$

Step 6: Let $Z(\omega)$ be the element in $\mathscr{H}(x)$ that corresponds to $g_\omega(\cdot)$ in $\mathscr{H}(F)$. It is easy to see that $Z(\omega)$ is a process with orthogonal increments and incremental variance given by $F(\omega)$:

$$\begin{aligned} &E\left((Z(\omega_4) - Z(\omega_3)) \cdot \overline{(Z(\omega_2) - Z(\omega_1))}\right) \\ &\quad = \int_{-\infty}^{\infty} (g_{\omega_4}(\omega) - g_{\omega_3}(\omega))(g_{\omega_2}(\omega) - g_{\omega_1}(\omega)) \, \mathrm{d}F(\omega) = 0, \\ &E(|Z(\omega_2) - Z(\omega_1)|^2) = F(\omega_2) - F(\omega_1), \end{aligned}$$

for $\omega_1 < \omega_2 < \omega_3 < \omega_4$.

Step 7: It remains to prove that $Z(\omega)$ is the spectral process to $x(t)$, i.e., that

$$x(t) = \int_{-\infty}^{\infty} e^{i\omega t} \, \mathrm{d}Z(\omega) = \lim_{n\to\infty} \sum e^{i\omega_k t} (Z(\omega_{k+1}) - Z(\omega_k)) = \lim_{n\to\infty} S_t^{(n)},$$

for an increasingly dense subdivision $\{\omega_k\}_1^n$ of $\mathbb{R}$ with $\omega_k < \omega_{k+1}$. But we have that $x(t) \in \mathscr{H}(x)$ and $e^{i \cdot t} \in \mathscr{H}(F)$ are corresponding elements. Further,

$$e^{i\omega t} = \lim_{n\to\infty} \sum e^{i\omega_k t} (g_{\omega_{k+1}}(\omega) - g_{\omega_k}(\omega)) = \lim_{n\to\infty} g_t^{(n)}(\omega),$$

where the difference of g-functions within parentheses is equal to one for $\omega_k < \omega \leq \omega_{k+1}$ and zero otherwise. The limits are in $\mathscr{H}(F)$, i.e., $g_t(\cdot) = e^{i \cdot t} = \lim g_t^{(n)}(\cdot)$. Since $S^{(n)}(t)$ corresponds to $g_t^{(n)}(\cdot)$ and limits are preserved under the isometry, we have that $x(t) = \lim S_t^{(n)}$.

$$
\begin{array}{ccc}
g_t^{(n)} & \xrightarrow{\ \mathscr{H}(F)\ } & g_t \\
\uparrow\downarrow & & \uparrow\downarrow \\
\int e^{i\omega t}\,dZ(\omega) & \xleftarrow{\hspace{1.5cm}} S_t^{(n)} \xrightarrow{\ \mathscr{H}(x)\ } & x(t)
\end{array}
$$

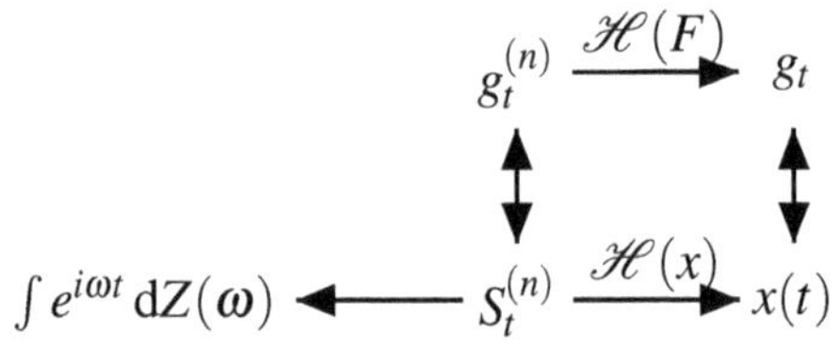

The diagram illustrates the convergence in $\mathscr{H}(F)$ and $\mathscr{H}(x)$, between the corresponding elements, and the equality of the limits of $S_t^{(n)}$. This ends the proof. $\qquad\square$

Corollary 3.1. *Every $y \in \mathscr{H}(x)$ can be expressed as*

$$
y = \int_{-\infty}^{\infty} g(\omega)\,dZ(\omega),
$$

for some function $g(\omega) \in \mathscr{H}(F)$, and

$$
E(|y|^2) = \int_{-\infty}^{\infty} |g(\omega)|^2\,dF(\omega).
$$

Proof. Every $y \in \mathscr{H}(x)$ is the limit of a sequence of linear combinations,

$$
y = \lim_{n} \sum_{k} \alpha_k^{(n)} x(t_k^{(n)}) = \lim_{n} \int_{\omega} \sum_{k} \alpha_k^{(n)} e^{i\omega t_k^{(n)}}\,dZ(\omega),
$$

and $g^{(n)}(\cdot) = \sum \alpha_k^{(n)} e^{i \cdot t_k^{(n)}}$ converges in $\mathscr{H}(F)$ to some function $g(\cdot)$, and then

$$
\int_{-\infty}^{\infty} g^{(n)}(\omega)\,dZ(\omega) \to \int_{-\infty}^{\infty} g(\omega)\,dZ(\omega)
$$

in $\mathscr{H}(x)$, which was to be shown. $\qquad\square$

3.3.3   Discrete and continuous spectrum

If the spectral distribution function $F(\omega)$ is piecewise constant with jumps of height ΔF_k at ω_k, then $Z(\omega)$ is also piecewise constant with jumps of random size ΔZ_k at ω_k, and $E(|\Delta Z_k|^2) = \Delta F_k$, so $x(t) = \sum \Delta Z_k e^{i\omega_k t}$. Note that the covariance function then has the corresponding form, $r(t) = \sum \Delta F_k e^{i\omega_k t}$.

In the general spectral representation, the complex $Z(\omega)$ defines a random amplitude and phase for the different components $e^{i\omega t}$. This fact is perhaps difficult to appreciate in the integral form, but is easily understood for processes with discrete spectrum, where the spectral process has jumps ΔZ_k. Take the polar form, $\Delta Z_k = |\Delta Z_k| e^{i \arg \Delta Z_k} = \rho_k e^{i\phi_k}$. Then,

$$x(t) = \sum \rho_k e^{i(\omega_k t + \phi_k)} = \sum \rho_k \cos(\omega_k t + \phi_k) + i \sum \rho_k \sin(\omega_k t + \phi_k).$$

For a real process, the imaginary part vanishes, and we have the now familiar representation (3.15), with amplitudes $A_k = \rho_k$,

$$x(t) = \sum A_k \cos(\omega_k t + \phi_k). \tag{3.25}$$

If the phases ϕ_k are independent and uniformly distributed between 0 and 2π, the process $x(t)$ is also strictly stationary.

For discrete spectrum we have the following important ways of recovering the discrete components of $F(\omega)$ and of $Z(\omega)$; the proof of the properties is part of the Fourier theory.

Theorem 3.8. *If $F(\omega)$ is a step function, with jumps of size ΔF_k at ω_k, then*

$$\lim_{T\to\infty} \frac{1}{T} \int_0^T r(t) e^{-i\omega_k t} \, \mathrm{d}t = \Delta F_k, \tag{3.26}$$

$$\lim_{T\to\infty} \frac{1}{T} \int_0^T r(t)^2 \, \mathrm{d}t = \sum_k (\Delta F_k)^2, \tag{3.27}$$

$$\lim_{T\to\infty} \frac{1}{T} \int_0^T x(t) e^{-i\omega_k t} \, \mathrm{d}t = \Delta Z_k. \tag{3.28}$$

In particular, from (3.27), if $r(t) \to 0$ as $t \to \infty$, the spectral distribution F is a continuous function without jumps.

If the spectrum is absolutely continuous, with $F(\omega) = \int_{-\infty}^{\omega} f(x)\,\mathrm{d}x$, then one can normalize the increments of $Z(\omega)$ by dividing by $\sqrt{f(\omega)}$, at least when $f(\omega) > 0$, and use the spectral representation in the form

$$x(t) = \int_{-\infty}^{\infty} e^{i\omega t} \sqrt{f(\omega)} \, \mathrm{d}\widetilde{Z}(\omega), \tag{3.29}$$

with

$$\widetilde{Z}(\omega) = \int_{\{x \le \omega; f(x)>0\}} \frac{dZ(x)}{\sqrt{f(x)}},$$

and

$$E\left(\left|d\widetilde{Z}(\omega)\right|^2\right) = \frac{dF(\omega)}{f(\omega)} = d\omega.$$

Even if $\widetilde{Z}(\omega)$ is not a true spectral process it is useful as a model for *white noise*. We will later meet the "constant spectral density" for the Poisson process in Section 3.5.4, and we shall use the formulation several times in later sections.

3.3.4 *One-sided spectral representation of a real process*

For a real process $x(t)$, the complex spectral representation has to produce a real integral. This of course requires $Z(\omega)$ to have certain symmetry properties, which we shall now investigate. Write ΔZ_0 for a possible Z-jump at $\omega = 0$. Then

$$\begin{aligned} x(t) = \int_{-\infty}^{\infty} e^{i\omega t}\, dZ(\omega) &= \Delta Z_0 + \int_{0+}^{\infty} e^{i\omega t}\, dZ(\omega) + \int_{0+}^{\infty} e^{-i\omega t}\, dZ(-\omega) \\ &= \Delta Z_0 + \int_{0+}^{\infty} \cos \omega t \cdot (dZ(\omega) + dZ(-\omega)) \\ &\quad + i \int_{0+}^{\infty} \sin \omega t \cdot (dZ(\omega) - dZ(-\omega)). \end{aligned}$$

For this to be real for all t it is necessary that ΔZ_0 is real, and also that $dZ(\omega) + dZ(-\omega)$ is real, and $dZ(\omega) - dZ(-\omega)$ is purely imaginary, which implies the important symmetry of the increments,

$$dZ(-\omega) = \overline{dZ(\omega)},$$

i.e., $\arg dZ(-\omega) = -\arg dZ(\omega)$ and $|dZ(-\omega)| = |dZ(\omega)|$. (These properties also imply that $x(t)$ is real.)

Now, introduce two real processes $\{u(\omega), 0 \le \omega < \infty\}$ and $\{v(\omega), 0 \le \omega < \infty\}$, with mean zero, and with $u(0-) = v(0-) = 0$, $du(0) = \Delta Z_0$, $v(0+) = 0$, and such that, for $\omega > 0$,

$$\begin{aligned} du(\omega) &= dZ(\omega) + dZ(-\omega) = 2\Re\, dZ(\omega) \\ dv(\omega) &= i(dZ(\omega) - dZ(-\omega)) = -2\Im\, dZ(\omega). \end{aligned}$$

The real spectral representation of $x(t)$ will then take the form

$$
\begin{aligned}
x(t) &= \int_0^\infty \cos \omega t \, \mathrm{d}u(\omega) + \int_0^\infty \sin \omega t \, \mathrm{d}v(\omega) \\
&= \int_{0+}^\infty \cos \omega t \, \mathrm{d}u(\omega) + \int_0^\infty \sin \omega t \, \mathrm{d}v(\omega) + \mathrm{d}u(0).
\end{aligned}
\tag{3.30}
$$

It is easily checked that with the one-sided spectral distribution function $G(\omega)$, defined by (3.11),

$$
E(\mathrm{d}u(\omega) \cdot \mathrm{d}v(\mu)) = 0, \quad \text{for all } \omega \text{ and } \mu, \tag{3.31}
$$

$$
E(\mathrm{d}u(\omega)^2) = \begin{cases} 2\mathrm{d}F(\omega) = \mathrm{d}G(\omega), & \omega > 0, \\ \mathrm{d}F(0) = \mathrm{d}G(0), & \omega = 0, \end{cases} \tag{3.32}
$$

$$
E(\mathrm{d}v(\omega)^2) = \begin{cases} 2\mathrm{d}F(\omega) = \mathrm{d}G(\omega), & \omega > 0, \\ \mathrm{d}F(0) = \mathrm{d}G(0), & \omega = 0. \end{cases} \tag{3.33}
$$

In almost all applications, when a spectral density for a time process $x(t)$ is presented, it is the one-sided density $g(\omega) = 2f(\omega) = \mathrm{d}G(\omega)/\mathrm{d}\omega$ that is given.

The representation (3.30) is useful for Monte Carlo simulation of a stationary process. We will deal with that application in the context of Gaussian processes in Section 3.4.1, with detailed recipes for simulation given in Appendix B.

3.3.5 *Why negative frequencies?*

One may ask why at all one should use negative frequencies, $\omega < 0$, in the spectral representation of a real process? Since the two complex functions $\mathrm{d}Z(\omega)\, e^{i\omega t}$ and $\mathrm{d}Z(-\omega)\, e^{-i\omega t} = \overline{\mathrm{d}Z(\omega)\, e^{i\omega t}}$, which build up the spectral representation, encircle the origin in the counterclockwise and clockwise directions with time their contribution to the total $x(t)$-process is real, and there seems to be no point in using the complex formulation.

One reason for the complex approach is, besides from some mathematical convenience, that negative frequencies are necessary when we want to build models for simultaneous *time and space* processes, for example a random water wave, which moves with time t. As described in Section 1.5.3, a random wave model, at time t and one-dimensional location s, can be built from elementary harmonics $A_\omega \cos(\omega t - \kappa s + \phi_\omega)$, cf. (7.27), where ω is the frequency in radians per time unit and κ is the wave number in radians per length unit. If ω and κ have the same sign the elementary wave moves to the right with increasing t and if they have opposite sign it moves to the left. In stochastic wave models for infinite water depth the dispersion relation states that $\kappa = \omega^2/g > 0$, with both positive and negative ω possible.

The (average) "energy" attached to the elementary wave $A_\omega \cos(\omega t - \kappa s + \phi_\omega)$ is $A_\omega^2/2$ or, in the random case, $E(A_\omega^2)/2$. If one observes the wave only at a single point $s = s_0$, it is not possible to determine its direction, and there is no way to decide, from the observations, how the wave energy should be divided between the two directions. One therefore chooses to divide the elementary energy equally between ω and $-\omega$.

If we have more than one observation point, perhaps a whole space interval of observations, we can determine wave direction and see how the energy should be divided between positive and negative ω. This can be done by splitting the process into two independent components, one $\{x_+(t), t \in \mathbb{R}\}$ with only positive frequencies, moving to the right, and one $\{x_-(t), t \in \mathbb{R}\}$ with only negative frequencies, moving to the left. The spectra on the positive and negative side need not be equal.

For a wave model with one time parameter and a two-dimensional space parameter (s_1, s_2), the wave direction is taken care of by a two-dimensional wave number $\kappa(\cos\theta, \sin\theta)$, and the spectrum is defined by one part that defines the energy distribution over frequencies and absolute wave numbers, and one directional spreading part that determines the energy for different wave directions. The spreading may depend on frequency and absolute wave number; see also Section 7.2.4, page 231.

For some practical aspects of negative frequencies in signal processing, see Remark 5.1 on page 157.

3.3.6 *Spectral representation of stationary sequences*

A stationary sequence $\{x(t), t \in \mathbb{Z}\}$ can be thought of as a stationary process which is observed only at integer times t. The spectral representation can then be restricted to ω-values only in $(-\pi, \pi]$ as for the spectral distribution. We may add all the spectral increments over $\omega + 2k\pi$, $k \in \mathbb{Z}$, to get the spectral formula

$$x(t) = \int_{-\pi+}^{\pi} e^{i\omega t}\, \mathrm{d}Z(\omega), \tag{3.34}$$

with the following explicit expression for the spectral process,

$$Z(\omega) = \frac{1}{2\pi}\left\{\omega x(0) - \sum_{k\neq 0} \frac{e^{-i\omega k}}{ik} x(k)\right\}.$$

Obviously, (3.34) defines a process $\{x(t), t \in \mathbb{R}\}$ also in continuous time with spectral process $Z(\omega)$ determined by $x(t)$ at integer points. Thus, it is possible to interpolate the process $x(t)$ exactly between the integer points. We return to this phenomenon in connection with the *Sampling theorem* in Section 5.2.

3.3.7 Fourier transformation of data

There is an obvious similarity, but also a fundamental difference, between the two Fourier representations

$$x(t) = \int_{-\pi+}^{\pi} e^{i\omega t}\, \mathrm{d}Z(\omega), \quad r(t) = \int_{-\pi+}^{\pi} e^{i\omega t}\, \mathrm{d}F(\omega),$$

of a stationary sequence $x(t), t \in \mathbb{Z}$, and its covariance function $r(t)$. The presentation in this chapter started from the spectral distribution $F(\omega)$, from which we derived the existence of the spectral process $Z(\omega)$. We have seen that under certain conditions the spectral density $f(\omega)$ can been obtained as the (inverse) Fourier transform

$$f(\omega) = \frac{\mathrm{d}F(\omega)}{\mathrm{d}\omega} = \frac{1}{2\pi} \sum_{t=-\infty}^{\infty} e^{-i\omega t} r(t), \tag{3.35}$$

and a natural question is if one can invert also $x(t)$ to find $\mathrm{d}Z(\omega)/\mathrm{d}\omega$. However, for reasons to be explained, this is not possible.

Historically, the development of the stationary process theory started, not from the statistical covariance concept, but from (Fourier) spectral analysis of data. It is probably fair to say that the theory of stationary processes as it is presented in this book has its roots in practical search for periodicities in physical and economic data; see the historical review by D. Brillinger [21].

From the late nineteenth century there are several examples of Fourier transformation of an *observed data series*, supposedly a realization of part of a stationary sequence, most notably Sir Arthur Schuster: *On the investigation of hidden periodicities with application to a supposed 26 day period of meteorological phenomena* (1898) [113], and A.A. Michelson and S.W. Stratton: *A new harmonic analyzer* (1898) [92]. These studies resulted in a number of false discoveries of apparent dominating frequencies in data series. It was not until 1929 that R.A. Fisher [51] developed a statistical theory in order to distinguish between an apparent and an established periodicity. In this section we shall investigate the statistical properties of such a transform of a finite data series from a stationary random sequence, and as we shall see, there is a fundamental difference between sequences with discrete and with continuous spectrum.

Suppose $x_0, x_1, \ldots, x_{n-1}$ come from a stationary sequence $\{x_n,\, n \in \mathbb{N}\}$ and define the Fourier transform

$$z_n(\omega) = \frac{1}{\sqrt{2\pi}} \sum_0^{n-1} x_k e^{-i\omega k}. \tag{3.36}$$

Definition 3.5 (Periodogram). *The function*

$$\begin{aligned} f_n^{\mathrm{per}}(\omega) &= \frac{1}{n}\left|z_n(\omega)\right|^2 \\ &= \frac{1}{2\pi n}\left\{\left(\sum_0^{n-1} x_k \cos \omega k\right)^2 + \left(\sum_0^{n-1} x_k \sin \omega k\right)^2\right\} \end{aligned}$$

is called the periodogram of the series $x_0, x_1, \ldots, x_{n-1}$.

Periodogram for sequences with discrete spectrum

We first consider a rather special discrete spectrum to see what happens. Suppose the observation series $x_0, \ldots, x_{N-1}$ comes from a stationary sequence with spectrum located at discrete frequencies $\omega_k = 2\pi k/N$, $k = 0, 1, \ldots, N/2-1$, the "Fourier frequencies." We know from Section 3.3.3, (3.25), that such a process has a spectral representation

$$x_t = \sum_{k=0}^{N/2-1} A_k \cos(\omega_k t + \phi_k), \tag{3.37}$$

with random amplitudes A_k and random, uniformly distributed phases ϕ_k.

The Fourier transform $z_N(\omega)$ of the data series is a function of frequency ω, defined for all ω. If we calculate its value for $\omega = 2\pi k_0/N$, $k_0 = 0, 1, \ldots, N-1$, we get, for $k_0 = 0, \ldots, N/2-1$,

$$\begin{aligned} z_N\left(\frac{2\pi k_0}{N}\right) &= \frac{1}{\sqrt{2\pi}} \sum_{t=0}^{N-1} x_t e^{-i2\pi(k_0/N)t} \\ &= \frac{1}{\sqrt{2\pi}} \sum_k \frac{A_k}{2}\left\{e^{i\phi_k} \sum_t e^{i2\pi(k-k_0)t/N} + e^{-i\phi_k} \sum_t e^{-i2\pi(k+k_0)t/N}\right\} \\ &= \frac{1}{\sqrt{2\pi}} \frac{N}{2} A_{k_0} e^{i\phi_{k_0}}, \end{aligned}$$

since the sums within brackets equal 0 for $k \neq \pm k_0, \mod N$.

We see that the normalized Fourier transform of the data series, calculated at frequencies $\omega_{k_0} = 2\pi k_0/N$, extracts the amplitude A_{k_0} and the phase ϕ_{k_0}. Taking absolute value, $|z_N(2\pi k_0/N)|^2$, one gets $(2\pi)^{-1}N^2A_{k_0}^2/4$. One should notice that

$$I = \frac{2\pi}{N}\sum_{k=0}^{N-1}\frac{1}{N}\left|z_N\left(\frac{2\pi k}{N}\right)\right|^2 = \sum_{k=0}^{N/2-1} A_k^2/2. \tag{3.38}$$

The expectation of I will be simply

$$E(I) = \sum_k \frac{E(A_k^2)}{2} = r(0) = V(x_t). \tag{3.39}$$

With statistical terminology one can interpret this relation as if we divided the total variation of x_t into contributions from the discrete frequencies in the spectrum, in a sort of *analysis of variance*.

Periodogram for sequences with continuous spectrum

What happens with the periodogram if the observations come from a random sequence with continuous spectrum? We start with an illustrative example.

Example 3.4. *We compare two types of periodograms. One is taken from a registration of the tone a, 220[Hz], played on a cello. The sampling rate is 8192 times per second. The other one is taken from a series of measurements of the ocean water level taken by a buoy anchored somewhere off the African coast, sampled four times per second.*

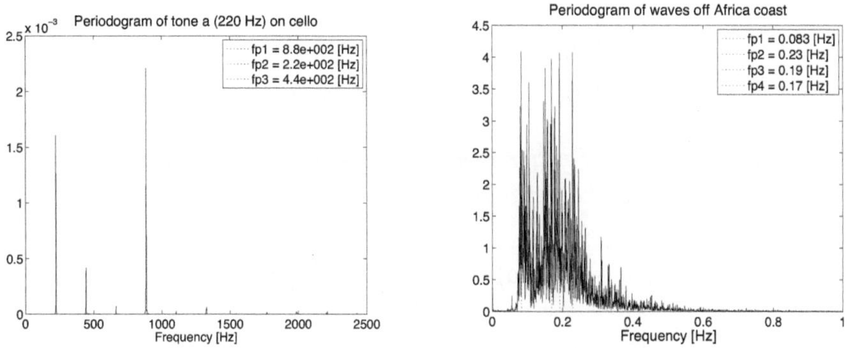

Figure 3.2 *Two types of periodogram. Left: The tone "a," 220[Hz], together with overtones, played on a cello, sampled 8192 times per second; $N = 9288$. Right: Water waves outside the African west coast, sampled four times per second; $N = 9524$.*

Figure 3.2 shows the two periodograms.[6] *To the left is shown the tone a. The tone is a pure tone together with overtones; the most prominent one is the third, 880[Hz]. One can imagine this as a stochastic process of the type (3.37), even if the phases in a cello tone in reality are not random and uniformly distributed, but rather quite characteristic for the instrument. The four most dominant frequencies are marked in the figure.*

The wave periodogram is quite different. The four most prominent frequencies are here 0.083[Hz], 0.17[Hz], 0.19[Hz], 0.23[Hz]. However, there is no reason why these particular harmonics should dominate the process; we recognize the spectrum as continuous and the spectral density to be a smooth function with a maximum somewhere near 0.2[Hz].

Why is the wave periodogram in the example so irregular? One way to understand the behavior is to imagine the continuous spectral density discretized, and its whole mass distributed to the discrete frequencies $\omega_k = 2\pi k/N$, $k = 0, 1, \ldots, N/2-1$. We have then to deal with a process of the type (3.37), and as we have seen, the periodogram reconstructs the squares of the amplitudes A_k at the discrete frequencies $2\pi k/N$. Since these are independent random variables, we get the irregular periodogram, regardless of the number of observations. If the sample size increases, the periodogram tries to estimate the squared amplitudes at more and more frequencies – it does not improve the individual estimates.

Remark 3.3 (Spectral density estimation)**.** *The periodogram is related to a statistical estimator of the covariance function $r(t) = E(x(s+t)x(s))$ for a stationary process with $E(x(t)) = 0$. Set*

$$r_N^*(t) = \begin{cases} \frac{1}{N}\sum_{s=0}^{N-1-t} x_{s+t}x_s, & \textit{for } t = 0, 1, \ldots, N-1, \\ r_N^*(-t), & \textit{for } t < 0, \\ 0, & \textit{otherwise}, \end{cases} \tag{3.40}$$

and insert it into the Fourier inversion formula (3.35) instead of $r(t)$, and simplify, and you get the periodogram,

$$f_n^{\text{per}}(\omega) = \frac{1}{2\pi}\sum_{t=-\infty}^{\infty} e^{-i\omega t} r_N^*(t). \tag{3.41}$$

This means that the periodogram is the natural spectrum to the covariance estimator (3.40).

One should note that $r_N^*(t)$ is a biased estimator of $r(t)$ since $E(r_N^*(t)) =$

[6] In the example we use the natural frequency unit "Hz," $f = \omega/2\pi$.

$\frac{N-t}{N} r(t)$. *However, the unbiased estimator, dividing with* $N-t$ *instead of* N, *is not a true covariance function and it is not non-negative definite, cf. Section 3.1.2, while* $r_N^*(t)$ *is.*

One should also note that the estimates are based on different numbers of observations, and that they are quite unreliable for large values of t. *There are several standard techniques to smooth the irregular periodogram to get a good estimate of the spectral density; see for example [22, 96] or [82, Ch. 9]. The three most common techniques are to split the data series into subseries and average the periodograms over the subseries, to take a weighted average of the periodogram values over neighboring frequencies, and to downweight those covariance estimates in (3.41) that are least reliable.*

Remark 3.4 (Uneven sampling intervals). *Most standard spectrum estimation techniques require evenly spaced samples at distance* d, *and then the highest identifiable frequency is the Nyquist frequency,* π/d. *It is an amazing fact that uneven sampling distances may considerably increase the range of estimable frequencies. An example where uneven sampling comes naturally is the estimation of periodicity of variable stars; the paper [47] contains a readable account of the technique.*

3.4 Gaussian processes

3.4.1 The Gaussian spectral process

For Gaussian processes $\{x(t), t \in \mathbb{R}\}$, the spectral process, $Z(\omega) = u(\omega) + iv(\omega)$, is complex Gaussian, with independent real and imaginary parts. Since $Z(\omega)$ is an element in the space $\mathscr{H}(x)$ of limits of linear combinations of x-variables, this is immediate from the characterization of Gaussian processes as those processes for which all linear combinations have a Gaussian distribution. Also the real spectral processes $\{u(\omega), 0 \le \omega < \infty\}$ and $\{v(\omega), 0 \le \omega < \infty\}$ are Gaussian with uncorrelated increments (3.31) and, hence, independent increments.

The sample paths of $u(\omega)$ and $v(\omega)$ can be continuous, or they could contain jump discontinuities, which then are normal random variables. In the continuous case, when there is a spectral density $f(\omega)$, they are almost like Wiener processes, and they can be transformed into Wiener processes by normalizing the incremental variance. In analogy with $\widetilde{Z}(\omega)$ in (3.29), define $w_1(\omega)$ and $w_2(\omega)$ from $u(\omega)$ and $v(\omega)$ by

$$w_1(\omega) = \int_{\{x \le \omega; f(x) > 0\}} \frac{\mathrm{d}u(x)}{\sqrt{2f(x)}}, \quad w_2(\omega) = \int_{\{x \le \omega; f(x) > 0\}} \frac{\mathrm{d}v(x)}{\sqrt{2f(x)}} \tag{3.42}$$

to get, from Theorem 2.18,

$$x(t) = \int_0^\infty \sqrt{2f(\omega)} \cos \omega t \, \mathrm{d}w_1(\omega) + \int_0^\infty \sqrt{2f(\omega)} \sin \omega t \, \mathrm{d}w_2(\omega). \quad (3.43)$$

Note that if $f(\omega) > 0$, then $E(\mathrm{d}w_1(\omega)^2) = E(\mathrm{d}w_2(\omega)^2) = \mathrm{d}\omega$. When $f(\omega) = 0$ we can choose $\mathrm{d}w_1(\omega), \mathrm{d}w_2(\omega)$ at will.

The representation (3.43) is particularly useful for simulation of stationary Gaussian processes, as described in detail in Appendix B. Then, the continuous spectrum is discretized to frequencies $\omega_k = k\Delta$, $k \in \mathbb{N}$, with the integrals (3.43) replaced by sums. Since the increments in the Wiener processes are independent normal variables, the approximative expressions become

$$x(t) = \sum U_k \sqrt{2\Delta F(\omega_k)} \cos \omega_k t + \sum V_k \sqrt{2\Delta F(\omega_k)} \sin \omega_k t, \quad (3.44)$$

$$= \sum A_k \sqrt{2\Delta F(\omega_k)} \cos(\omega_k t + \phi_k), \quad (3.45)$$

where U_k and V_k are independent standard normal variables, and

$$A_k = \sqrt{U_k^2 + V_k^2}, \quad \phi_k = -\arg(U_k + iV_k).$$

The distribution of the amplitudes A_k is standard Rayleigh, with mean $\sqrt{\pi/2}$, variance $(4-\pi)/2$, and probability density $f_A(x) = xe^{-x^2/2}, x \geq 0$, and the phases ϕ_k are uniformly distributed in $[0, 2\pi]$ and independent of the amplitudes.

The representation (3.44) was used explicitly already by Lord Rayleigh in connection with heat radiation and by Einstein (1910) and others to introduce Gaussian randomness. The form (3.45) appears to have come later, at least according to S.O. Rice [101], who cites work written by W.R. Bennett in the 1930's.

3.4.2 *Gaussian white noise in continuous time*

The differentials $\mathrm{d}w_1(\omega)$ and $\mathrm{d}w_2(\omega)$ in (3.43) are examples of *Gaussian white noise*. White noise in general is a common notion in stochastic process theory when one needs a process in continuous time where all process values are virtually independent, regardless of how close they are in time. Complete independence would require $r_x(t) = 0$ for all t except $t = 0$, i.e., the covariance function is not continuous and Bochner's theorem, Theorem 3.3, is of no use to find a corresponding spectrum. Fourier's inversion formula (3.9) hints that the spectrum should be constant, independent of ω, but $f(\omega) > 0$

is not a spectral density, since its integral is infinite. On the other hand, the δ-distribution, $\delta_0(\omega)$ (Dirac delta function), forms a Fourier transform pair together with the constant function $f(\omega) = 1/2\pi$. It is in fact possible to formulate a theory for "distribution valued" stationary processes and covariance functions, but that theory is little used in practical work and we do not go into any details on this; for a brief introduction, see [125, Appendix I].

Instead we will use the two Wiener processes defined by (3.42) to illustrate the common way to go around the problem with constant spectral density. We have already used them as spectral processes in (3.43) without any difficulty; we only noted that $E(\mathrm{d}w_1(\omega)^2) = E(\mathrm{d}w_2(\omega)^2) = \mathrm{d}\omega$.

In the theory of stochastic differential equations, one often uses the notation $w'(t)$ or $\mathrm{d}w(t)$ with the understanding that it is shorthand used in a stochastic integral of the form $\int_{t_0}^{t} g(u)w'(u)\,\mathrm{d}u = \int_{t_0}^{t} g(u)\,\mathrm{d}w(u)$, for $\int_{t_0}^{t} g(u)^2\,\mathrm{d}u < \infty$, as introduced in (2.37) on page 64. We will illustrate this with the previously mentioned Langevin equation (1.13) and deal with these more in detail in Section 4.3.

Example 3.5 (Ornstein-Uhlenbeck process)**.** *The Ornstein-Uhlenbeck process is a Gaussian stationary process with covariance function* $r(t) = \sigma^2 e^{-\alpha|t|}$, *and spectral density*

$$f(\omega) = \frac{\sigma^2}{\pi} \cdot \frac{\alpha}{\alpha^2 + \omega^2}.$$

It follows from the criteria in Chapter 2 that a Gaussian process with this covariance function and spectrum is continuous but, similar to the Wiener process, not differentiable.

The process can be realized as a stochastic integral

$$x(t) = \sqrt{2\alpha\sigma^2} \int_{-\infty}^{t} e^{-\alpha(t-\tau)}\,\mathrm{d}w(\tau), \tag{3.46}$$

which is the solution to the linear stochastic differential (Langevin) equation

$$\alpha x(t) + x'(t) = \sqrt{2\alpha}\,\sigma\, w'(t) \tag{3.47}$$

with Gaussian white noise $w'(t)$.

For large α ($\alpha \to \infty$)*, the covariance function falls off very rapidly around* $t = 0$ *and the correlation between* $x(s)$ *and* $x(t)$ *becomes negligible when* $s \neq t$. *In the integral (3.46) each* $x(t)$ *depends asymptotically only on the increment* $\mathrm{d}w(t)$*, implying approximate independence. With increasing* α*, the spectral density becomes increasingly flatter at the same time as* $f(\omega) \to 0$. *In order to keep the variance of the process constant, not going to* 0 *or*

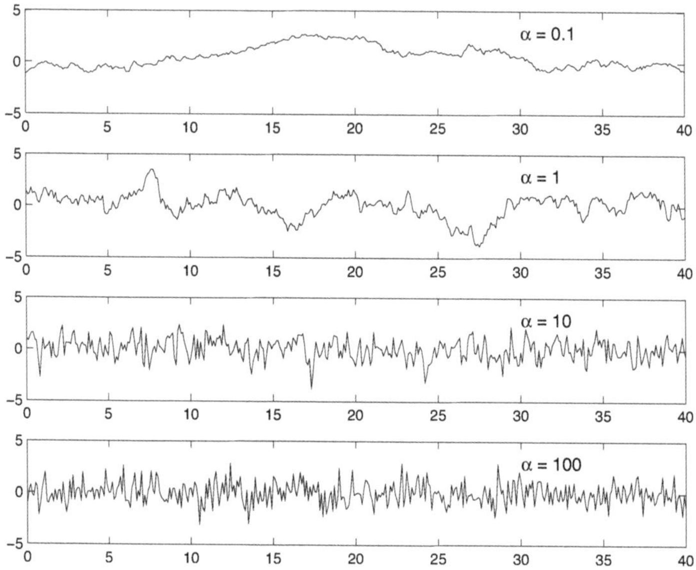

Figure 3.3 *Realization of the Ornstein-Uhlenbeck process with different values of* α.

∞, we can let $\sigma^2 \to \infty$ in such a way that $\sigma^2/\alpha \to C > 0$. Therefore, the Ornstein-Uhlenbeck process with large α and $\sigma^2/\alpha = C$ can be used as an approximation to Gaussian white noise; see Figure 3.3 for realizations with different α-values.

As a stationary process the Ornstein-Uhlenbeck has a spectral representation of the type (3.43), and one may ask what connection there is between the two integral representations. To see the analogy, take $w_1(\omega)$ and $w_2(\omega)$ from (3.43) and define $w_C(\omega)$ for $-\infty < \omega < \infty$, as

$$w_C(\omega) = \begin{cases} w_1(\omega) + iw_2(\omega), & \omega > 0, \\ w_1(-\omega) - iw_2(-\omega), & \omega < 0, \end{cases}$$

to get $x(t) = \int_{-\infty}^{\infty} e^{i\omega t} \sqrt{f(\omega)}\, \mathrm{d}w_C(\omega)$.

We then do some formal calculation with the derivative of the Wiener process, $w'(t) = \frac{1}{\sqrt{2\pi}} \int_{-\infty}^{\infty} e^{i\omega t}\, \mathrm{d}w_C(\omega)$, for some complex Wiener process $w_C(\omega)$. Inserting this in (3.46), we obtain

$$\begin{aligned} x(t) &= \sqrt{2\alpha\sigma^2} \int_{-\infty}^{t} e^{-\alpha(t-\tau)} w'(\tau)\, \mathrm{d}\tau \\ &= \frac{\sqrt{2\alpha\sigma^2}}{\sqrt{2\pi}} \int_{\tau=-\infty}^{t} e^{-\alpha(t-\tau)} \left\{ \int_{\omega=-\infty}^{\infty} e^{i\omega\tau}\, \mathrm{d}w_C(\omega) \right\} \mathrm{d}\tau \end{aligned}$$

$$= \frac{\sqrt{2\alpha\sigma^2}}{\sqrt{2\pi}} \int_{\omega=-\infty}^{\infty} \left\{ \int_{\tau=-\infty}^{t} e^{-(\alpha+i\omega)(t-\tau)}\,\mathrm{d}\tau \right\} e^{i\omega t}\,\mathrm{d}w_C(\omega)$$

$$= \frac{\sqrt{2\alpha\sigma^2}}{\sqrt{2\pi}} \int_{\omega=-\infty}^{\infty} \frac{1}{\alpha+i\omega} e^{i\omega t}\,\mathrm{d}w_C(\omega)$$

$$= \frac{\sqrt{2\alpha\sigma^2}}{\sqrt{2\pi}} \int_{\omega=-\infty}^{\infty} \frac{1}{\sqrt{\alpha^2+\omega^2}} e^{i(-\arg(\alpha+i\omega)+\omega t)}\,\mathrm{d}w_C(\omega)$$

$$= \int_{\omega=-\infty}^{\infty} e^{i(\omega t+\gamma(\omega))} \sqrt{f(\omega)}\,\mathrm{d}w_C(\omega),$$

with $\gamma(-\omega) = -\gamma(\omega)$. Thus, the same Wiener process $w_C(\omega)$ that works in the spectral representation of the white noise in (3.46) can be used as spectral process after correction of the phase.

Example 3.6 (Low-frequency white noise)**.** *In Example 3.2 we met the low-frequency white noise with the sinc covariance function $r(t) = \sigma^2 \frac{\sin at}{at}$, and its box spectral density $f(\omega) = \frac{\sigma^2}{2a}$ for $|\omega| < a$. Note, that a Gaussian process with this covariance function is differentiable infinitely many times; for a similar process, solve Exercise 4.13.*

3.5 Stationary counting processes

3.5.1 *Counting processes*

Counting processes are part of the vast and diversified area of point processes. Even if not a central theme in this book, it may be appropriate to mention a few of the second order properties of stationary point processes on the real line, namely the covariance (intensity) function and its Fourier transform, the spectrum, introduced by M.S. Bartlett [7]. These notions are general, in the sense that they do not depend on any special mechanism for the point generation, and they have found wide applications in the statistical analysis and characterization of random point patterns in space and time. For a comprehensive account of point process theory and applications, see [37], and also [39].

A *counting process* is a stochastic process $\{N\}$ that "counts" the number of events that happen at random instances in $\mathbb{R}$. The number of events in an interval A is $N(A)$. We will use also the notation $N(t) = N((0,t])$ for the number of events in $(0,t]$; it is a piecewise constant function that increases in (positive) integer steps at random time points. The following terminology is common for counting processes; cf. [37].

Terminology: A counting point process N is/has

stationary: if the joint distribution of increments over disjoint intervals does not change with a common shift of time;

mean density of events: $m = E(N((0,1]) \leq \infty$;

intensity: $\lambda = \lim_{h \downarrow 0} P(N((0,h]) \geq 1)/h \leq \infty$;

orderly: if $\lim_{h \downarrow 0} P(N((0,h]) \geq 2)/h = 0$;

simple: if N increases in unit steps, i.e., multiple events do not occur.

The following relations can be shown to hold for a stationary counting process; see, for example [35, Sect. 3.8] and [37, Sect. 3.3].

- If N is simple then $m = \lambda \leq \infty$;
- If N is simple and $\lambda < \infty$ then N is orderly;
- If N is orderly then it is simple.

3.5.2 The Poisson process

The Poisson process is introduced rather informally in most probability and statistics textbooks, and we have already used some of its basic properties. In Remark 1.1 we discussed one constructive definition and one definition directly referring to Kolmogorov's existence theorem. We shall now formulate a third equivalent definition, summarizing its most basic property as a *counting process*, and use that as definition, thereby following Breiman [19, Sect. 14.6].

Definition 3.6 (Poisson process). *A stochastic process $\{N(t), 0 \leq t < \infty\}$ is a Poisson process if $N(0) = 0$, its sample paths are piecewise constant and change at discrete time points with upwards jumps of size one, and*

a) its increments over disjoint intervals are independent random variables, and

b) the distribution of the increment over an interval I of length $\|I\|$ depends only on the length of the interval and not on its location.

It is shown in [19] that this completely characterizes the Poisson process, and that the increment over an interval I is Poisson distributed with parameter $\lambda \|I\|$, for some constant $\lambda \geq 0$, called the *intensity*.

The Poisson process is sometimes defined via the inter-event times. If $T_1 < T_2 < \ldots$ denote the time of the events, then T_1, $T_2 - T_1$, $T_3 - T_2, \ldots$, is a sequence of independent random variables with exponential distribution with mean $1/\lambda$. Also this property uniquely characterizes the Poisson process.

We give now some more aspects of the Poisson process. If a Poisson process $\{N(t), 0 \leq t < \infty\}$ is used for counting a stationary stream of events, the intensity λ is equal to the expected number of events per time unit. But λ can also be interpreted as proportional to the probability of an event occurring during a very small time interval of width $h > 0$,

$$P(N(t+h) = N(t) + 1) \approx \lambda h. \tag{3.48}$$

This local probability is constant over time, in agreement with the stationarity.

A (two-sided) Poisson process over the whole real line is defined by taking two independent Poisson processes $N^+(t)$ and $N^-(t)$, with $t \geq 0$, and defining

$$N(t) = \begin{cases} N^+(t), & \text{for } t \geq 0, \\ -N^-(-t-0), & \text{for } t < 0. \end{cases} \tag{3.49}$$

To prepare for a more general notation, we write $N((a,b]) = N(b) - N(a)$ for the random number of events in the interval $(a,b]$, and observe that its distribution depends only on the interval length $b - a$, not on the location.

Remark 3.5 (The inspection paradox). *In the two-sided Poisson process, $\{N(t), t \in \mathbb{R}\}$, the first event on the positive side occurs at a time T_1 with an exponential distribution with mean $1/\lambda$. The last event on the negative side occurs at a time T_0, say, with $-T_0$ exponential, also with mean $1/\lambda$ and independent of T_1. Thus, the interval that contains the origin is on average $2/\lambda$, twice as long as all other inter-event times. Nevertheless, the stream of events is stationary. We will meet this type of "biased sampling" in Section 8.3.2 on crossings and wave patterns.*

Inhomogeneous Poisson process

The inhomogeneous Poisson process is a building block for more general counting processes. In an inhomogeneous Poisson process the number of events in disjoint intervals are independent and Poisson distributed, but the intensity of events is a function of time, $\lambda(t)$. The expected number of events in the interval $(a,b]$ is the integrated intensity, $\int_a^b \lambda(t)\,\mathrm{d}t$.

Rényi's characterization of the Poisson process

We end this section by a characterization of the stationary Poisson process by the probability of no event. The characterization follows from a theorem by A. Rényi, 1967 [100]; see [37, Thm. 2.3.I].

Theorem 3.9 (Rényi's characterization of the Poisson process). *Let N be an orderly counting process on $\mathbb{R}$. Then, for N to be a stationary Poisson process it is necessary and sufficient that for all sets A that can be represented as the union of a finite number of finite intervals,*

$$P(N(A) = 0) = e^{-\lambda \|A\|}.$$

The theorem says that it suffices to specify the probability of no events in sets in a sufficiently rich family in order to get the entire Poisson distribution, including the independent for disjoint sets.

3.5.3 Correlation intensity function

Now, one could go a step further and ask for the probability that in a Poisson process there will be an event in each of two small intervals, separated by a distance τ. For the Poisson process, increments over disjoint intervals are independent (and therefore also uncorrelated), and hence, for $\tau \neq 0$,

$$\lim_{h \downarrow 0} \frac{P(N(t+h) = N(t)+1 \text{ and } N(t+\tau+h) = N(t+\tau)+1)}{h^2} = \lambda^2,$$

$$\lim_{h \downarrow 0} \frac{E((N(t+h)-N(t))(N(t+\tau+h)-N(t+\tau))}{h^2} = \lambda^2, \tag{3.50}$$

$$\lim_{h \downarrow 0} \frac{C(N(t+h)-N(t), N(t+\tau+h)-N(t+\tau))}{h^2} = 0. \tag{3.51}$$

One could say that, in a Poisson process, the *correlation intensity* is zero; there is no correlation between events, whatsoever. We will next discuss counting processes with dependent events, for which the covariance limit in (3.51) is not zero.

Counting processes with correlated increments

Consider now a counting process $\{N(t), 0 \leq t < \infty\}$, for which the increments may be dependent. This means that events in disjoint intervals can influence each other. The influence can take many different forms. It may be deterministic, for example, if in a Poisson process each event is followed by exactly one new event after a fixed time. Or it may be stochastic, as is the case in a Poisson with a *cascade* of aftereffects, or as in a *renewal process*, where the sequence of events occur with independent, but non-exponential, inter-arrival times.

Definition 3.7. *If $\{N(t), 0 \le t < \infty\}$ is a regular counting process with intensity λ, the function, defined for $\tau \neq 0$,*

$$w(\tau) = \lim_{h \downarrow 0} \frac{P(N(t+h) = N(t)+1, N(t+\tau+h) = N(t+\tau)+1)}{h^2} - \lambda^2$$
$$= \lim_{h \downarrow 0} \frac{C(N(t+h) - N(t), N(t+\tau+h) - N(t+\tau))}{h^2}$$

is called the covariance intensity function (when the limit exists). The normalized function $w(\tau)/\lambda^2 \ge -1$ is called the correlation intensity function.

The correlation intensity function is a simple descriptor of the dependence between events. The interpretation is that if there is an event at time t, then the conditional probability for an event also in the small interval $[t+\tau, t+\tau+h)$ is approximately $h\lambda(1 + w(\tau)/\lambda^2)$. A negative $w(\tau)$ decreases the chance that two events occur at distance τ; a positive value increases the chance. Note that nothing is assumed on the number of events *between t* and $t+\tau$; there may be none, one, or many.

The definitions of intensity and covariance intensity are often written in differential form,

$$\lambda = \frac{E(\mathrm{d}N(t))}{\mathrm{d}t}, \tag{3.52}$$

$$w(\tau) = \frac{E(\mathrm{d}N(t)\,\mathrm{d}N(t+\tau))}{(\mathrm{d}t)^2} - \lambda^2, \quad \tau \neq 0, \tag{3.53}$$

with $\mathrm{d}N(t) = N(t+\mathrm{d}t) - N(t)$, and with $w(-\tau) = w(\tau)$.

For a regular process, $E((\mathrm{d}N(t))^2) = E(\mathrm{d}N(t))$, so $w(0)$ is of larger order than $w(\tau)$, $\tau \neq 0$. To accommodate $\tau = 0$ and $\tau \neq 0$ in one expression one can use the *Dirac delta function*, $\delta_0(t)$, with the interpretation that $\int f(t)\delta_0(t)\,\mathrm{d}t =$

$f(0)$ for every continuous function f. Using this formalism, the function

$$w_c(\tau) = \lambda \delta_0(\tau) + w(\tau) \tag{3.54}$$

(with $w(0) = \lim_{\tau \to 0} w(\tau)$) is called the *complete covariance intensity function.*

3.5.4 Spectral distribution for counting processes

If the covariance intensity $w(\tau)$ is absolutely integrable, we can apply the inversion formula (3.9) on (3.54), and use the constant $1/2\pi$ as the Fourier transform of the delta function. We then obtain the *complete spectral density function* for $\{N(t), t \in \mathbb{R}\}$, also called the *Bartlett spectrum,*

$$\begin{aligned} f_c(\omega) &= \frac{1}{2\pi}\int_{-\infty}^{\infty} e^{-i\omega t} w_c(t)\,\mathrm{d}t = \frac{\lambda}{2\pi} + \frac{1}{2\pi}\int_{-\infty}^{\infty} e^{-i\omega t} w(t)\,\mathrm{d}t \\ &= \frac{\lambda}{2\pi} + f(\omega), \end{aligned} \tag{3.55}$$

for $-\infty < \omega < \infty$. This is not a proper spectral density in the sense of Section 3.2, but it still provides useful information about the periodicity of events.

First, one can note that the first term $\lambda/2\pi$ is the Bartlett spectrum of a Poisson process, for which $w(\tau) \equiv 0$. As we shall see in Section 4.3.2, a constant spectral density for a stationary process in continuous time corresponds to complete independence, also called *white noise,* at least in a formal sense. (For discrete time this is Theorem 3.6.) Thus, the Poisson process is the point process equivalent to white noise.

Alternative definition of Bartlett spectrum

Until now we have used the interval $[0,t]$ as a reference interval, and defined $N(t)$ as the number of events in $[0,t]$. Actually, there is no specific reason to start counting at time 0, and therefore one should rather consider a counting process, $\{N(C)\}$, that counts the number of events in any specified interval $C = [a,b]$, or, more generally, in any Borel set $A \in \mathscr{B}$.[7] Stationarity is defined as usual – distributions have to be independent of a time shift – and the process is simple if $P(N([x]) > 1) = 0$, for all x, and we denote the intensity $\lambda = E(N([0,1])) \leq \infty$.

A simple counting process $\{N(C)\}$ with events at times $\{T_n\}_{n=-\infty}^{\infty}$, $T_0 \leq$

[7]This is a special type of *random measure.*

$0 < T_1$, defines a *counting measure*, $N(C) = \#\{T_n \in C\}$, and one can use the notation,

$$N(\phi) = \int_{-\infty}^{\infty} \phi(t)\,\mathrm{d}N(t) = \sum_n \phi(T_n), \tag{3.56}$$

when ϕ is a real or complex-valued function. The distribution of the counting process $N(C)$ is specified by the distributions of the random variables $N(\phi)$ for a sufficiently rich class of functions ϕ.

We can now give a more general definition of the Bartlett spectrum. Consider the family of functions ϕ that are both absolutely and square integrable ($\phi \in \mathscr{L}^1 \cap \mathscr{L}^2$), and denote the (inverse) Fourier transform of ϕ by $\hat{\phi}(\omega) = \int_{-\infty}^{\infty} e^{-i\omega t}\phi(t)\,\mathrm{d}t$.

Definition 3.8 (Bartlett spectrum). *The Bartlett spectrum to a stationary, simple and bounded counting process $\{N(C)\}$ is any measure $\mu_N(\omega)$ with the property that, for every $\phi, \phi_1, \phi_2 \in \mathscr{L}^1 \cap \mathscr{L}^2$,*

$$C(N(\phi_1), N(\phi_2)) = \int_{-\infty}^{\infty} \widehat{\phi_1}(\omega)\overline{\widehat{\phi_2}}(\omega)\,\mathrm{d}\mu_N(\omega), \tag{3.57}$$

$$V(N(\phi)) = \int_{-\infty}^{\infty} |\widehat{\phi}(\omega)|^2\,\mathrm{d}\mu_N(\omega). \tag{3.58}$$

It is shown in [37, Sect. 11.2] that a necessary and sufficient condition for the existence of μ_N is that

$$E(N([a,b])^2) < \infty,$$

for all bounded intervals $[a,b]$. As for the spectrum of a stationary process, μ_N can be expressed by means of a real, non-decreasing symmetric spectral function, $F_N(\omega)$, such that $F_N(b) - F_N(a) = \int_a^b \mathrm{d}\mu_N(\omega)$, with the difference that it is not bounded.

Example 3.7 (Poisson process). *We carry out the arguments for a Poisson process over the whole real line, as defined by* (3.49). *To find $V(N(\phi))$, assume ϕ is continuous, and reason heuristically. Imagine the real line divided into intervals $\{I_k\}_{-\infty}^{\infty}$, of length Δt. Denote by t_k the left endpoint of interval I_k. If Δt is small, the sum in* (3.56) *is approximately the sum of $\phi(t_k)$ over those intervals where there is at least one event; since when Δt is small there will be at most one event in each interval. Since the number of events in disjoint* *intervals is independent, and has variance $\lambda\,\Delta t$, we have*

$$V(N(\phi)) \approx \lambda \sum_k |\phi(t_k)|^2\,\Delta t \approx \lambda \int_{-\infty}^{\infty} |\phi(t)|^2\,\mathrm{d}t.$$

By Parseval's relation, applied to

$$\hat{\phi}(\omega) = \int_{-\infty}^{\infty} e^{-i\omega t}\phi(t)\,\mathrm{d}t, \text{ and } \phi(t) = (2\pi)^{-1}\int_{-\infty}^{\infty} e^{i\omega t}\hat{\phi}(\omega)\,\mathrm{d}\omega,$$

one has

$$\int_{-\infty}^{\infty} |\phi(t)|^2\,\mathrm{d}t = \frac{1}{2\pi}\int_{-\infty}^{\infty} |\hat{\phi}(\omega)|^2\,\mathrm{d}\omega,$$

which implies that $\mathrm{d}\mu_N(\omega) = (\lambda/2\pi)\,\mathrm{d}\omega$. *We see again that the Bartlett spectrum for the Poisson process is constant,* $\lambda/2\pi$.

Example 3.8 (Cox process, doubly stochastic Poisson process). *The standard Poisson process has constant event intensity. If the event probability is time-dependent, the intensity λ is a function of time, $\lambda = \lambda(t)$. The Poisson character still requires that the number of events in disjoint intervals is Poisson distributed, but now with mean equal to the intensity function integrated over the interval. Such a process is called an inhomogeneous Poisson process.*

A Cox process is a type of inhomogeneous Poisson, where the intensity is itself a (non-negative) stationary process $\{\lambda(t), t \in \mathbb{R}\}$. Conditioned on the realization of $\{\lambda(t)\}$, the Cox process is an inhomogeneous Poisson process with intensity function $\lambda(t)$. An older name for the Cox process is doubly stochastic Poisson process.

Now, consider a Cox process $\{N(t), t \in \mathbb{R}\}$, for which the non-negative stationary intensity process $\{\lambda(t)\}$ has mean $E(\lambda(t)) = \lambda_0$ and assume its spectrum is absolutely continuous with spectral density $f_\lambda(\omega)$. To find the variance $V(N(\phi))$ we first find the conditional moments, for a certain realization of $\lambda(t)$. By the same heuristics as for the Poisson process,

$$E(N(\phi) \mid \lambda) = \int_{-\infty}^{\infty} \phi(t)\lambda(t)\,\mathrm{d}t, \tag{3.59}$$

$$V(N(\phi) \mid \lambda) = \int_{-\infty}^{\infty} |\phi(t)|^2\lambda(t)\,\mathrm{d}t. \tag{3.60}$$

Combining these expressions in the rule (A.7) *for total variance, in Appendix A, we get*

$$\begin{aligned} V(N(\phi)) &= V(E(N(\phi) \mid \lambda)) + E(V(N(\phi) \mid \lambda)) \\ &= V\left(\int_{-\infty}^{\infty} \phi(t)\lambda(t)\,\mathrm{d}t\right) + E\left(\int_{-\infty}^{\infty} |\phi(t)|^2\lambda(t)\,\mathrm{d}t\right). \end{aligned}$$

To find the variance of the integral, we could use Theorem 2.17. However, we rather anticipate Theorem 4.1 in Chapter 4, and regard $E(N(\phi) \mid \lambda)$ as the output of a linear filter with impulse response $\phi^*(u) = \phi(-u)$. With filter terminology, the frequency response is, by (4.5),

$$g^*(\omega) = \int_{-\infty}^{\infty} e^{-i\omega u} \phi^*(u)\,\mathrm{d}u = \int_{-\infty}^{\infty} e^{i\omega u} \phi(u)\,\mathrm{d}u = \hat{\phi}(-\omega).$$

The rule (4.9) *for variance of a filter output, together with Parseval's formula, then gives (since f_λ is symmetric)*

$$\begin{aligned} V(N(\phi)) &= \int_{-\infty}^{\infty} |g^*(\omega)|^2 f_\lambda(\omega)\,\mathrm{d}\omega + \lambda_0 \int_{-\infty}^{\infty} |\phi(t)|^2\,\mathrm{d}t \\ &= \int_{-\infty}^{\infty} |\hat{\phi}(\omega)|^2 f_\lambda(\omega)\,\mathrm{d}\omega + \frac{\lambda_0}{2\pi} \int_{-\infty}^{\infty} |\hat{\phi}(\omega)|^2\,\mathrm{d}\omega. \end{aligned}$$

We have found that the Bartlett spectrum of the Cox process has density

$$\frac{\lambda_0}{2\pi} + f_\lambda(\omega).$$

Remark 3.6 (Random measure)**.** *The counting measure is a special case of a random measure with unit point masses at the times of events. Any non-negative continuous process, $\{\lambda(t), t \in \mathbb{R}\}$, defines a random measure Λ with density* $\mathrm{d}\Lambda(t) = \lambda(t)\,\mathrm{d}t$. *In analogy with* (3.56) *one can then define $\Lambda(\phi) = \int_{-\infty}^{\infty} \phi(t)\,\mathrm{d}\Lambda(t)$, and define its Bartlett spectrum as in* (3.58). *When you have read Chapter 4, you should be able to prove that the Bartlett spectrum to Λ is equal to the regular spectrum $F_\lambda(\omega)$ of the stationary process $\{\lambda(t)\}$.*

Exercises

3:1. A function $r(t), t \in \mathbb{Z}$, takes values $r(0) = a$, $r(\pm 1) = b$, $r(t) = 0$, otherwise. For what combinations of a and b is $r(t)$ a covariance function?

3:2. If r_1 and r_2 are covariance functions, when are the product $r_1 r_2$ and the convolution $r_1 * r_2$ covariance functions?

3:3. Prove "Polya's criterion": if $r(t)$ is symmetric, continuous, and non-increasing and convex on $(0, \infty)$, then it is a covariance function.

3:4. Show that $\mu(\omega/2M)$ is the Fourier transform of the sinc function.

3:5. Show that the spectral density to the squared sinc function $r(t) = \sin^2 t/t^2$ is triangular.

3:6. Present two different proofs of the fact that $r(t) = \max(0, 1 - |t|)$ is a covariance function with spectral density $f(\omega) = \frac{1-\cos\omega}{\pi\omega^2}$.

3:7. Generalize the Random Telegraphy Signal by assuming that the times the process stays in position are independent and exponential, but with different mean values, α when it is in position -1 and β when it is in position $+1$. Find the covariance function and spectrum of the modified process, which is the general two-stage Markov process.

3:8. A random process has covariance function $r(t) = B\cos^2(\omega_0 t)\, e^{-W|t|}$, where B, ω_0, and W are positive constants. Find and sketch the spectrum when ω_0 is much larger than W.

3:9. Let $x(t)$ be a stationary Gaussian process with $E(x(t)) = 0$, covariance function $r_x(t)$, and spectral density $f_x(\omega)$. Find the covariance function for

$$y(t) = x^2(t) - r_x(0),$$

and show that it has the spectral density

$$f_y(\omega) = 2\int_{-\infty}^{\infty} f_x(\mu) f_x(\omega - \mu)\, \mathrm{d}\mu.$$

3:10. Derive the spectral density for $u(t) = 2x(t)x'(t)$ if $x(t)$ is a differentiable stationary Gaussian process with spectral density $f_x(\omega)$.

3:11. Let $e_t, t = 0, \pm 1, \pm 2, \ldots$, be independent standard normal variables and define, for $|\theta| < 1$, the stationary processes

$$x_t = \theta x_{t-1} + e_t = \sum_{n=-\infty}^{t} \theta^{t-n} e_n, \quad y_t = e_t + \psi e_{t-1}.$$

a) Find the covariances and spectral densities of x_t and y_t.

b) Express the spectral processes $Z_x(\omega)$ and $Z_y(\omega)$ in terms of the spectral process $Z_e(\omega)$, and derive the cross-spectrum between x_t and y_t.

3:12. Let u_n and v_n be two sequences of independent, identically distributed variables with zero mean and let the stationary sequences x_n and y_n be defined by

$$y_n = a_1 + b_1 x_{n-1} + u_n, \quad x_n = a_2 - b_2 y_n + v_n.$$

Express the spectral processes $\mathrm{d}Z_x$ and $\mathrm{d}Z_y$ as functions of u_n and v_n, and derive the spectral densities for x_n and y_n and their cross-spectrum.

3:13. Show the following form of the *law of large numbers*: If $\{x_n, n \in \mathbb{N}\}$ is a stationary sequence with spectral process $Z(\omega)$, then

$$\frac{1}{n}\sum_{k=1}^{n} x_n \overset{q.m.}{\to} Z(0+) - Z(0-);$$

cf. the limit (3.28) in Theorem 3.8.

3:14. Let Y, Γ, ϕ be independent random variables with $E(Y) = 0$, $E(Y^2) = \sigma^2$, and ϕ uniform in $(0, 2\pi)$. Write $f(u)$ for the probability density function of Γ, and assume it to be symmetric around 0. Define the process $\{x(t), t \in \mathbb{R}\}$ as $x(t) = Y \cos(\Gamma t + \phi)$.

a) Show that $x(t)$ is stationary and has covariance function

$$r_x(t) = \sigma^2 \int_{-\infty}^{\infty} \cos(\omega t) f(\omega)\, d\omega.$$

b) Argue that to every covariance function in continuous time, there is a differentiable process that has the given function as its covariance function. What distribution $f(u)$ makes $r_x(t) = e^{-\alpha|t|}$?

3:15. Take a stationary process $\{x(t), t \in \mathbb{R}\}$ with $E(x(t)) = m$ and spectral density $f(\omega)$, and use it to amplitude modulate a pure randomly shifted cosine function with constant frequency Γ. Find the covariance function, spectral density, and spectral representation of the resulting process

$$y(t) = x(t) \cos(\Gamma t + \phi),$$

when ϕ is uniformly distributed in $(0, 2\pi)$.

3:16. Use the limit

$$\lim_{T \to \infty} \frac{1}{\pi} \int_{-T}^{T} \frac{\sin \omega t}{t}\, dt = \begin{cases} 1 & \text{for } \omega > 0, \\ 0 & \text{for } \omega = 0, \\ -1 & \text{for } \omega < 0 \end{cases}$$

for the following alternative derivation of the spectral representation of a stationary process $x(t)$ with spectral distribution function $F(\omega)$.

a) First show that the following integral and limit exists in quadratic mean:

$$Z(\omega) = \lim_{T \to \infty} \frac{1}{2\pi} \int_{-T}^{T} \frac{e^{-i\omega t} - 1}{-it} x(t)\, dt.$$

b) Then show that the process $Z(\omega)$, $-\infty < \omega < \infty$, has orthogonal increments and that, for $\omega_1 < \omega_2$, $E(|Z(\omega_2) - Z(\omega_1)|^2) = F(\omega_2) - F(\omega_1)$.

c) Finally, show that the integral

$$\int_{-\infty}^{\infty} e^{i\omega t}\, dZ(\omega) = \lim \sum e^{i\omega_k t} (Z(\omega_{k+1}) - Z(\omega_k))$$

exists, and that $E(|x(t) - \int_{-\infty}^{\infty} e^{i\omega t}\, dZ(\omega)|^2) = 0$.

3:17. A stochastic process $x(t), t \geq 0$, is called a *diffusion process* if the conditional density $g_u(\tau, v) = f_{x(t+\tau)|x(t)=v}(u)$ satisfies the equation

$$\frac{\partial g_u(\tau,v)}{\partial \tau} = \frac{1}{2}a(\tau,v)\frac{\partial^2 g_u(\tau,v)}{\partial v^2} + b(\tau,v)\frac{\partial g_u(\tau,v)}{\partial v},$$

for some $a(\tau, v) > 0$, $b(\tau, v)$. Show that the Wiener process is a diffusion.

3:18. Show that the Ornstein-Uhlenbeck process is a diffusion.

3:19. ("Fractional Brownian motion") Let $w(t)$ be a Wiener process. For what values of the constant κ is the following process well defined,

$$x(t) = \int_{-\infty}^{\infty} \max(0,(t-s))^{\kappa-1/2} - \max(0,-s)^{\kappa-1/2})\,\mathrm{d}w(s).$$

3:20. The two stationary processes $x(t)$ and $y(t)$ have mean zero and spectral densities $f_x(\omega)$ and $f_y(\omega)$, respectively.

a) Show that, if $f_x(\omega) \leq f_y(\omega)$ for all ω, then $V\left(\sum_k a_k x(t_k)\right) \leq V\left(\sum_k a_k y(t_k)\right)$ for all choices of $t_1, \ldots, t_n$ and real constants $a_1, \ldots, a_n$.

b) If $f_x(\omega) > f_y(\omega)$ in some interval $\omega_1 < \omega < \omega_2$, then there exist some t_k:s and a_k:s so the opposite inequality holds. Prove this, as formally correct as you can.

3:21. Consider a Cox process whose random (but constant in time) intensity has density $f_\lambda(x) = x^{-2}$, for $x \geq 1$. Show that it is simple, with infinite intensity, and find $\lim_{h \downarrow 0} h^{-1} P(N([0,h]) = 2)$. Is the process orderly?

3:22. Let $\mu_N(\omega)$ be the Bartlett spectrum of a simple counting process $\{N(t), t \in \mathbb{R}\}$. Construct a new counting process $\{M(t), t \in \mathbb{R}\}$ by shifting all events to new positions by independent random amounts. Express the Bartlett spectrum $\mu_M(\omega)$ in terms of the distribution/characteristic function of the shifts. What happens if the shifts are normal with variance σ^2 and $\sigma \to \infty$?

3:23. (Hawkes process) The classical Hawkes process [60] is generated by a basic Poisson process with constant intensity function λ, in which each event changes the intensity of future events in a random or deterministic way. Denote the basic Poisson process by $\{N(t), t \in \mathbb{R}\}$ with events at $\{T_n\}$, and let $h(t) > 0$, $t > 0$, be the additional intensity at time t after an original Poisson event. The total intensity, conditional on $\{N(s), s < t\}$, is thus

$$\lambda(t) = \lambda + \sum_{n; T \leq t} h(t - T_n).$$

Find an expression for the Bartlett spectrum of a Hawkes process.

Chapter 4

Linear filters – general properties

Derivatives, integrals, and weighted averages are examples of linear operations on random as well as on deterministic functions and data sequences. A practical implementation of a linear filter works in the time domain, changing the time course of a function. The theoretical analysis is however most conveniently performed in the frequency domain, where the effects of the filter is split on the different frequency components in the signal. It is perhaps fair to say that the spectral theory of stationary processes proves to be most useful in connection with linear filters. The formal and informal treatment of the spectral representation introduced in Chapter 3 considerably facilitates the derivation of the characteristics of a linear filter. This chapter deals with the general tools to handle covariance and spectral properties of linear filters. Special emphasis is given to the use of white noise in linear filters. Some special topics are treated in the next chapter.

4.1 Linear time invariant filters

4.1.1 Linear filter on a stationary process

One of the most useful instruments in the theory of stochastic processes is the linear prediction device, by which we "predict" or approximate a random variable x by a linear combination of a set of observed random variables, or by a limit of such linear combinations. One might, for example, want to predict the value of a stochastic process as some future time $t+h$ by a linear combination of values observed up till time t:

$$\widehat{x}(t+h) = \sum_k a_k x(t-s_k), \text{ some } a_k, s_k > 0.$$

Filters may also be constructed to reduce noise in an audio or video signal or to fill in missing data in a time series. The simple operations "differentiation" and integration are examples of linear filter operations.

One class of filters is of the *convolution type*, defined as a convolution between an *impulse response function* $h(u)$ and an *input function* $x(t)$. The *output function* $y(t)$ is then

$$y(t) = \int_{-\infty}^{\infty} h(u)x(t-u)\,du = \int_{-\infty}^{\infty} h(t-u)x(u)\,du. \tag{4.1}$$

The impulse response may contain delta functions, δ_{τ_k}, which act as time delays; for example $y(t) = \int \delta_{\tau_0}(u)x(t-u)\,du = x(t-\tau_0)$. The term impulse response is explained by the fact that if the input function $x(u)$ is an "impulse", i.e., $x(u) = \delta_0(u)$, then $y(t) = h(t)$.

If the impulse response function $h(u)$ is zero for $u < 0$, the value $y(t)$ can be calculated at time t from available values $x(s), s \leq t$. Such a filter is called *causal* or *physically realizable*, indicating that the output from the filter at time t depends on the past and not on the future. Otherwise $y(t)$ is not available until some further time has passed. We can talk about *linear prediction* or *linear reconstruction*, respectively.

After these introductory examples we are ready for a definition. There are several formal definitions of linear filters, depending on the type of application at hand. In our context of stationary processes with finite variance we shall concentrate on *linear time-invariant filters* that transform an input stationary process $x(t)$ into an output stationary process $y(t)$ through action on the spectral components in the following way.

Definition 4.1 (Definition F[1]). *A linear time-invariant filter is a transformation $\mathscr{S}$ that takes a stationary process $\{x(t), t \in \mathbb{R}\}$ with spectral representation $x(t) = \int_{-\infty}^{\infty} e^{i\omega t}\,dZ(\omega)$ into a new stationary process $\{y(t), t \in \mathbb{R}\}$ so that*

$$y(t) = \int_{-\infty}^{\infty} g(\omega)e^{i\omega t}\,dZ(\omega). \tag{4.2}$$

The function $g(\omega)$ is called the frequency function or frequency response (also transfer function), and it has to satisfy

$$\int_{-\infty}^{\infty} |g(\omega)|^2\,dF(\omega) < \infty.$$

[1] "Definition F" as in Frequency; this definition was suggested by J.L. Doob [41, Section XI.9]. "Definition L" as in Linear.

In a more general setting one could use the following, less operational characterization of a linear time-invariant filter.

Definition 4.2 (Definition L[1]). *A linear time-invariant filter on a function space Ξ is a transformation $\mathscr{S}$, $\Xi \ni x \mapsto \mathscr{S}(x) \in \Xi$, such that for any constants a_1, a_2 and time delay τ,*

$$\mathscr{S}(a_1 x_1 + a_2 x_2) = a_1 \mathscr{S}(x_1) + a_2 \mathscr{S}(x_2),$$
$$\mathscr{S}(x(\cdot + \tau)) = \mathscr{S}(x)(\cdot + \tau).$$

4.1.2 Impulse response/frequency response relations

Suppose a linear filter is defined by its impulse response function $h(t)$, as in (4.1),

$$y(t) = \int_{-\infty}^{\infty} h(u)x(t-u)\,\mathrm{d}u = \int_{-\infty}^{\infty} h(t-u)x(u)\,\mathrm{d}u.$$

Inserting the spectral representation of $x(t)$ and changing the order of integration, we obtain a filter in frequency response form,

$$y(t) = \int_{\omega=-\infty}^{\infty} \left\{ \int_{u=-\infty}^{\infty} e^{i\omega u} h(t-u)\,\mathrm{d}u \right\} \mathrm{d}Z(\omega) = \int_{-\infty}^{\infty} g(\omega)\, e^{i\omega t}\,\mathrm{d}Z(\omega),$$

where

$$g(\omega) = \int_{-\infty}^{\infty} e^{-i\omega u} h(u)\,\mathrm{d}u \tag{4.3}$$

is well defined, e.g., if $\int_{-\infty}^{\infty} |h(u)|\,\mathrm{d}u < \infty$.

Conversely, we can start with the frequency response formulation,

$$y(t) = \int_{-\infty}^{\infty} g(\omega)\, e^{i\omega t}\,\mathrm{d}Z(\omega),$$

with $\int |g(\omega)|^2\,\mathrm{d}F(\omega) < \infty$. Then, if $h(u) = \frac{1}{2\pi}\int e^{i\omega u} g(\omega)\,\mathrm{d}\omega$ is absolutely integrable, i.e., $\int |h(u)|\,\mathrm{d}u < \infty$, the inversion (4.3) holds.

Inserting the expression for $g(\omega)$ and changing the order of integration we get the filter in impulse response form,

$$\begin{aligned} y(t) &= \int_{-\infty}^{\infty} e^{i\omega t} \left\{ \int_{-\infty}^{\infty} e^{-i\omega u} h(u)\,\mathrm{d}u \right\} \mathrm{d}Z(\omega) \\ &= \int_{-\infty}^{\infty} h(u) \left\{ \int_{-\infty}^{\infty} e^{i\omega(t-u)}\,\mathrm{d}Z(\omega) \right\} \mathrm{d}u = \int_{-\infty}^{\infty} h(u)x(t-u)\,\mathrm{d}u. \end{aligned}$$

We summarize these relations.

The impulse response and frequency response function form a Fourier transform pair,

$$h(u) = \frac{1}{2\pi}\int_{-\infty}^{\infty} e^{i\omega u} g(\omega)\,\mathrm{d}\omega, \text{ if } \int_{-\infty}^{\infty} |g(\omega)|^2\,\mathrm{d}F(\omega) < \infty, \quad (4.4)$$

$$g(\omega) = \int_{-\infty}^{\infty} e^{-i\omega u} h(u)\,\mathrm{d}u, \text{ if } \int_{-\infty}^{\infty} |h(u)|\,\mathrm{d}u < \infty. \quad (4.5)$$

As noted in Section 3.2.2, there are natural situations where the impulse response function is not absolutely integrable. An example is the low-pass filter with $g(\omega) = 1$ when $|\omega| < 1$, and 0 otherwise, which has $h(u) = \frac{\sin u}{2\pi u}$. One then has to take special care of possible discrete spectral components at the filter endpoints. We do not go deeper into these special cases.

Common linear filters: Here is a summary of three important linear filters and their impulse and frequency response functions.

Operation	Symbolic	Impulse response	Frequency response
time shift	$y(t) = x(t-\tau_0)$	$\delta_{\tau_0}(u)$	$e^{-i\omega\tau_0}$
derivation	$y(t) = x'(t)$	$\delta_0'(u)$	$i\omega$
exponential smoothing	$y(t) = \int_{-\infty}^{t} e^{-\alpha(t-u)} x(u)\,\mathrm{d}u$	$e^{-\alpha u}$, for $u > 0$	$\frac{1}{\alpha+i\omega}$

Gaussian input gives Gaussian output

The convolution representation

$$y(t) = \int_{-\infty}^{\infty} h(t-u)x(u)\,\mathrm{d}u$$

defines $y(t)$ as the limit of a sequence of approximating sums. This means that if the input stationary process $x(t)$ is Gaussian then the output process is also a stationary Gaussian process.

Linear filters in discrete time

We have formulated most of the filter theory for continuous time processes. No new difficulties appear in discrete time, where the integrals are replaced by summation in analogy with the relation between covariance function and spectrum in discrete time. The frequency function may be restricted to a finite frequency band, depending on the sampling interval.

A discrete filter of the form $y(t) = \sum_k h_k x(t - t_k)$ for constants h_k, t_k can be written in integral form $y(t) = \int_{-\pi+}^{\pi} g(\omega)e^{i\omega t}\,\mathrm{d}Z(\omega)$, with impulse and frequency response functions

$$h(u) = \sum_k h_k \delta_{t_k}(u), \quad g(\omega) = \sum_k h_k e^{-i\omega t_k}. \tag{4.6}$$

4.1.3 Correlation and spectral relations

The spectral formulation of linear filters gives us an easy-to-use tool to find covariance and spectral relations in processes generated by linear filtering. If the process $x(t) = \int_{-\infty}^{\infty} e^{i\omega t}\,\mathrm{d}Z(\omega)$ is stationary with spectral distribution function $F_x(\omega)$, and the two processes $\{u(t), t \in \mathbb{R}\}$ and $\{v(t), t \in \mathbb{R}\}$ are generated by linear filtering,

$$u(t) = \int_{-\infty}^{\infty} g(\omega)e^{i\omega t}\,\mathrm{d}Z(\omega), \quad v(t) = \int_{-\infty}^{\infty} h(\omega)e^{i\omega t}\,\mathrm{d}Z(\omega),$$

then, by (3.20),

$$C(u(s), v(t)) = \int_{-\infty}^{\infty} g(\omega)\overline{h(\omega)}\,e^{i(s-t)\omega}\,\mathrm{d}F_x(\omega). \tag{4.7}$$

Applying these relations to a filter output, we get the following theorem.

Theorem 4.1. *If the input $\{x(t), t \in \mathbb{R}\}$ to a linear filter is a stationary process with mean zero and spectral distribution $F(\omega)$, then the output process $y(t) = \int_\omega g(\omega)\,e^{i\omega t}\,\mathrm{d}Z(\omega)$ is also stationary with mean zero, and its covariance function is given by*

$$r_y(t) = E\left(y(s+t)\cdot\overline{y(s)}\right) = \int_{-\infty}^{\infty} |g(\omega)|^2\,e^{i\omega t}\,\mathrm{d}F(\omega). \tag{4.8}$$

In particular, if $x(t)$ has spectral density $f_x(\omega)$, the spectral density of $y(t)$ is

$$f_y(\omega) = |g(\omega)|^2\,f_x(\omega). \tag{4.9}$$

The simplicity of (4.8) contrasts with the following cumbersome covariance formula via the impulse response, obtained from Theorem 2.17.

Theorem 4.2 (Covariance function from impulse response). *If the input process $\{x(t), t \in \mathbb{R}\}$ has mean m_x and covariance function $r_x(t)$, then the output process has mean and covariance*

$$E(y(t)) = m_y = m_x \int_{-\infty}^{\infty} h(u)\,\mathrm{d}u = m_x g(0),$$

$$r_y(t) = \int_u \int_v h(u)\overline{h(v)}\, r_x(t+u-v)\,\mathrm{d}u\,\mathrm{d}v,$$

where $g(0) = \int h(u)\,\mathrm{d}u$ is called the gain of the filter.

Many of the interesting processes we have studied in previous sections were obtained as linear combinations of $x(t)$-variables, or, more commonly, as limits of linear combinations. To formulate the spectral forms of these operations, we need the following property, cf. Step 3 in the proof of Theorem 3.7.

Lemma 4.1. *If $g_n \to g$ in $\mathcal{H}(F)$, i.e., $\int_{-\infty}^{\infty} |g_n(\omega) - g(\omega)|^2\,\mathrm{d}F(\omega) \to 0$, then*

$$\int_{-\infty}^{\infty} g_n(\omega)\, e^{i\omega t}\,\mathrm{d}Z(\omega) \to \int_{-\infty}^{\infty} g(\omega)\, e^{i\omega t}\,\mathrm{d}Z(\omega)$$

in $\mathcal{H}(x)$, i.e., in quadratic mean.

Proof. Use the isometry from Corollary 3.1, page 92,

$$\left\| \int_{-\infty}^{\infty} g_n(\omega)\, e^{i\omega t}\,\mathrm{d}Z(\omega) - \int_{-\infty}^{\infty} g(\omega)\, e^{i\omega t}\,\mathrm{d}Z(\omega) \right\|^2_{\mathcal{H}(x)}$$

$$= \int_{-\infty}^{\infty} \left| g_n(\omega) e^{i\omega t} - g(\omega) e^{i\omega t} \right|^2 \mathrm{d}F(\omega) = \|g_n - g\|^2_{\mathcal{H}(F)},$$

to get the convergence equivalence. □

Example 4.1 (Derivation). *The derivation operation is the limit of*

$$\frac{x(t+h)-x(t)}{h} = \int_{-\infty}^{\infty} \frac{e^{ih\omega}-1}{h} e^{i\omega t}\, \mathrm{d}Z(\omega).$$

If the spectrum of $x(t)$ satisfies the condition $\int \omega^2 \, \mathrm{d}F(\omega) < \infty$ for quadratic mean differentiability, then $(e^{i\omega h}-1)/h \to i\omega$ in $\mathcal{H}(F)$ as $h \to 0$, and hence

$$x'(t) = \int_{-\infty}^{\infty} i\omega\, e^{i\omega t}\, \mathrm{d}Z(\omega) = \int_{-\infty}^{\infty} \omega\, e^{i(\omega t+\pi/2)}\, \mathrm{d}Z(\omega).$$

The frequency function for derivation is therefore $g(\omega) = i\omega = \omega e^{i\pi/2}$, and the spectral density of the derivative is $f_{x'}(\omega) = \omega^2 f_x(\omega)$.

In general, writing $y(t) = \int |g(\omega)|\, e^{i(\omega t + \arg g(\omega))}\, \mathrm{d}Z(\omega)$, we see how the filter amplifies the amplitude of $\mathrm{d}Z(\omega)$ by a factor $|g(\omega)|$ and adds $\arg g(\omega)$ to the phase. For the derivative, the phase increases by $\pi/2$, while the amplitude increases by the frequency dependent factor ω.

Example 4.2 (Bandpass filter). *A filter with frequency response $g(\omega) = 1$ when $0 < \omega_1 \le |\omega| \le \omega_2$ and zero otherwise acts a a bandpass filter, and removes all frequency components with frequency outside $[\omega_1, \omega_2]$. The filtered process will have variance*

$$V(u(t)) = 2\int_{\omega_1}^{\omega_2} \mathrm{d}F(\omega),$$

which again illustrates how the spectrum distributes the total variance (= energy) over frequencies.

4.1.4 Linear processes

A stationary sequence $\{x_n, n \in \mathbb{Z}\}$ or a stationary process $\{x(t), t \in \mathbb{R}\}$ is called *linear* if it is the output of a linear time invariant filter acting on a sequence of orthogonal random variables, i.e.,

$$x_n = \sum_{k=-\infty}^{\infty} h_{n-k} y_k, \quad x(t) = \int_{u=-\infty}^{\infty} h(t-u)\, \mathrm{d}Y(u),$$

where the variables y_k are uncorrelated with mean 0, and $E(|y_k|^2) = 1$, and $Y(t)$ is a stationary process with orthogonal increments,

$$E(\mathrm{d}Y(u) \cdot \overline{\mathrm{d}Y(v)}) = \begin{cases} \mathrm{d}u, & \text{for } u = v, \\ 0, & \text{for } u \neq v. \end{cases}$$

The term *infinite moving average* is also used for processes of this type.

Theorem 4.3. *a) A stationary sequence $\{x_n, n \in \mathbb{Z}\}$ is an infinite moving average $x_n = \sum_{k=-\infty}^{\infty} h_{n-k} y_k$, with orthonormal y_k and $\sum_k |h_k|^2 < \infty$, if and only if its spectrum is absolutely continuous, $F(\omega) = \int_{-\pi}^{\omega} f(x)\,\mathrm{d}x$. The spectral density is $f(\omega) = |g(\omega)|^2/2\pi$, with $g(\omega) = \sum_k h_k e^{-i\omega k}$.*

b) A stationary process $\{x(t), t \in \mathbb{R}\}$ is an infinite continuous moving average $x(t) = \int_{-\infty}^{\infty} h(t-u)\,\mathrm{d}Y(u)$, with an orthogonal increment process $Y(u)$, and $\int_{-\infty}^{\infty} |h(u)|^2\,\mathrm{d}u < \infty$, if and only if its spectrum is absolutely continuous, $F(\omega) = \int_{-\infty}^{\omega} f(x)\,\mathrm{d}x$. The spectral density is $f(\omega) = |g(\omega)|^2/2\pi$, with $g(\omega) = \int_u h(u)e^{-i\omega u}\,\mathrm{d}u$.

Proof. We show part a). Part b) is quite similar. For the “only if” part, use that $y_k = \int_{-\pi}^{\pi} e^{i\omega k}\,\mathrm{d}Z(\omega)$, where $E(|\mathrm{d}Z(\omega)|^2) = \frac{\mathrm{d}\omega}{2\pi}$. Then

$$x_t = \sum_k h_{t-k} \int_{-\pi}^{\pi} e^{i\omega k}\,\mathrm{d}Z(\omega) = \int_{-\pi}^{\pi} e^{i\omega t} \left\{ \sum_k h_{t-k} e^{-i\omega(t-k)} \right\} \mathrm{d}Z(\omega)$$

$$= \int_{-\pi}^{\pi} e^{i\omega t}\, g(\omega)\,\mathrm{d}Z(\omega),$$

with $g(\omega) = \sum_k h_k e^{-i\omega k}$. Thus, the spectral distribution of x_k has increment

$$\mathrm{d}F(\omega) = E(|g(\omega)\,\mathrm{d}Z(\omega)|^2) = |g(\omega)|^2 \frac{\mathrm{d}\omega}{2\pi},$$

and hence is absolutely continuous with spectral density, $f(\omega) = |g(\omega)|^2/2\pi$.

For the “if” part, when $F(\omega) = \int_{-\infty}^{\omega} f(x)\,\mathrm{d}x$, write $f(\omega) = |g(\omega)|^2/2\pi$, and expand $|g(\omega)|$ in a Fourier series, $|g(\omega)| = \sum_k c_k e^{i\omega k}$. From the normalized spectral representation (3.29),

$$x_t = \int_{-\pi}^{\pi} e^{i\omega t} \sqrt{f(\omega)}\,\mathrm{d}\tilde{Z}(\omega), \quad \text{with} \quad E(|\mathrm{d}\tilde{Z}(\omega)|^2) = \mathrm{d}\omega,$$

we then get

$$x_t = \int_{-\pi}^{\pi} e^{i\omega t} \left\{ \frac{1}{\sqrt{2\pi}} \sum_k c_k e^{i\omega k} \right\} \mathrm{d}\widetilde{Z}(\omega)$$

$$= \sum_k \frac{c_k}{\sqrt{2\pi}} \int_{-\pi}^{\pi} e^{i\omega(t+k)} \,\mathrm{d}\widetilde{Z}(\omega) = \sum_k c_k\, e_{t+k} = \sum_k h_{t-k}\, e_k,$$

with $e_k = \frac{1}{\sqrt{2\pi}} \int_{-\pi}^{\pi} e^{i\omega k}\, \mathrm{d}\widetilde{Z}(\omega)$, and $h_k = c_{k+t}$. Since $\widetilde{Z}(\omega)$ has constant incremental variance, the e_k-variables are uncorrelated and normalized as required. □

4.1.5 *Lévy processes*

The linear processes treated in the previous section form a very flexible class of stationary processes with second order moments. The increments in the process $Y(u)$ in Theorem 4.3(b) are uncorrelated. With the stronger requirement that the increments are independent, one gets a new class of processes that is subject to special interest.

Definition 4.3 (Lévy process). *A Lévy process is a right-continuous process $\{x(t), t \in \mathbb{R}\}$ such that the distribution of the increment $x(t+h) - x(t)$ is independent of t, and the increments $x(t_2) - x(t_1), x(t_3) - x(t_2), \ldots, x(t_n) - x(t_{n-1})$ over disjoint intervals are independent. A Lévy process has stationary independent increments.*

The Poisson and Wiener processes are common representatives of Lévy processes. These processes share another characteristic property, of classical interest in probability theory; they both have infinitely divisible distributions.

Definition 4.4 (Infinitely divisible distribution). *A distribution of a random variable is said to be infinitely divisible if, for every integer n, there are independent and identically distributed random variables $x_1^{(n)}, \ldots, x_n^{(n)}$, such that x and $x_1^{(n)} + \ldots + x_n^{(n)}$ have the same distribution.*

The Poisson and the normal distribution are both infinitely divisible, and they have finite mean and finite variance. A general Lévy process consists of a continuous Wiener component varying around a drift term with constant slope, and on top of this random jumps of different sizes that occur according to independent Poisson processes. This somewhat vague description is made precise in the theory of infinitely divisible processes; see for example [19, Ch. 14] or [111].

There is an important class of infinitely divisible distributions that do not have finite variance, and they play an important role in connection with the stochastic integrals in Theorem 4.3. For example, in stochastic models for financial processes, the impact on the economy is often modeled by such *heavy tailed* distributions.

The standard Wiener process is a Lévy process with stationary independent and normal increments. One characteristic property is that $x(\lambda t)$ and $\sqrt{\lambda}x(t)$ have the same normal distribution, with mean zero and variance λt.

Definition 4.5 (Self similarity and stability). *A process $\{x(t), t \in \mathbb{R}\}$ is called self-similar with index H if, for all $\lambda > 0$, the process $\{y(t), t \in \mathbb{R}\}$ with $y(t) = x(\lambda t)$ has the same distributions as $\{\lambda^H x(t), t \in \mathbb{R}\}$. A self-similar Lévy process is called a stable process.*

The following characterization of self-similar processes is due to J. Lamperti [72].

Theorem 4.4 (Lamperti). *A process $\{x(t), t \geq 0\}$ is self-similar with index H if and only if $y(t) = e^{-H\alpha t}x(e^{\alpha t}), t \in \mathbb{R}$, makes a stationary process for all $\alpha > 0$.*

Proof. We show the statements only in one dimension; the n-dimensional statements are trivial reformulations. If $\{x(t), t \in \mathbb{R}\}$ is self-similar, then $x(e^{\alpha(t+h)})$ has the same distribution as $e^{H\alpha h}x(e^{\alpha t})$, so

$$y(t+h) = e^{-H\alpha t}(e^{-H\alpha h}x(e^{\alpha(t+h)})) \stackrel{\mathscr{L}}{=} e^{-H\alpha t}x(e^{\alpha t}) = y(t).$$

Similarly, $x(\lambda x) = (\lambda t)^H y(\alpha^{-1}\log \lambda t)) \stackrel{\mathscr{L}}{=} (\lambda t)^H y(\alpha^{-1}\log t) = \lambda^H x(t)$ gives the converse. □

4.1.6 Fractional Brownian motion

The Wiener process is self-similar with index $1/2$. It has stationary independent increments with $V(x(t+h)-x(t)) = hV(x(1))$. Due to the independence,

$$V(x(t)) = V(x(t_1)) + V(x(t_2)-x(t_1)) + \ldots + V(x(t)-x(t_n)),$$

when $0 < t_1 \ldots < t_n < t$. One can think of this as a way to produce independent normal variables – just take increments in a Wiener process.

Fractional Brownian motion is a class of Gaussian processes, with the Wiener process as one of its members, with some useful properties.

Definition 4.6 (Fractional Brownian motion). *a) A zero mean Gaussian process $\{x(t), 0 \le t < \infty\}$ with $x(0) = 0$ is called fractional Brownian motion (fBm) if it is self-similar with index $H \in (0,1)$ and has stationary increments. It is simple to see that its covariance function and variance are, respectively,*

$$r(s,t) = C(x(s),x(t)) = \frac{\sigma^2}{2}\left(s^{2H} + t^{2H} - |s-t|^{2H}\right), \tag{4.10}$$

$$V(x(t)) = \sigma^2 t^{2H}. \tag{4.11}$$

b) The sequential increments $y_n = x(n) - x(n-1)$, $n = 1,2,\ldots$, in fractional Brownian motion form a stationary sequence called fractional Gaussian noise. The covariance function is

$$r_y(t) = C(y_{n+t}, y_n) = \frac{\sigma^2}{2}\left((t-1)^{2H} - 2t^{2H} + (t+1)^{2H}\right)$$

$$\sim \sigma^2 H(2H-1)t^{2H-2}, \text{ as } t \to \infty.$$

The index H has dramatic influence on the covariance structure of the fractional Brownian motion and on its increments $y_n = x(n) - x(n-1)$. If $H = 1/2$, then the process is regular Brownian motion with independent increments and linearly increasing variance, while if $1/2 < H < 1$ the variance increases at a faster rate than linear, and the covariances of the increment decrease slowly. For $0 < H < 1/2$, the variance of the fractional Brownian motion increases slower than as $\sqrt{n}$, indicating an excess of negative dependence. More on this will come in Section 4.4.1.

4.2 Linear filters and differential equations

Linear filters expressed in terms of differential equations are common in the engineering sciences. The linear oscillator, also called the harmonic oscillator, is the basic element in lightly damped mechanical systems which exhibit resonant random vibrations. Its counterpart in electronic systems is the resonance circuit. We shall describe both of these as examples of a general technique, common in the theory of ordinary differential equations, here translated into stochastic language. To illustrate the general ideas we start with two practically important filters, the exponential smoothing filter, also called the RC-filter, with a term borrowed from electrical engineering, and the linear resonant oscillator.

4.2.1 The RC-filter and exponential smoothing

Consider the electrical circuit in Figure 4.1 with an input potential difference $x(t)$ on the left hand side and an output potential difference $y(t)$ on the right hand side. The circuit consists of a resistance R and a capacitance C. Regarding $x(t)$ as the driving process and $y(t)$ as the resulting process we will see that this device acts as a *smoother* that reduces rapid high frequency variations in $x(t)$. The relation between the input $x(t)$ and the output $y(t)$ is

$$RCy'(t) + y(t) = x(t), \tag{4.12}$$

and the equation has the (deterministic) solution

$$y(t) = \frac{1}{RC} \int_{-\infty}^{t} e^{-(t-u)/(RC)} x(u)\, du.$$

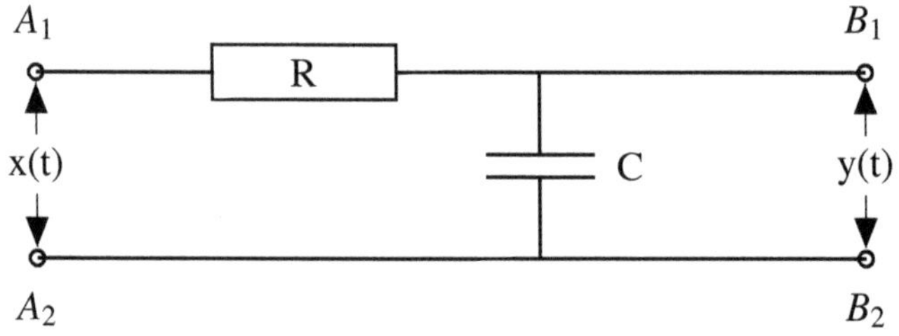

Figure 4.1: *Input $x(t)$ and output $y(t)$ in an exponential smoother (RC-filter).*

This is a relation of the convolution type, and the impulse response of the RC-filter is

$$h(u) = \frac{1}{RC} e^{-u/(RC)}, \quad \text{for } u > 0,$$

with frequency response

$$g(\omega) = \int_0^\infty e^{-i\omega u} \frac{1}{RC} e^{-u/(RC)} \, du = \frac{1}{1 + i\omega RC}.$$

Applying the relation (4.9) we get the spectral density relation between input and output,

$$f_y(\omega) = \frac{f_x(\omega)}{(\omega RC)^2 + 1}. \tag{4.13}$$

The depreciation of high frequencies in the spectrum explains the use of the RC-filter as a smoother.

To anticipate the general results for covariance function relations we also make the following elementary observation about the covariance functions, where we use the cross-covariances from Theorem 2.3,

$$\begin{aligned} r_x(\tau) &= C(RCy'(t) + y(t), RCy'(t+\tau) + y(t+\tau)) \\ &= (RC)^2 r_{y'}(\tau) + RC\, r_{y,y'}(t, t+\tau) + RC\, r_{y',y}(t, t+\tau) + r_y(\tau) \\ &= (RC)^2 r_{y'}(\tau) + RC\, r_y'(\tau) + RC\, r_y'(-\tau) + r_y(\tau) = (RC)^2 r_{y'}(\tau) + r_y(\tau). \end{aligned}$$

Using the spectral density $\omega^2 f_y(\omega)$ for $\{y'(t), t \in \mathbb{R}\}$, according to Example 4.1, we find

$$\begin{aligned} r_x(t) &= (RC)^2 \int_{-\infty}^{\infty} e^{i\omega t} \omega^2 f_y(\omega) \, d\omega + \int_{-\infty}^{\infty} e^{i\omega t} f_y(\omega) \, d\omega \\ &= \int_{-\infty}^{\infty} e^{i\omega t} \{(\omega RC)^2 + 1\} f_y(\omega) \, d\omega, \end{aligned}$$

and get the spectral density for $\{x(t), t \in \mathbb{R}\}$,

$$f_x(\omega) = \{(\omega RC)^2 + 1\} f_y(\omega), \tag{4.14}$$

in accordance with (4.13).

As a final observation we note that the impulse response function satisfies the differential equation

$$RCh'(u) + h(u) = 0, \tag{4.15}$$

for $u > 0$, with the initial condition $h(0) = 1/(RC)$.

4.2.2 *The linear oscillator*

The linear random oscillator is the basic ingredient in many engineering applications of stationary processes, in particular in mechanical and electrical engineering. The simple model in the following example is then duplicated into a complex structure, giving a large, but "simple," in the sense that it is linear, structure. Such models are used as first approximations in, for example, safety analysis of buildings, aircraft, cars, etc.

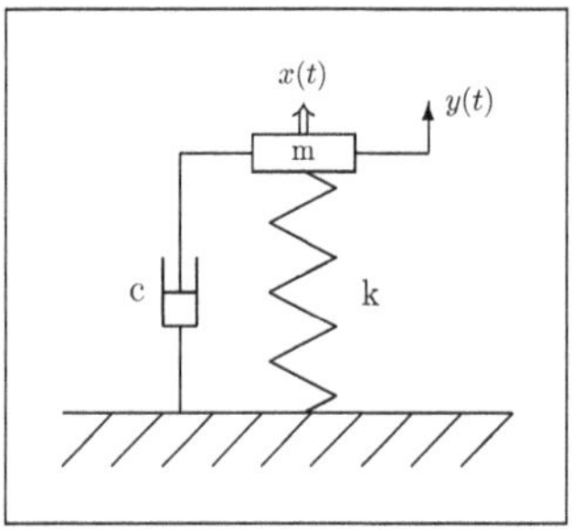

Figure 4.2: *A linear oscillator.*

Example 4.3 (Linear oscillator). *First consider a spring-and-damper system as in Figure 4.2, with mass m, stiffness k, and damping coefficient c. When the mass is subject to a regular or irregular varying force $x(t)$ it moves more or less periodically, and we denote the displacement from the equilibrium by $y(t)$. The relation between the force $x(t)$ and the resulting displacement is described by the differential equation,*

$$my''(t) + cy'(t) + ky(t) = x(t). \tag{4.16}$$

Here $\omega_0 = \sqrt{k/m}$ is called the resonance frequency *or* eigenfrequency, *and $\zeta = \frac{c}{2\sqrt{mk}}$ the* relative damping, *which we here assume satisfies $0 < \zeta < 1$. Expressed in terms of the damping and eigenfrequency, the fundamental equation is*

$$y''(t) + 2\zeta\omega_0 y'(t) + \omega_0^2 y(t) = m^{-1}x(t).$$

This equation can be solved just like an ordinary differential equation with a continuous $x(t)$ and, from Theorem 4.5 later in this section, it has the solution

$$y(t) = \int_{-\infty}^{t} h(t-u)x(u)\,\mathrm{d}u,$$

expressed with the impulse response function

$$h(u) = m^{-1}\widetilde{\omega}_0^{-1}e^{-\alpha u}\sin(\widetilde{\omega}_0 u), \quad u \geq 0,$$

with the constants

$$\alpha = \zeta\omega_0, \quad \widetilde{\omega}_0 = \omega_0(1-\zeta^2)^{1/2}.$$

To find the frequency function $g(\omega)$ for the linear oscillator we consider each term on the left hand side in (4.16). Since differentiation has frequency function $i\omega$, and, hence, repeated differentiation has frequency function $-\omega^2$, we see that $g(\omega)$ satisfies the equation

$$\{-m\omega^2 + ci\omega + k\} \cdot g(\omega) = 1, \tag{4.17}$$

and hence

$$g(\omega) = \frac{1}{-m\omega^2 + ic\omega + k}. \tag{4.18}$$

Since

$$|g(\omega)|^2 = \frac{1}{(k - m\omega^2)^2 + c^2\omega^2},$$

the spectral density for the output signal $y(t)$ is

$$f_y(\omega) = \frac{f_x(\omega)}{(k - m\omega^2)^2 + c^2\omega^2} = \frac{f_x(\omega)/m^2}{(\omega_0^2 - \omega^2)^2 + 4\alpha^2\omega^2}. \tag{4.19}$$

When the relative damping ζ is small, the filter accentuates the relative importance of frequencies near ω_0. Small amplitudes in $x(t)$ can cause very large displacements in the output.

Example 4.4 (Resonance circuit). *A resonance circuit with one inductance, one resistance, and one capacitance in series is an electronic counterpart to the harmonic mechanical oscillator; see Figure 4.3.*

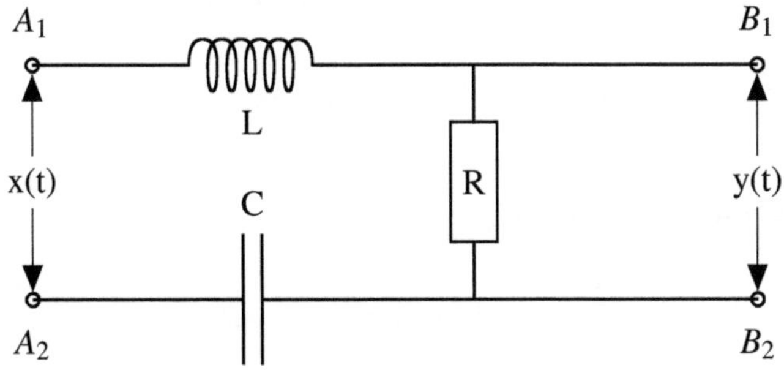

Figure 4.3: *A resonance circuit.*

If the input potential between A_1 and A_2 is $x(t)$, the electric current $I(t)$ through the circuit obeys the equation

$$LI'(t) + RI(t) + \frac{1}{C}\int_{-\infty}^{t} I(s)\,\mathrm{d}s = x(t).$$

The output potential between B_1 and B_2, which is $y(t) = RI(t)$, therefore follows the same equation (4.16) as the linear mechanical oscillator,

$$Ly''(t) + Ry'(t) + \frac{1}{C}y(t) = Rx'(t), \tag{4.20}$$

but this time with $x'(t)$ as driving force. In parallel to (4.17) we have now

$$\{-(L/R)\omega^2 + i\omega + 1/(RC)\} \cdot g(\omega) = i\omega, \tag{4.21}$$

which gives the frequency function for the filter between $x(t)$ and $y(t)$,

$$g(\omega) = \frac{i\omega}{-(L/R)\omega^2 + i\omega + 1/(RC)}.$$

The resonance frequency is $\omega_0 = 1/\sqrt{LC}$ and the relative damping ζ corresponds to the relative bandwidth $1/Q = 2\zeta$, where

$$1/Q = \Delta\omega/\omega_0 = R\sqrt{C/L},$$

and $\Delta\omega = \omega_2 - \omega_1$ is such that $|g(\omega_1)| = |g(\omega_2)| = |g(\omega_0)|/\sqrt{2}$.

4.2.3 *Linear differential equations driven by a stochastic process*

Suppose we have a stationary process $\{x(t), t \in \mathbb{R}\}$, sufficiently differentiable, and assume that the process $\{y(t), t \in \mathbb{R}\}$ is a solution to an ordinary linear differential equation with constant coefficients,

$$\sum_{k=0}^{p} a_{p-k} y^{(k)}(t) = x(t), \tag{4.22}$$

or, seemingly more generally,

$$\sum_{k=0}^{p} a_{p-k} y^{(k)}(t) = \sum_{j=0}^{q} b_{q-j} x^{(j)}(t). \tag{4.23}$$

By "solution" we mean either that (almost all) sample functions satisfy the equations or that there exists a process $\{y(t), t \in \mathbb{R}\}$ such that the two sides are equivalent. Note that (4.23) is only marginally more general than (4.22), since the right hand sides in both equations are stationary processes without any further assumption.

What can then be said about the solution to these equations: when does a solution exist and when is it a stationary process; and in that case, what is its spectrum and covariance function?

For the linear differential equation (4.22),

$$a_0 y^{(p)}(t) + a_1 y^{(p-1)}(t) + \ldots + a_{p-1} y'(t) + a_p y(t) = x(t), \tag{4.24}$$

we define the *generating polynomial*,

$$A(z) = a_0 + a_1 z + \ldots + a_p z^p,$$

and the corresponding *characteristic equation*

$$z^p A(z^{-1}) = a_0 z^p + a_1 z^{p-1} + \ldots + a_{p-1} z + a_p = 0. \tag{4.25}$$

The existence of a stationary process solution depends on the solutions to the characteristic equation.

Definition 4.7 (Stable equation). *The differential equation (4.24) is called stable if the roots of the characteristic equation all have negative real part. The solutions of a stable equation tend to* 0 *as* $t \to \infty$, *when the right hand side is constant and equal to* 0.

One can work with (4.24) as a special case of a multivariate first order differential equation. Assuming for a moment that $a_0 = 1$, the form is

$$\mathbf{y}' = \mathbf{A}\mathbf{y} + \mathbf{x}, \tag{4.26}$$

with $\mathbf{y}(t) = (y(t), y'(t), \cdots, y^{(p-1)}(t))'$, $\mathbf{x}(t) = (0, 0, \cdots, x(t))'$, and

$$\mathbf{A} = \begin{pmatrix} 0 & 1 & 0 & \ldots & 0 \\ 0 & 0 & 1 & \cdots & 0 \\ \vdots & \vdots & \vdots & \ddots & 1 \\ -a_p & -a_{p-1} & a_{p-2} & \cdots & -a_1 \end{pmatrix}.$$

This formulation is common in linear and non-linear systems theory; cf. for example [65, Ch. 8], to which we refer for part of the following Theorem 4.5.

Remark 4.1. *For discrete time versions of linear stochastic differential equations and their stability, we refer to any good textbook in Time series analysis; for example [22].*

Theorem 4.5. *If the differential equation (4.24) is stable, and the right hand side* $\{x(t), t \in \mathbb{R}\}$ *is a stationary process, then there exists a stationary process* $\{y(t), t \in \mathbb{R}\}$ *that solves the equation. The solution can be written as the output of a linear filter*

$$y(t) = \int_{-\infty}^{t} h(t-u)\,x(u)\,\mathrm{d}u, \tag{4.27}$$

where the impulse function $h(u)$*, with* $\int_0^\infty |h(u)|\,\mathrm{d}u < \infty$*, is a solution to the equation*

$$a_0 h^{(p)}(t) + a_1 h^{(p-1)}(t) + \ldots + a_p h(t) = 0,\ t \geq 0, \tag{4.28}$$

with initial conditions $h(0) = \ldots = h^{(p-2)}(0) = 0, h^{(p-1)}(0) = a_p/a_0$.

Proof. If $\{x(t), t \in \mathbb{R}\}$ has q times differentiable sample paths (with probability one), we use a standard result in ordinary differential equations to get a solution for almost every sample path; see [65, Ch. 8].

If we work only with second order properties one can take (4.27) as the definition of a process $\{y(t), t \in \mathbb{R}\}$ and then show that it is p times differentiable (in quadratic mean) and that the two sides of (4.24) are equivalent. □

Theorem 4.6. *If* $\{x(t), t \in \mathbb{R}\}$ *and* $\{y(t), t \in \mathbb{R}\}$ *are two stationary processes that solve the differential equation (4.23), then their spectral densities obey the relation*

$$\left|\sum_{k=0}^{p} a_k (i\omega)^{p-k}\right|^2 f_y(\omega) = \left|\sum_{j=0}^{q} b_j (i\omega)^{q-j}\right|^2 f_x(\omega). \tag{4.29}$$

Proof. The theorem is an easy consequence of the spectral process property (3.20). Just write the differential form in the sides in (4.23) by means of the spectral representation and carry out the integration. □

4.3 White noise in linear systems

4.3.1 White noise in linear differential equations

The Wiener process was constructed as a mathematical model for the Brownian motion of particles suspended in a viscous fluid in which the erratic particle movements are the results of bombardment by the fluid molecules. The Wiener process model requires that the particles have zero mass. A more realistic model gives room also for the particle mass. If $x(t)$ denotes the force acting on the particle and $y(t)$ is the velocity, we get the Ornstein-Uhlenbeck differential equation (3.47) from Example 3.5, page 103,

$$a_0 y'(t) + a_1 y(t) = x(t),$$

where a_1 depends on the viscosity and a_0 is the particle mass. If the force $x(t)$ is caused by collisions from independent molecules it is reasonable that different $x(t)$ be independent. Adding the assumption that they are Gaussian leads us to take $x(t) = \sigma w'(t)$ as the "derivative of a Wiener process," i.e., Gaussian white noise,

$$a_0 y'(t) + a_1 y(t) = \sigma w'(t). \tag{4.30}$$

This equation can be solved as an ordinary differential equation, when $\alpha = a_1/a_0 > 0$,

$$y(t) = \frac{\sigma}{a_0} \int_{-\infty}^{t} e^{-\alpha(t-u)} w'(u)\, \mathrm{d}u = \frac{\sigma}{a_0} \int_{-\infty}^{t} e^{-\alpha(t-u)}\, \mathrm{d}w(u). \tag{4.31}$$

Here the last integral is well defined, Example 2.12 on page 63 and the covariance formula (2.37), even if the differential equation we started out from is not.

Interpretation as integral equation

By carrying out the integration it is easy to see that the process $y(t)$ defined by (4.31) satisfies

$$a_1 \int_{t_0}^{t} y(u)\, \mathrm{d}u = -a_0(y(t) - y(t_0)) + \sigma(w(t) - w(t_0)),$$

which means that, instead of equation (4.30), we could have used the integral equation

$$a_0(y(t) - y(t_0)) + a_1 \int_{t_0}^{t} y(u)\, \mathrm{d}u = \sigma\,(w(t) - w(t_0)), \tag{4.32}$$

to describe the increments of $y(t)$.

The general differential equation,

$$a_0 y^{(p)}(t) + a_1 y^{(p-1)}(t) + \ldots + a_{p-1} y'(t) + a_p y(t) = \sigma w'(t), \tag{4.33}$$

can be solved in a similar way, and expressed as a stochastic integral,

$$y(t) = \sigma \int_{-\infty}^{t} h(t-u)\,\mathrm{d}w(u). \tag{4.34}$$

The impulse response function $h(u)$ is the solution to

$$a_0 h^{(p)}(t) + a_1 h^{(p-1)}(t) + \ldots + a_{p-1} h'(t) + a_p h(t) = 0,$$

as in Theorem 4.5. The formal differential equation can be replaced by the well defined differential-integral equation

$$\begin{aligned} a_0(y^{(p-1)}(t) - y^{(p-1)}(t_0)) + a_1(y^{(p-2)}(t) - y^{(p-2)}(t_0)) + \\ + \ldots + a_{p-1}(y(t) - y(t_0)) + a_p \int_{t_0}^{t} y(u)\,\mathrm{d}u = \sigma(w(t) - w(t_0)). \end{aligned} \tag{4.35}$$

Stochastic calculus

Stochastic differential equations involving Gaussian white noise are often written in differential form, $a_0 \mathrm{d}y(t) + a_1 y(t)\,\mathrm{d}t = \sigma\,\mathrm{d}w(t)$, or, more generally, allowing the parameters to vary with time, as

$$\mathrm{d}y(t) = a(t)y(t)\,\mathrm{d}t + \sigma(t)\,\mathrm{d}w(t),$$

with variable deterministic coefficients. For the most general form, with random coefficients,

$$\mathrm{d}y(t) = a(y(t),t)\,y(t)\,\mathrm{d}t + \sigma(y(t),t)\,\mathrm{d}w(t), \tag{4.36}$$

a completely new theory is needed, namely *stochastic calculus*; the reader is referred to [95] for a good introduction. A special case of (4.36) has been used in *stochastic mechanics* under the name *parametric excitation*, meaning that the parameters in a linear oscillator themselves are random processes. The idea is to allow for variability in environment, material properties, etc.

Strictly speaking, the term "stochastic differential equation" should be reserved for equations like (4.36), where the equation itself is random.

4.3.2 White noise and constant spectral density

A linear system acts as a frequency dependent amplifier and phase modifier on the input. Of special importance is the case when the input is *white noise*. This idealized type of process is defined only in the context of linear systems.

The characteristic feature of white noise, which may be denoted $n(t)$ or, if it is Gaussian, $w'(t)$, is that all frequencies are represented in equal amount, i.e., it has *constant spectral density*, $f_n(\omega) = \frac{\sigma^2}{2\pi}$, $-\infty < \omega < \infty$. This is not a proper spectral density of a stationary process, but used in connection with a linear system that satisfies $\int_{-\infty}^{\infty} |h(u)|^2 \, du < \infty$, it produces the correct spectral density for the stationary output process. This property is a parallel to the interpretation (4.35) of the differential equation (4.33).

Theorem 4.7. *The stationary process* $\{x(t), t \in \mathbb{R}\}$, *defined as a stochastic integral from a Wiener process* $\{w(t), t \in \mathbb{R}\}$ *by* $x(t) = \int_{-\infty}^{\infty} h(t-u) \, dw(u)$, *with* $\int_{-\infty}^{\infty} |h(u)|^2 \, du < \infty$, *has the covariance function* $r_x(t) = \int_{-\infty}^{\infty} h(t-u)h(-u) \, du$ *and spectral density* $f_x(\omega) = \frac{|g(\omega)|^2}{2\pi}$, *where* $g(\omega) = \int_{-\infty}^{\infty} e^{-i\omega u} h(u) \, du$ *is the frequency response function corresponding to the impulse response* $h(u)$.

Proof. The covariance function follows from Theorem 2.18. Now the right hand side of the integral expression for $r_x(t)$ is the convolution of $h(u)$ with $h(-v)$ and their Fourier transforms are $g(\omega)$ and $\overline{g(\omega)}$, respectively. Since convolution corresponds to multiplication of the Fourier transforms, the spectral density of $r_x(t)$ is, as stated, $f_x(\omega) = \frac{1}{2\pi} g(\omega)\overline{g(\omega)} = \frac{1}{2\pi}|g(\omega)|^2$. □

Theorem 4.8. *If the process* $x(t) = \int_{-\infty}^{t} h(t-u) \, dw(u)$ *is stationary and solves the stable stochastic differential equation*

$$a_0 y^{(p)}(t) + a_1 y^{(p-1)}(t) + \ldots + a_{p-1} y'(t) + a_p y(t) = \sigma w'(t), \tag{4.37}$$

then its spectral density is

$$f_x(\omega) = \frac{\sigma^2}{2\pi} \cdot \frac{1}{\left|\sum_{k=0}^{p} a_k (i\omega)^{p-k}\right|^2}.$$

The theorem finally confirms our claim that the Gaussian white noise $\sigma w'(t)$ can be treated as if it has constant spectral density $\sigma^2/(2\pi)$.

Proof. The relation between the impulse response and the frequency response function $g(\omega)$ is a property of the systems equation (4.37) and does not depend on any stochastic property. One can therefore use the established relation

$$g(\omega) = \frac{1}{\sum_{k=0}^{p} a_k (i\omega)^{p-k}}$$

to get the result; see [65]. □

A stationary process with spectral density of the form $\frac{C}{|P(i\omega)|^2}$ where $P(\omega)$ is a complex polynomial can be generated as the output from a stable linear system with white noise input; this is a very convenient way to produce a stationary process with suitable spectral properties.

Example 4.5 (Stable linear oscillator). *The linear oscillator driven by white noise,*

$$y''(t) + 2\zeta\omega_0 y'(t) + \omega_0^2 y(t) = \sigma w'(t),$$

with relative damping $0 < \zeta < 1$, *has spectral density (cf. (4.19)),*

$$f_y(\omega) = \frac{\sigma^2}{2\pi} \cdot \frac{1}{(\omega^2 - \omega_0{}^2)^2 + 4\alpha^2\omega^2},$$

with $\alpha = \zeta\omega_0$. *The covariance function is found by residue calculus from* $r_y(t) = \int_{-\infty}^{\infty} e^{i\omega t} f_y(\omega)\,\mathrm{d}\omega$. *With* $\beta = \omega_0(1 - \zeta^2)^{1/2}$ *one gets the covariance function*

$$r_y(t) = \frac{\sigma^2}{4\alpha\omega_0^2} e^{-\alpha|t|} \left(\cos\beta t + \frac{\alpha}{\beta} \sin\beta|t| \right). \tag{4.38}$$

This covariance function is of the oscillating type often encountered in practice,

$$r(t) = Ce^{-\alpha|t|} \cos(\beta|t| - \psi), \tag{4.39}$$

with constant β *and* ψ. *In (4.38) the shift* $\psi = \arctan(\alpha/\beta)$. *When* $\psi = 0$ *the covariance function is that of an amplitude modulated cosine function; we met such a covariance function in Exercise 3.15. For non-oscillating alternatives, see the table in Appendix C.*

Remark 4.2 (Non-stable equations). *In the entire section we have considered stable linear equations, with characteristic roots in the left complex half plane. Solutions were constructed from a Wiener process as in Theorem 4.8 and were found to be stationary processes, stable in time, in the sense that new disturbances eventually died out, often exponentially fast, when time went from past to future. One might want to know what can be said about equations which are not stable, for example,* $y(t) - y''(t) = \sigma w'(t)$. *In fact, such equations, and their generalizations to random fields, are important in many applications, and we will deal with some examples in Section 4.4.2 and Section 7.2.3.*

4.3.3 Shot noise

The term *shot noise* is historically associated with the physical phenomenon of an aggregation of electric pulses triggered by random events and is nowadays also used as a general statistical term for *filtered point processes* with deterministic or random impulse response. In this section we shall combine the theory from Section 3.5 with the filter techniques in this chapter, to obtain covariance and spectral properties of the general shot noise process.

The simplest shot noise process is based on the two-sided Poisson process $\{N(t), t \in \mathbb{R}\}$ with constant intensity λ. Each event in the Poisson process triggers a response of fixed shape $h(u)$, and these responses are added to form the shot noise process.

Campbell's formulas

We shall derive the correlation and spectral properties of a shot noise process driven by a stationary orderly stream of events, $\{N(t), t \in \mathbb{R}\}$, with Bartlett spectrum $\mathrm{d}\mu_N(\omega)$; see Definition 3.8. As special cases we get the properties of shot noise driven by a Poisson process and by a Cox process.

Let the times of the driving point process be denoted by $\{T_n\}_{-\infty}^{\infty}$, and let the impulse response be $h(u)$. In the special case when the response is acting only in the future, $h(u) = 0$ for $u < 0$, but that restriction is not used in the coming derivations. Now, the shot noise process $\{x(t), t \in \mathbb{R}\}$ is defined as

$$x(t) = \sum_n h(t - T_n) = \int_{-\infty}^{\infty} h(t-s)\,\mathrm{d}N(s), \tag{4.40}$$

with summation over all events, or, in the special case, over the events that have happened up till time t.

Taking $h_\tau(u) = h(\tau + u)$ and $h^*(u) = h(-u)$, $h_\tau^*(u) = h_\tau(-u) = h(\tau - u)$, we get, with the notation (3.56),

$$x(0) = \int h_0^*(t)\,\mathrm{d}N(t) = N(h_0^*), \quad x(\tau) = \int h_\tau^*(t)\,\mathrm{d}N(t) = N(h_\tau^*), \tag{4.41}$$

and, by the defnition of the Bartlett spectrum $\mathrm{d}\mu_N(\omega)$, for $\{N(t), t \in \mathbb{R}\}$,

$$\begin{aligned} C(x(\tau), x(0)) &= C(N(h_\tau^*), N(h_0^*)) \\ &= \int_{-\infty}^{\infty} \widehat{h_\tau^*}(\omega)\,\overline{\widehat{h_0^*}(\omega)}\,\mathrm{d}\mu_N(\omega). \end{aligned} \tag{4.42}$$

Now, it is readily found that the Fourier transforms in (4.42) can be expressed in term of the frequency function $g(\omega)$ for the filter with impulse response $h(u)$,

$$\widehat{h^*}(\omega) = \int_{-\infty}^{\infty} e^{i\omega u} h^*(u)\,\mathrm{d}u = \int_{-\infty}^{\infty} e^{-i\omega u} h(u)\,\mathrm{d}u = g(\omega),$$

and that

$$\widehat{h_\tau^*}(\omega) = \int_{-\infty}^{\infty} e^{i\omega u} h_\tau^*(u)\,\mathrm{d}u = \int_{-\infty}^{\infty} e^{-i\omega u} h_\tau(u)\,\mathrm{d}u = e^{-i\omega\tau} g(\omega).$$

Inserting these expressions into (4.42), we get

$$C(x(\tau), x(0)) = \int_{-\infty}^{\infty} |g(\omega)|^2 e^{i\omega\tau}\,\mathrm{d}\mu_N(\omega), \tag{4.43}$$

relating the covariance function for the shot noise process to the Bartlett spectrum for the driving point process and the filter frequency function. This relation is exactly the same as the one between input and output spectrum in a common linear filter.

Taking $\tau = 0$, we get the variance of the shot noise process,

$$V(x(0)) = V(N(h^*)) = \int_{-\infty}^{\infty} |\widehat{h^*}(\omega)|^2\,\mathrm{d}\mu_N(\omega) = \int_{-\infty}^{\infty} |g(\omega)|^2\,\mathrm{d}\mu_N(\omega). \tag{4.44}$$

We formulate our findings in a theorem.

Theorem 4.9. *The covariance function of a shot noise process*

$$x(t) = \int_{-\infty}^{\infty} h(t-u)\,\mathrm{d}N(u),$$

driven by a counting process $\{N(t), t \in \mathbb{R}\}$, has a spectral representation

$$r_x(\tau) = C(x(\tau), x(0)) = \int_{-\infty}^{\infty} |g(\omega)|^2 e^{i\omega\tau}\,\mathrm{d}\mu_N(\omega),$$

where $\mu_N(\omega)$ is the Bartlett spectrum of the counting process.

Example 4.6 (Shot noise driven by a Poisson process). *The covariance function for a Poisson shot noise process $x(t) = \int h(t-s)\,\mathrm{d}N(s)$, with $g(\omega) = \int_{-\infty}^{\infty} e^{-i\omega u} h(u)\,\mathrm{d}u$, follows from* (4.42) *and the Plancherel's relation* (A.21). *With the Bartlett spectral density $\lambda/2\pi$ for the Poisson process, the covariance function is given by the simple expression*

$$r(\tau) = \lambda \int_{-\infty}^{\infty} h(u)\,h(u-\tau)\,\mathrm{d}u, \tag{4.45}$$

and its spectral density is

$$f_x(\omega) = \frac{\lambda}{2\pi} |g(\omega)|^2, \tag{4.46}$$

where $g(\omega)$ is the frequency response to the impulse response $h(u)$. The relation (4.45) *is called* Campbell's formula, *after N. Campbell, who found the mean and variance of shot noise [24]. The covariance formula was derived by S.O. Rice [101]; see also [112].*

Example 4.7 (Shot noise driven by a Cox process). *A Cox process (Example 3.8) has Bartlett spectrum with density $\frac{\lambda_0}{2\pi} + f_\lambda(\omega)$. The resulting shot noise process has spectral density*

$$f_x(\omega) = |g(\omega)|^2 \left(\frac{\lambda_0}{2\pi} + f_\lambda(\omega) \right).$$

If one includes a random element in the impulse response $h(u)$, so it can take different shape and/or amplitude, one obtains a very flexible class of processes. For examples of different shot noise processes with randomly varying impulse response and for other types of counting processes, see [20].

4.4 Long range dependence, non-integrable spectra, and unstable systems

This section deals with three related topics, which extend the simplistic view on stationary processes that has been dominating the presentation so far. A stationary process has been defined by a consistent family of distributions, from which we derived all properties of the process (in the σ-field of events). The view adopted in this section is a more uncertain one – we don't pretend to know the distribution family in full details, but only those part of it that can be observed after the process has passed through a certain linear filter. We start the presentation with what is called *long range dependence*, a concept that is related to the two next topics, *non-integrable spectra* and *unstable systems*. Common for the three themes is the fact that, from an applied viewpoint, it is difficult, or even impossible, to distinguish between deterministic and stochastic properties of an observed random sequence or function; only some local structure can be observed and modeled.

In the previous section we studied processes with a high degree of independence over time. As we have seen, this independence is coupled to the content of high frequency components in the process, as reflected in the frequency spectrum. The number of finite spectral moments determines the number of derivatives. In the extreme case, with constant spectral density we have the white noise of independent variables. As noted in Section 4.3.2 the continuous time white noise makes sense when used as input to a linear filter, which acts as a smoother of high frequencies if the impulse response is a squared integrable function.

Too much high frequency content in the spectrum makes the variance break down. Another type of degenerate spectrum gives rise to a long range dependence, where there is an excess of dependence in terms of correlation at far distances. This dependence makes it difficult, almost impossible, to evaluate the stationarity of the process from any finite series of observations. What looks like a stable mean level may well be part of a very long cycle. For such processes it may be sensible to analyze local variability, and leave the average behavior to be inferred by other means, for example from a physical model.

The term long range dependence, or long-memory processes, has been used in a number of different ways in the literature, and there is not one unique accepted definition. The survey tract [110] by G. Samorodnitsky contains an illuminative discussion of different theoretical and practical sides of long range dependence behavior, and their relations to other stochastic phenomena; for more on statistical issues, see also [13]. In the definition adopted here, long range dependent processes have a spectral density, which is unbounded, but still integrable, near the zero frequency.

A special operation to remove a slowly changing mean is simple differencing. A stationary process is called *intrinsically (weakly) stationary* if the variance of any difference is finite, $V(x(t)-x(s)) < \infty$, and only depends on the difference $|t-s|$. Any weakly stationary process is also intrinsically stationary. The term was introduced for random fields by G. Matheron in 1973 [90]. We discuss some generalizations later in this section, and also in Section 7.2.3, where also other types of non-stationarities are treated, for example linear, quadratic, or exponential trends. As discussed in [110], processes with long memory lie on the boundary between stationary and non-stationary processes, and one could claim that the intrinsically stationary lie on the other side of that unclear boundary.

The third topic treated in this section is unstable systems, which may generate non-stationary (random) elements, with local stationary variations. Also these processes can be called intrinsically stationary.

4.4.1 *Long range dependence*

There are several definitions of what can be meant by long range dependence for a stationary process. The following definition is most closely related to the main themes in this book. It is based on the shape of the spectrum near zero frequency, or the related covariance decay at long distances; see [13, 110]. We formulate the definition for continuous spectrum with spectral density $f(\omega)$.

Definition 4.8 (Long range dependence - slow correlation decay). *A stationary stochastic process $\{x(t)\}$ is said to exhibit long range dependence, or have long memory, if there is a constant H, $1/2 < H < 1$, and a slowly varying function $L_1(\omega)$ such that*[2]

$$f(\omega) = L_1(|\omega|)\,|\omega|^{1-2H}, \quad \text{as } \omega \to 0, \tag{4.47}$$

or equivalently (under mild conditions, given in [110]), if the covariance function satisfies

$$r(t) = L_2(t)\,t^{2H-2}, \quad \text{as } t \to \infty, \tag{4.48}$$

for some function $L_2(t)$ that varies slowly at $+\infty$.

[2] A function $L(t), t > 0$, is slowly varying at 0 (∞) if the limit $\lim_{t\to 0(\infty)} L(at)/L(t) = 1$ for every $a > 0$.

The two conditions are often simplified to $f(\omega) \sim C_1 |\omega|^{1-2H}$ near $\omega = 0$, and $r(t) \sim C_2 t^{2H-2}$ for large t. Note that the restriction on H implies that the spectrum becomes unbounded near the zero frequency, but still integrable, and that the covariance function is not absolutely integrable (or summable, if time is discrete).

The slow covariance decay for a long memory process has serious consequences for statistical inference, in particular for the assessment of the accuracy of parameter estimates. The naive estimate of the mean value for a stationary sequence is the arithmetic mean, $\overline{x}_n = n^{-1} \sum_1^n x_k$, of a sequence of n successive observations. It has variance

$$V(\overline{x}_n) = \frac{1}{n^2} \left\{ nr(0) + 2 \sum_1^{n-1} (n-k) r(k) \right\},$$

and if the covariance function is absolutely summable,

$$\lim_{n \to \infty} n V(\overline{x}_n) = r(0) + 2 \sum_1^{\infty} r(k),$$

and the variance of $\overline{x}_n$ is of the order n^{-1} for large n.

For a long memory process, the variance is of higher order, and it can be shown, see [110], that

$$\lim_{n \to \infty} n V(\overline{x}_n) \frac{1}{nr(n)} = \frac{2}{2H(2H-1)}.$$

Thus, the factor $nr(n)$ and the parameter H are important for statistical inference, and estimation of H has been a topic in statistics since the early 1950's; see [13, 110].

Remark 4.3 (The Hurst effect). *The use of the symbol H for the index of fractional Brownian motion has historical reasons. H. Hurst, British hydrologist and Nile expert, analyzed in 1951 the fluctuations in ancient water flow data from the river Nile over long periods of time [61]. He observed that the maximal fluctuations around the expected accumulated flow increased with the length, n, of the studied period at a rate* $n^{0.74}$, *which is larger than the expected rate* $\sqrt{n}$.

B. Mandelbrot suggested in 1965 the self-similar fractional Brownian motion as a possible stochastic model for the Hurst phenomenon [87]. The important property of the fBm in that application is the long range memory, not the self-similarity.

The Hurst effect with the long range dependence has been observed in many different empirical studies, in climatology, economics, internet traffic, to mention a few. It should be emphasized, however, that a statistical model for long memory, like the fBm, is only a model for an empirical fact, and it does not explain the mechanism behind it.

4.4.2 Non-integrable spectral density - intrinsic stationarity

The spectral density for a long memory process goes to infinity near $\omega = 0$, while keeping the integral, i.e., the process variance, finite. As mentioned in the introduction to this section, a high content of very low frequency components makes it questionable to imagine a stable average level for a stationary process. In the extreme case, when the spectral integral diverges near zero, the process has no finite variance, and it may be more fruitful to study the local variability, instead of the variability around a global average value. This can be accomplished by removing the local mean by a real *contrast filter*, i.e., a filter with zero gain, characterized by $\sum h_k = 0$; cf. Theorem 4.2.

Example 4.8. *To see how it works, and to prepare for a serious application in Section 7.2 on intrinsic random fields, we explore what properties can be extracted from a non-integrable spectral density, following [117, Sect. 2.9]. Consider the function* $f(\omega) = |\omega|^{-\alpha}$, $\omega \in \mathbb{R}$, *with* $1 < \alpha < 3$, *which is obviously not a valid spectral density. Let* $h(u) = \sum_{k=1}^n h_k\, \delta_{t_k}(u)$ *be the impulse response function to a zero gain filter,* $\sum_1^n h_k = 0$, *and with frequency response function* $g(\omega) = \sum_1^n h_k e^{-i\omega t_k} \sim C\omega$ *as* $\omega \to 0$. *If* $f(\omega)$ *had been the spectral density of a real stationary input process* $\{x(t), t \in \mathbb{R}\}$, *then, for* $1 < \alpha < 3$,

$$V(y(t)) = V\left(\sum_1^n h_k x(t-t_k)\right) = \int_{-\infty}^{\infty} |g(\omega)|^2\, |\omega|^{-\alpha}\, d\omega < \infty, \tag{4.49}$$

although $V(x(t)) = \infty$. *Thus, the filter output is well behaved, even if the input is degenerated.*

In fact, we can extract an even more interesting feature from the ill-behaved "spectral density" $f(\omega)$, *namely a parallel to the alternative variance formula in linear filters,*

$$V(y(t)) = \int_{-\infty}^{\infty} |g(\omega)|^2 f_x(\omega)\, d\omega = \sum_{j,k} h_j h_k\, r_x(t_j - t_k). \tag{4.50}$$

With $\sum h_k = 0$, *we evaluate (4.49) to obtain an expression for* $V(y(t))$:

$$\int_{-\infty}^{\infty} \left|\sum_k h_k e^{-i\omega t_k}\right|^2 |\omega|^{-\alpha}\, d\omega = \int_{-\infty}^{\infty} \sum_{j,k} h_j h_k \cos(\omega(t_j - t_k))\, |\omega|^{-\alpha}\, d\omega$$

$$= \int_{-\infty}^{\infty} \sum_{j,k} h_j h_k \left\{\cos(\omega(t_j - t_k)) - 1\right\} |\omega|^{-\alpha} \, d\omega$$
$$= \sum_{j,k} h_j h_k \int_{-\infty}^{\infty} \left\{\cos(\omega(t_j - t_k)) - 1\right\} |\omega|^{-\alpha} \, d\omega$$
$$= -4 \sum_{j,k} h_j h_k \int_0^{\infty} \sin^2(\omega(t_j - t_k)/2) \, |\omega|^{-\alpha} \, d\omega.$$

The integral in the right hand side can be evaluated for $\alpha \in (1,3)$, [55, 3.823, 8.334(3)],

$$\int_0^{\infty} \frac{\sin^2(\omega d/2)}{\omega^\alpha} \, d\omega = \frac{\pi}{4\Gamma(\alpha)\sin(\pi(\alpha-1)/2)} |d|^{\alpha-1}.$$

Hence, if we define

$$G(t) = -\frac{\pi}{\Gamma(\alpha)\sin(\pi(\alpha-1)/2)} |d|^{\alpha-1},$$

we get

$$0 \le \sum_{j,k} h_j h_k G(t_j - t_k) = \int_{-\infty}^{\infty} |g(\omega)|^2 \, |\omega|^{-\alpha} \, d\omega, \tag{4.51}$$

i.e., $f(\omega)$ and $G(t)$ act in a similar way as a spectral density/covariance function pair when used as input to a contrast filter. Note, however, that $\sum_{j,k} h_j h_k G(t_j - t_k)$ is necessarily non-negative only for $\sum_k h_k = 0$. A function of this type is called by Matheron [90] a generalized covariance function.

Example 4.9 (Brownian motion)**.** *The Wiener process (fractional Brownian motion with index $H = 1/2$) has covariance function $r_w(s,t) = \sigma^2 \min(s,t)$. Its generalized covariance function is $G(t) = -\frac{\sigma^2}{2}|t|$, with $f(\omega) = \frac{\sigma^2}{2\pi}\omega^{-2}$ as "spectral density," satisfying* (4.51).

Intrinsic stationarity

Definition 4.9 (Intrinsic stationarity)**.** *A (possibly non-stationary) stochastic process $\{x(t), t \in \mathbb{R}\}$ is called intrinsically stationary if there exists a time invariant contrast filter $\mathscr{S}$ such that $y(t) = \mathscr{S}(x)(t)$ is (weakly) stationary.*

The idea behind intrinsic stationarity is to remove an unknown structure of specific type from a stochastic process and get a new process that is stationary, in some well defined sense. For example, taking differences, $y(t) = x(t) - x(t-h)$, with a fixed $h > 0$, removes not only any constant mean value function, but also periodic components with period h. A typical example is an ARIMA(p,d,q)-process, i.e., a random sequence for which the dth order differences $(1-U^{-1})^d x_t$ form a causal ARMA(p,q)-process. Another example is the seasonal ARIMA-process (SARIMA), for which a season of period D is removed to make an ARIMA-model.

To remove a linear trend one needs a second order filter with specific constraints, and so on. In general, to remove a polynomial trend of degree p with a discrete filter in continuous time, $h(u) = \sum_1^n h_k \, \delta_{t_k}(u)$, one needs $p+1$ constraints,

$$\sum_{k=1}^{n} h_k t_k^r = 0, \quad r = 0, \ldots, p. \tag{4.52}$$

Then, $\{x(t), t \in \mathbb{R}\}$ is called *intrinsic of order (at most)* $p+1$ if

$$y(t) = \sum_{k=1}^{n} h_k x(t - t_k) \tag{4.53}$$

defines a stationary process when $\{h_k, t_k\}_1^n$ satisfy (4.52); cf. [107, Ch. 3].

It should be noted that (4.53) removes not only polynomials, but also many functions with higher order periodicities. If time is discrete, and the t_k are chosen as $p+1$ consecutive integers, it removes exactly all polynomials of degree at most p. In continuous time, it suffices to differentiate $p+1$ times to do the same thing.

Intrinsic stationarity is a useful concept in hierarchial models of random fields, and in Bayesian analysis of such fields; see the discussion on generalized random fields in Appendix B.2 in [76]. We return to some of these issues in Chapter 7.

Unstable systems

In our study of linear stochastic systems in Sections 4.2 and 4.3 we assumed throughout that the governing differential equations (4.24) and (4.33) were stable, with the roots of the characteristic equation (4.25) all having negative real part. Then we could easily find the characteristics of any stationary process that solves the equation, and assume that any impact that could disturb the statistical stationarity would eventually die out in the future. The direction of time from $-\infty$ to $+\infty$ was important.

Now, it is obvious that there are many other solutions to the differential equations, namely all solutions to the homogeneous equation with right hand side identically zero. For example, the process

$$y(t) = \frac{\sigma}{a_0} \int_{-\infty}^{t} e^{-\alpha(t-u)} \, \mathrm{d}w(u),$$

with $\alpha = a_1/a_0 > 0$, is stationary and solves the equation $a_0 y'(t) + a_1 y(t) = \sigma w'(t)$, in the sense that the left hand side is Gaussian white noise. But then also

$$x(t) = y(t) + Ce^{-\alpha t}$$

solves the equation, and C can be any constant or random variable.

Definition 4.10 (Intrinsic stationarity with respect to a linear filter). *A family $\mathscr{X}$ of stochastic process is called intrinsically stationary with covariance function $r(t)$ if there is a linear filter $\mathscr{S}$ such that for all processes $x \in \mathscr{X}$ the process $y(t) = \mathscr{S}(x)(t)$ is stationary with covariance function $r(t)$.*

Example 4.10 (Whittle-Matérn family). *Consider the linear differential equation*

$$y(t) - y''(t) = w'(t). \tag{4.54}$$

Its characteristic equation, $1 - r^2 = 0$, has two roots, $r_{1,2} = \pm 1$, so the equation is not stable. The left hand side represents a linear filter $\mathscr{S}$ applied to the process $\{y(t), t \in \mathbb{R}\}$. The standard rule for derivation implies that the frequency function of the filter is

$$g(\omega) = 1 + \omega^2,$$

so a possible solution process should satisfy

$$|g(\omega)|^2 f_y(\omega) = (1+\omega^2)^2 f_y(\omega) = f_{w'}(\omega) = \frac{1}{2\pi},$$

and have spectral density

$$f_y(\omega) = \frac{1}{2\pi} \frac{1}{(1+\omega^2)^2}. \tag{4.55}$$

Any stationary process with this spectrum is a solution.

A spectral density of the form $f(\omega) = C(1+\omega^2)^{-\alpha}$ *can be inverted to a covariance function involving a modified Bessel function of the second kind,* K_ν*. Using, for example, [55, 8.432.5] with* $\nu = 3/2$*, we obtain the covariance function for the spectrum* (4.55),

$$r_y(t) = \frac{1}{2\sqrt{2\pi}} |t|^{3/2} K_{3/2}(|t|), \tag{4.56}$$

with variance $r_y(0) = 1/4$*. The spectral density obviously fulfills the differentiability condition, as it should, and the derivative is continuous but not differentiable.*

It is obvious that any process that differs from $y(t)$ *by* $C_1 e^{-t} + C_2 e^t$ *with random* C_1, C_2 *is also a solution to* (4.54)*. The family* $\mathscr{X}$ *of all such processes is intrinsically stationary.*

Example 4.11 (Whittle-Matérn family, cont.). *The spectral density* (4.55) *is a member of a family of very useful spectra with parameters* a, ν, σ^2*, of the form*

$$f(\omega) = \frac{\sigma^2}{2\pi} \frac{1}{(a^2+\omega^2)^{\nu+1/2}}. \tag{4.57}$$

The corresponding covariance function is, in terms of the Bessel function K_ν*,*

$$r(t) = \frac{\sigma^2}{A} \cdot \frac{1}{2^{\nu-1}\Gamma(\nu)} (at)^\nu K_\nu(at), \tag{4.58}$$

where $A = \frac{\Gamma(\nu+1/2)}{\Gamma(\nu)} \sqrt{4\pi}\, a^{2\nu}$*. Of the parameters,* a *is related to the correlation length of the process and* $\sigma^2/A = r(0)$ *is the variance, while* ν *defines the smoothness. From the form of the spectral density it is clear that a process with this covariance function is* m *times (q.m.) differentiable, whenever* $\nu > m$*. When* $\nu - 1/2 = m$ *is a non-negative integer, the covariance function is of the form* $e^{-a|t|}$ *times a polynomial in* $|t|$ *of degree* m*. We will meet this covariance/spectrum pair in Chapter 7,* (7.24)-(7.25)*, as a very flexible covariance structure for random fields.*

4.5 The ARMA-family

In statistics, the ARMA-models, autoregressive-moving average models, are by far the most used discrete time stationary processes, both in their own right and as parametric approximations to general sequences and processes.

They are not the main focus in this book, but for easy reference, and for use in examples, we state the definitions and their main spectral properties. In the definitions, $\{e_n, n \in \mathbb{Z}\}$ is a white noise sequence of independent variables with mean zero and variance σ^2.

Definition 4.11 (Autoregressive, moving average, and mixed, AR, MA , ARMA processes). *A random sequence $\{x_n, n \in \mathbb{Z}\}$ is called an* ARMA*(p,q)-process if it is stationary, and there exist real constants $\{a_k\}_{k=0}^p$, $\{c_k\}_{k=0}^q$, and a white noise sequence $\{e_n, n \in \mathbb{Z}\}$, such that*

$$x_t + a_1 x_{t-1} + \ldots + a_p x_{t-p} = e_t + c_1 e_{t-1} + \ldots + c_q e_{t-q}. \tag{4.59}$$

The sequence $\{x_n, n \in \mathbb{Z}\}$ is an AR*(p)-process if $c_1 = \ldots, c_q = 0$ and it is an* MA*(q)-process if $a_1 = \ldots, a_p = 0$.*

The polynomials, $A(z) = \sum_{k=0}^p a_k z^k$ and $C(z) = \sum_{k=0}^q c_k z^k$ (with $a_0 = c_0 = 1$), are called the *generating polynomials*, and the equations

$$z^p A(z^{-1}) = z^p + a_1 z^{p-1} + \ldots + a_p = 0, \tag{4.60}$$
$$z^q C(z^{-1}) = z^q + c_1 z^{q-1} + \ldots + c_q = 0, \tag{4.61}$$

are the *characteristic equations*.[3] Writing U^{-1} for the backward shift operator, i.e., $U^{-j} x_t = x_{t-j}$, the ARMA-equation can be briefly stated as

$$A(U^{-1}) x_t = C(U^{-1}) e_t. \tag{4.62}$$

Causal and invertible sequences

When ARMA-models are used to model stationary random sequences that develop in time, it is common practice to assume that e_t acts as an *innovation* to the x-sequence at time t, independent of, or at least uncorrelated with, x_s for $s < t$, and write the ARMA-equation

$$x_t = -a_1 x_{t-1} - \ldots - a_p x_{t-p} + e_t + c_1 e_{t-1} + \ldots + c_q e_{t-q}.$$

An ARMA-process is said to be *causal* if x_t can be expressed as a sum $x_t = \sum_{k=0}^{\infty} d_k e_{t-k}$ with $\sum_{k=0}^{\infty} |d_k| < \infty$. This means in particular that e_{t+1} is uncorrelated with all x_s for $s \leq t$. On the other hand, for statistical inference,

[3]The reason for choosing the letter C for the MA-polynomial is that in some literature the letter B is used for a deterministic part of the right hand side of (4.59).

it would be desirable that one could reconstruct all e_t linearly from a series of observations of x_s up till time t. When that is possible the ARMA-process is called *invertible*.

Theorem 4.10. *Let $\{x_n, n \in \mathbb{Z}\}$ be an ARMA(p,q)-process (4.59) for which the polynomials $A(z)$ and $C(z)$ have no common zeroes. Then it is*

a) causal if and only if $A(z) \neq 0$ for all z with $|z| \leq 1$. There then exist $\{d_k\}_{k=0}^{\infty}$ with $\sum_0^\infty |d_k| < \infty$, and

$$x_t = \sum_{k=0}^{\infty} d_k e_{t-k}; \tag{4.63}$$

b) invertible if and only if all roots of the characteristic equation $z^q C(z^{-1}) = 0$ lie inside the unit circle.

Proof. For (a) we consider the formal expression $x_t = \frac{C(U^{-1})}{A(U^{-1})} e_t$, and seek an expansion of the right hand side in terms of e_{t-k} with $k \geq 0$, and conversely for (b).

First assume that $A(z) \neq 0$ for all complex z with $|z| \leq 1$. Then, $1/A(z)$ can be expanded in a power series in z that converges inside a disk with radius $z_0 = 1 + \varepsilon > 1$,

$$1/A(z) = \sum_{k=0}^{\infty} \widetilde{a}_k z^k = \widetilde{A}(z), \text{ for } |z| \leq 1 + \varepsilon.$$

This implies that $\widetilde{A}(z)A(z) \equiv 1$ and $\widetilde{A}(z)C(z) = D(z) = \sum_{k=0}^{\infty} d_k z^k$ for $|z| \leq 1$, and $\sum_{k=0}^{\infty} |d_k| < \infty$. Thus x_t is the output of a causal filter with input $\{e_n, n \in \mathbb{Z}\}$ and impulse response sequence $d_k, k \geq 0$, showing the "if" part,

$$x_t = \sum_{k=0}^{\infty} d_k e_{t-k}.$$

For the "only if" part, assume $x_t = \sum_{k=0}^{\infty} d_k e_{t-k} = D(U^{-1})e_t$, for some sequence $\{d_k\}$ with $\sum_k |d_k| < \infty$. Inserting this into $A(U^{-1})x_t = C(U^{-1})e_t$ and using that the e_t-variables are uncorrelated, one can conclude that $A(z)D(z) = C(z)$. Since $|D(z)| < \infty$ for $|z| \leq 1$, and $A(z)$ and $C(z)$ have no common zero, $A(z) = C(z)/D(z)$ cannot be zero for $|z| \leq 1$.

Part (b) is analogous; see also [22, Sect. 3.1]. □

The spectrum – poles and zeroes

The following theorem is a direct consequence of (4.6) and the time discrete version of Theorem 4.1.

Theorem 4.11. *The spectral density for an* ARMA(p,q)*-process* $\{x_n, n \in \mathbb{Z}\}$ *with generating polynomials* $A(z), C(z)$ *is*

$$f_x(\omega) = \sigma^2 \frac{|C(e^{-i\omega})|^2}{|A(e^{-i\omega})|^2}. \tag{4.64}$$

The rational form (4.64) of the spectrum makes the ARMA-family a very flexible class of processes that can approximate many empirical spectra. To see how, consider the characteristic equations $z^p A(z^{-1}) = 0$, $z^q C(z^{-1}) = 0$, which have all their roots inside the unit circle. If we denote the roots $z_1^a, \dots, z_p^a$ and $c_1^c, \dots, z_q^c$, respectively, we can factorize the polynomials and write

$$A(z^{-1}) = z^{-p}(z - z_1^a) \cdot \dots \cdot (z - z_p^a),$$
$$C(z^{-1}) = z^{-q}(z - z_1^c) \cdot \dots \cdot (z - z_p^c),$$

and insert into (4.64) to obtain the spectrum in useful form,

$$f_x(\omega) = \sigma^2 \frac{\prod_1^q |e^{i\omega} - z_k^c|^2}{\prod_1^p |e^{i\omega} - z_k^a|^2}, \qquad -\pi < \omega \leq \pi.$$

The zeroes z_k^c in the numerator are simply called the *zeroes* of the process, while the zeroes z_k^a in the denominator are called the *poles*. Note that zeroes and poles are either real or appear in complex conjugate pairs. For more on pole/zero analysis, see [82, Section 7.2.3].

Exercises

4:1. a) Use the Taylor expansion of $E(e^{i(t_1x_1+t_2x_2+t_3x_3+t_4x_4)})$ to prove

$$\begin{aligned} E(x_1x_2x_3x_4) &= E(x_1x_2)E(x_3x_4) + E(x_1x_3)E(x_2x_4) \\ &\quad + E(x_1x_4)E(x_2x_3) - 2E(x_1)E(x_2)E(x_3)E(x_4), \end{aligned}$$

for Gaussian x_1, x_2, x_3, x_4. For normal variables with mean zero, this is a special case of Isserlis' theorem [64].

b) Let $\{x(t), t \in \mathbb{R}\}$ be a stationary Gaussian process with mean zero, covariance function $r_x(t)$, and spectral density $f_x(\omega)$, and consider the process

$$y(t) = x(t) \cdot x(t-T)$$

where T is a constant delay. Derive expressions for the covariance function and spectral density of $y(t)$ in terms of r_x and f_x.

4:2. If $r_x(t)$ is a covariance function, then also the linear interpolation is a covariance function,

$$r^*(t) = \begin{cases} r_x(t), & \text{for } t = 0, \pm 1, \pm 2, \ldots, \\ \text{linearly interpolated,} & \text{otherwise.} \end{cases}$$

Prove this statement.

4:3. Complete the proof of Lemma 4.1 on page 122.

4:4. Find the covariance function of the (non-stationary) process $y(t) = \int_0^t e^{-(t-u)}\,\mathrm{d}w(u)$, where $\{w(t), 0 \le t < \infty\}$ is a standard Wiener process.

4:5. Define the process $y(t) = x(t) + \int_0^\infty e^{-u} x(t-u)\,du$ as the output from a linear filter with $x(t)$ as input. What are the impulse and frequency functions of the filter? Derive the covariance function and spectrum of $y(t)$ if $x(t)$ is stationary with covariance function $r_x(t) = e^{-2|t|}$.

4:6. Consider the covariance function

$$r_y(t) = \frac{\sigma^2}{4\alpha\omega_0^2} e^{-\alpha|t|} \left(\cos\beta t + \frac{\alpha}{\beta} \sin\beta|t| \right)$$

of the linear oscillator in Example 4.5 on page 138.

The covariance function contains some non-differentiable $|t|$; show that it still fulfills a condition for sample function differentiability, but not for twice differentiability.

Find the relation between the relative damping ζ and the spectral width parameter $\alpha = \omega_2/\sqrt{\omega_0\omega_4}$.

4:7. Define

$$v(T) = V\left(\frac{1}{T} \int_0^T x(t)\,\mathrm{d}t \right),$$

and assume $\{x(t), t \in \mathbb{R}\}$ has a bounded spectral density $f(\omega)$. Find the relation between $f(0)$ and the asymtotic behavior of $v(T)$ as $T \to \infty$.

4:8. Let $\{x(t), t \in \mathbb{R}\}$ be stationary with mean zero and covariance function $r(t) = e^{-|t|}$. Find functions $f(t)$ and $g(t)$ such that one can express $x(t)$ as

$$x(t) = f(t)w(g(t)/f(t)),$$

where $w(s)$ is a standard Wiener process.

4:9. Prove that fractional Brownian motion, Definition 4.6, has (an equivalent version with) continuous sample paths.

4:10. Generalizing Example 4.9, find the generalized covariance function for fractional Brownian motion.

4:11. Define the process $x(t) = A_0 + A_1 t$, with uncorrelated random coefficients, A_0, A_1. Is it intrinsically stationary (Definition 4.9)? Show that

$$G(t) = -\frac{1}{2}E(A_1^2)t^2$$

can be used as a generalized covariance function in analogy with (4.51) and find the corresponding "spectral density" $f(\omega)$.

4:12. (Cont.) Analyze the properties of $y(t) = x(t) + x_0(t)$, when $x_0(t)$ is a stationary process, independent of A_0, A_1, with covariance function $r_0(t)$ and spectral density $f_0(\omega)$.

4:13. Let $\{x(t), t \in \mathbb{R}\}$ be a stationary process with covariance function $r(t) = e^{-t^2}$. Show that the process is infinitely differentiable, and that one can predict future values $x(t+h)$ perfectly by observing the process without error in an arbitrarily small interval around t. Hint: calculate

$$E\left(\left|x(t+h) - \sum_{k=0}^{n} \frac{h^k}{k!} \frac{d^k x(t)}{dt^k}\right|^2\right).$$

You should return to this example when you have studied Chapter 6 and considered Exercise 6.14.

4:14. Take a spectral distribution $F(\omega)$ with unit mass over $(-\pi, \pi]$, with $F(\pi) - F(-\pi+) = 1$, and let the random variable v be distributed according to F. Also, let u be uniformly distributed over $[-\pi, \pi)$, and independent of v. Prove that the random sequence $x_n = e^{i(vn+u)}$ is stationary, compute its covariance function, and show that its spectrum is F.

Chapter 5

Linear filters – special topics

In this chapter we present three special applications of the general filter theory, which have played important roles in the theoretical development of stationary process theory in information science and in engineering: the *envelope* representation of the local amplitude, used for example in AM radio, the *sampling theorem* for continuous reconstruction of a sampled process, and the *Karhunen-Loève expansion*, which can be used to make explicit inference in continuously observed processes.

5.1 The Hilbert transform and the envelope

5.1.1 The Hilbert transform

The spectral representation presents a stationary process $\{x(t), t \in \mathbb{R}\}$ in terms of random cosine and sine functions with positive and negative frequencies. The complex form of the spectral representation,

$$x(t) = \int_{-\infty}^{\infty} e^{i\omega t}\, \mathrm{d}Z(\omega),$$

is real under the Hermitian symmetry requirement $\mathrm{d}Z(-\omega) = \overline{\mathrm{d}Z(\omega)}$. In fact, $x(t)$ is then expressed as the sum of two complex processes, of which one is the complex conjugate of the other. If we take only half of the spectral representation, with $\omega \geq 0$, we obtain a complex process,

$$x^*(t) = 2\int_{0+}^{\infty} e^{i\omega t}\, \mathrm{d}Z(\omega) + \Delta Z(0),$$

where $\Delta Z(0)$ is the jump of $Z(\omega)$ at the origin. Due to the Hermitian symmetry we obtain a linear transform of $x(t)$ whose spectral representation contains the same information as that of $x(t)$.

Definition 5.1 (Hilbert transform). *The Hilbert transform of a stationary process* $\{x(t), t \in \mathbb{R}\}$ *is the process* $\{\widehat{x}(t), t \in \mathbb{R}\}$, *in*

$$x^*(t) = x(t) + i\widehat{x}(t),$$

and it is the result of a linear filter on $x(t)$ *with frequency function*

$$g(\omega) = \begin{cases} i, & \text{for } \omega < 0, \\ 0, & \text{for } \omega = 0, \\ -i, & \text{for } \omega > 0. \end{cases}$$

One can obtain $x^*(t)$ as the limit, as $h \downarrow 0$ through continuity points of $F(\omega)$, of the linear operation with frequency function

$$g_h(\omega) = \begin{cases} 0, & \text{for } \omega < -h, \\ 1, & \text{for } |\omega| \le h, \\ 2, & \text{for } \omega > h. \end{cases}$$

When $x(t)$ is real, with $\mathrm{d}Z(-\omega) = \overline{\mathrm{d}Z(\omega)}$, it follows from the real spectral representation (3.30),

$$x(t) = \int_{0+}^{\infty} \cos \omega t \, \mathrm{d}u(\omega) + \int_0^{\infty} \sin \omega t \, \mathrm{d}v(\omega) + \mathrm{d}u(0), \tag{5.1}$$

that also $\widehat{x}(t) = i(x(t) - x^*(t))$ is real, and that it is given by

$$\widehat{x}(t) = \int_0^{\infty} \sin \omega t \, \mathrm{d}u(\omega) - \int_{0+}^{\infty} \cos \omega t \, \mathrm{d}v(\omega). \tag{5.2}$$

Thus, $x^*(t)$ is a complex process with $x(t)$ as real part and $\widehat{x}(t)$ as imaginary part. All involved processes can be generated by the same real spectral processes $\{u(\lambda), 0 \le \lambda < \infty\}$ and $\{v(\lambda), 0 \le \lambda < \infty\}$ from Section 3.2.3. Note that for a pure cosine process, $\cos \omega t = (e^{i\omega t} + e^{-i\omega t})/2$, one gets $x^*(t) = e^{i\omega t} = \cos \omega t + i \sin \omega t$, so the Hilbert transform of a cosine function is a sine function with the same frequency; similarly, the Hilbert transform of a sine function is a cosine with sign reversed, as in (5.2).

Theorem 5.1. *Let $\{x(t), t \in \mathbb{R}\}$ be stationary and real, with mean 0, covariance function $r(t)$, and spectral distribution function $F(\omega)$, with a possible jump $\Delta F(0)$ at $\omega = 0$. Denote the Hilbert transform of $x(t)$ by $\widehat{x}(t)$. Then, with $G(\omega)$ denoting the one-sided spectrum, the following statements hold.*

a) $\{\widehat{x}(t), t \in \mathbb{R}\}$ is stationary and real, with mean 0, and covariance function

$$\widehat{r}(t) = r(t) - \Delta F(0) = \int_{-\infty}^{\infty} e^{i\omega t}\, \mathrm{d}F(\omega) - \Delta F(0)$$
$$= \int_{0+}^{\infty} \cos \omega t\, \mathrm{d}G(\omega).$$

b) $\{\widehat{x}(t), t \in \mathbb{R}\}$ has the same spectrum $F(\omega)$ as $\{x(t)\}$, except that any jump at $\omega = 0$ has been removed.

c) The cross-covariance function between $\{x(t)\}$ and $\{\widehat{x}(t)\}$ is

$$r^*(t) = E(x(s+t) \cdot \widehat{x}(s)) = -\int_0^{\infty} \sin \omega t\, \mathrm{d}G(\omega).$$

In particular, $x(t)$ and $\widehat{x}(t)$ are uncorrelated, for Gaussian processes even independent, when taken at the same time instant. Of course, the processes $\{x(t), t \in \mathbb{R}\}$ and $\{\widehat{x}(t), t \in \mathbb{R}\}$ are dependent.

Proof. Part (a) and (c) follow from (3.30), (5.2), and the correlation properties (3.31-3.33) of the real spectral processes. Part (b) is immediate. □

Remark 5.1. *We have defined the Hilbert transform of a stationary process in the spectral plane by means of the frequency function $g(\omega)$, noting that $\int_{-\infty}^{\infty} |g(\omega)|^2\, \mathrm{d}F(\omega) < \infty$, so the transform is well defined. In signal processing, and also in mathematics, the Hilbert transform of a function $f(t)$ is usually defined as the Cauchy principal value of the convolution of f with the impulse function $h(u) = 1/(\pi u)$.*

In signal processing the function $f^* = f + i\widehat{f}$ is called "the analytic signal" corresponding to the real signal f. It has several mathematical advantages in the design of signal processing algorithms. For more on the use of the Hilbert transform in signal processing, see [59].

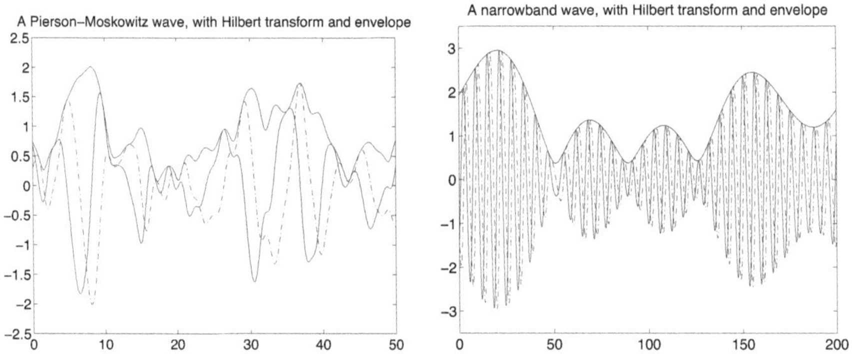

Figure 5.1 *Gaussian processes (——), their Hilbert transforms (– – –), and envelopes (——). Left: A process with Pierson-Moskowitz broad band spectrum. Right: A process with narrow triangular spectrum over* $(0.8, 1.2)$.

5.1.2 The envelope

The envelope is a tool to extract essential information from a random signal. Take a look at the two processes in Figure 5.1. It is obvious that the solid curve on top of the random curves has quite a different message to convey about the character of the signals. In the right diagram the solid curve at the top of the oscillating curve follows quite well the slowly changing amplitude of the rather regular oscillations, while in the left diagram it is difficult to see any systematic relation between the upper curve and the random processes below.

The difference behavior in the two examples is caused by different spectral width. The narrow Pierson-Moskowitz spectrum implies that the process is "almost," but not exactly, a cosine function with a sine function as Hilbert transform. Instead, the process is a mixture of cosine functions, with almost the same frequency, and the envelope describes the *fading phenomenon*, with a slowly changing amplitude, like a piano, slightly out of tune.

The envelope curve encloses a function (signal, random process) in a way that catches some important feature of the function. For example, in AM (amplitude modulated) radio transmission, the radio station sends out a high frequency "carrier wave" whose amplitude is changed (modulated) according to the audible signal – the "message." The simplest way to recover the audio signal is by means of an *envelope detector* that extracts the envelope of the radio signal.

We will now define the envelope of the sample functions of a stationary stochastic process and investigate its statistical properties expressed in terms of covariance and spectral properties.

Consider a stationary (Gaussian) process $\{x(t), t \in \mathbb{R}\}$ with covariance function $r(t)$ and spectral distribution function $F(\omega)$. Assume that $F(\omega)$ is continuous at $\omega = 0$ so there is no "constant" random component in $x(t)$, and consider the joint behavior of $x(t)$ and its Hilbert transform $\widehat{x}(t)$. Since $x(t)$ is a Gaussian process, also $\widehat{x}(t)$ is Gaussian, and both have covariance function $r(t)$. Consider the complex process

$$x^*(t) = x(t) + i\widehat{x}(t) = 2\int_{0+}^{\infty} e^{i\omega t}\,\mathrm{d}Z(\omega).$$

Definition 5.2 (Envelope [35]). *The envelope $R(t)$ of a stationary process $\{x(t), t \in \mathbb{R}\}$ is the absolute value of $x^*(t)$,*

$$R(t) = \sqrt{x(t)^2 + \widehat{x}(t)^2}.$$

In particular, $|x(t)| \leq R(t)$, with equality when $\widehat{x}(t) = 0$.

We have assumed $F(\omega)$ to be continuous at $\omega = 0$, so the random variables $x(t)$ and $\widehat{x}(t)$ have the same Gaussian distribution with variance $\sigma^2 = r(0)$, and since they are independent the sum of squares is distributed as $\sigma^2\chi^2(2)$, which is exponential.[1] Therefore the envelope has a Rayleigh distribution with density

$$f_R(r) = \frac{r}{\sigma^2} e^{-r^2/2\sigma^2}, \quad r \geq 0.$$

The envelope, as defined here, does always exist. Its physical meaning is however not clear from the mathematical definition, and we must turn to special types of processes before we can identify any particularly interesting properties of $R(t)$.

5.1.3 The envelope of a narrow band process

For processes with spectrum concentrated to a narrow frequency band, the sample functions have a characteristic "fading" appearance, as a wave with

[1] The sum of squares of n independent standard normal variables has a $\chi^2(n)$-distribution, a "Chi-squared distribution" with n degrees of freedom.

one dominating frequency and a slowly changing amplitude, described by the envelope. The narrow band process to the right in Figure 5.1 shows how the envelope follows the crests of the process. For the wide band process to the left, the envelope has little to say about the process behavior. The traditional treatment of the envelope in signal processing assumes that the process is "band limited" to a narrow spectral band around a center frequency ω_0. In that setting the Hilbert transform is almost like a smoothed differentiation, but without the frequency dependent factor ω.[2]

We consider a stationary process $x(t)$ with spectral density $f(\omega)$ concentrated around some frequency ω_0,

$$f(\omega) = \frac{1}{2} f_0(\omega - \omega_0) + \frac{1}{2} f_0(\omega + \omega_0),$$

for some function $f_0(\omega)$ such that $f_0(\omega) = f_0(-\omega)$, and $f_0(\omega) = 0$ for $|\omega| > d$, some $d < \omega_0$. Express $x(t)$ in spectral form, using the normalized form (3.29), to obtain

$$\begin{aligned} x(t) &= \int_{-\infty}^{\infty} e^{i\omega t} \sqrt{f(\omega)}\, \mathrm{d}w_C(\omega) \\ &= \int_{\omega_0 - d}^{\omega_0 + d} \sqrt{f_0(\omega - \omega_0)/2}\, e^{i\omega t}\, \mathrm{d}w_C(\omega) \\ &\qquad + \int_{-\omega_0 - d}^{-\omega_0 + d} \sqrt{f_0(\omega + \omega_0)/2}\, e^{i\omega t}\, \mathrm{d}w_C(\omega) \\ &= I_1(t) + I_2(t), \quad \text{say.} \end{aligned}$$

By change of variables in $I_1(t)$ and $I_2(t)$, this gives

$$\begin{aligned} I_1(t) &= e^{i\omega_0 t} \int_{-d}^{d} \sqrt{f_0(\omega)/2}\, e^{i\omega t}\, \mathrm{d}_\omega w_C(\omega + \omega_0) = e^{i\omega_0 t} y(t), \\ I_2(t) &= e^{-i\omega_0 t} \int_{-d}^{d} \sqrt{f_0(\omega)/2}\, e^{i\omega t}\, \mathrm{d}_\omega w_C(\omega - \omega_0) \\ &= e^{-i\omega_0 t} \int_{-d}^{d} \sqrt{f_0(\omega)/2}\, e^{-i\omega t}\, \mathrm{d}_\omega w_C(-\omega - \omega_0) = \overline{I_1(t)}, \end{aligned}$$

and, in combination, $x(t) = 2\Re\, y(t)\, e^{i\omega_0 t}$. Here, $y(t)$ is a complex process,

$$y(t) = y_1(t) + i y_2(t) = \int_{-d}^{d} \sqrt{f_0(\omega)/2}\, e^{i\omega t}\, \mathrm{d}_\omega w_C(\omega + \omega_0)$$

[2] An approximate envelope is then obtained as $\sqrt{x(t)^2 + x'(t)^2/\omega_0^2}$.

with only low frequencies. With $R(t) = 2|y(t)|$ and $\Theta(t) = \arg y(t)$, we obtain

$$x(t) = R(t)\,\Re\, e^{i(\omega_0 t+\Theta(t))} = R(t)\cos(\omega_0 t + \Theta(t)). \tag{5.3}$$

The current amplitude/phase representation (5.3) is a convenient way to represent the local character of a stationary process contrary to the global spectral representation.

The envelope $R(t)$ has here a real physical meaning as the slowly varying amplitude of the process, while $\Theta(t)$ is a time-varying random phase; this is in contrast to the random phases in the elementary representation (3.15) with randomly chosen phases that do not change over time.

5.2 The sampling theorem

The spectral form expresses a stationary process as an integral of elementary cosine functions with random amplitude and phase. If the spectrum is discrete and all amplitudes and phases are known, the process can be reconstructed completely from this discrete set of data. More surprising is that also with a continuous spectrum, it may sometimes be possible to reconstruct the entire process from discrete observations. The following "sampling theorem" is named after the Swedish-American engineer Harry Nyquist, cf. the Nyquist frequency, and the American mathematician Claude Shannon, called "the father of information theory."

For *band-limited* processes, the reconstruction can be achieved by sampling at equidistant points. A process is band-limited to frequency ω_0 if

$$F(-\omega_0+) - F(-\infty) = F(\infty) - F(\omega_0-) = 0,$$

i.e., its spectrum is restricted to the interval $(-\omega_0, \omega_0)$. We require that there is no spectral mass at the points $\pm\omega_0$.

Theorem 5.2 (The Nyquist-Shannon sampling theorem). *If the stationary process $\{x(t), t \in \mathbb{R}\}$ is band-limited to ω_0, then it is perfectly specified by its values at discrete time points spaced $t_0 = \pi/\omega_0$ apart. More specifically, for any arbitrary constant α,*

$$x(t) = \sum_{k=-\infty}^{\infty} x(\alpha + kt_0) \cdot \frac{\sin \omega_0(t - \alpha - kt_0)}{\omega_0(t - \alpha - kt_0)}, \tag{5.4}$$

with convergence in quadratic mean, uniformly on every bounded interval.

Proof. The spectral representation says that

$$x(t) = \int_{-\omega_0+}^{\omega_0-} e^{i\omega t}\,\mathrm{d}Z(\omega). \tag{5.5}$$

For a fixed t, the function $g_t(\omega) = e^{i\omega t} \cdot I_{[-\omega_0,\omega_0]}$ is square integrable over $(-\omega_0, \omega_0)$, and from the theory of Fourier series, it can be expanded as

$$e^{i\omega t} = \lim_{N\to\infty} \sum_{k=-N}^{N} e^{i\omega k t_0} \cdot \frac{\sin \omega_0(t-kt_0)}{\omega_0(t-kt_0)},$$

with convergence in $\mathscr{H}(F)$, i.e.,

$$\int_{-\omega_0}^{\omega_0} \left| e^{i\omega t} - \sum_{k=-N}^{N} e^{i\omega k t_0} \cdot \frac{\sin \omega_0(t-kt_0)}{\omega_0(t-kt_0)} \right|^2 \mathrm{d}F(\omega) \to 0.$$

The convergence is also uniform for $|\omega| < \omega_0 = \pi/t_0$. For $\omega = \pm\omega_0$ it converges to

$$\frac{e^{i\omega_0 t} + e^{-i\omega_0 t}}{2} = \cos \omega_0 t.$$

Therefore, when $\mathrm{d}F(\pm\omega_0) = 0$, then

$$\sum_{k=-N}^{N} e^{i\omega k t_0} \cdot \frac{\sin \omega_0(t-kt_0)}{\omega_0(t-kt_0)} \to e^{i\omega t}, \tag{5.6}$$

in $\mathscr{H}(F)$ as $N \to \infty$.

Inserting this expansion into the spectral representation of $\{x(t), t \in \mathbb{R}\}$, we obtain, using Lemma 4.1,

$$E\left(\left| x(t) - \sum_{k=-N}^{N} x(kt_0) \cdot \frac{\sin \omega_0(t-kt_0)}{\omega_0(t-kt_0)} \right|^2 \right) \tag{5.7}$$

$$= E\left(\left| \int_{-\omega_0}^{\omega_0} \left(e^{i\omega t} - \sum_{k=-N}^{N} e^{i\omega k t_0} \cdot \frac{\sin \omega_0(t-kt_0)}{\omega_0(t-kt_0)} \right) \mathrm{d}Z(\omega) \right|^2 \right)$$

$$= \int_{-\omega_0}^{\omega_0} \left| e^{i\omega t} - \sum_{k=-N}^{N} e^{i\omega k t_0} \cdot \frac{\sin \omega_0(t-kt_0)}{\omega_0(t-kt_0)} \right|^2 \mathrm{d}F(\omega).$$

By (5.6), this goes to 0 as $N \to \infty$, i.e.,

$$x(t) = \lim_{N\to\infty} \sum_{k=-N}^{N} x(kt_0) \cdot \frac{\sin \omega_0(t-kt_0)}{\omega_0(t-kt_0)},$$

in $\mathscr{H}(x)$, which is the statement of the theorem for $\alpha = 0$. For arbitrary α, apply the just proved result to $y(t) = x(t+\alpha)$. □

Remark 5.2. *If there is spectral mass* $F^+ = \mathrm{d}F(\omega_0)$, $F^- = \mathrm{d}F(\omega_0)$, *at the end points* $\pm\omega_0$, *then (5.7) would tend to*

$$\sin^2 \omega_0 t (F^+ + F^-),$$

failing the sampling representation. An example of this is the simple random cosine process, $x(t) = \cos(\omega_0 t + \phi)$, *which has covariance function* $r(t) = \frac{1}{2}\cos \omega_0 t$, *and spectrum concentrated at* $\pm\omega_0$. *Then*

$$x(\alpha + kt_0) = (-1)^k x(\alpha),$$

which means that for every t, the sum

$$\sum_k x(\alpha + kt_0) \cdot \frac{\sin \omega_0(t-\alpha-kt_0)}{\omega_0(t-\alpha-kt_0)}$$

is proportional to $x(\alpha)$. *On the other hand* $x(\alpha + t_0/2)$ *is uncorrelated with* $x(\alpha)$ *and cannot be represented by the sampling theorem.*

Remark 5.3. *The sampling theorem is the basis for efficient signal and image compression techniques. However, used in its original form the reconstruction requires an infinite sequence of evenly spaced samples and this is not possible in practice. The problem of reconstruction from irregular sampling has received much attention, both in the theoretical and applied literature; see [12, 48].*

5.3 Karhunen-Loève expansion

5.3.1 *Principal components*

In a multivariate distribution of a random vector the components may be more or less statistically dependent. If there is strong dependence between the components it is possible that it suffices to specify a few values in order to reconstruct almost the entire outcome of the full vector. The formal tool to generate such a common behavior is the concept of *principal components*, which is a tool to reduce the dimensionality of a variation space.

Let $\mathbf{x} = (x_1, \ldots, x_n)'$ be a vector of n real random variables with mean zero and a covariance matrix $\mathbf{\Sigma}$, which is symmetric and non-negative definite by construction. The covariance matrix $\mathbf{\Sigma} = (\sigma_{jk}) = E(\mathbf{x}\mathbf{x}')$, $\sigma_{jk} = C(x_j, x_k)$, has n non-negative eigenvalues λ_k with corresponding orthonormal eigenvectors $\mathbf{p}_k$ ($|\mathbf{p}_k|^2 = 1, \mathbf{p}'_j\mathbf{p}_k = 0$, for $j \neq k$), decreasingly ordered as $\lambda_1 \geq \lambda_2 \geq \ldots \geq \lambda_n$, such that

$$\mathbf{\Sigma}\mathbf{p}_k = \lambda_k \mathbf{p}_k.$$

The transformation

$$z_k = \frac{1}{\sqrt{\lambda_k}} \mathbf{p}'_k \mathbf{x}$$

gives us n new random variables, $z_1, \ldots, z_n$, with mean zero and unit variance,

$$V(z_k) = \frac{\mathbf{p}'_k \mathbf{\Sigma} \mathbf{p}_k}{\lambda_k} = \frac{\mathbf{p}'_k \mathbf{p}_k \lambda_k}{\lambda_k} = 1.$$

Furthermore, they are uncorrelated, for $j \neq k$,

$$C(z_j, z_k) = E(z_j z_k) = \frac{1}{\sqrt{\lambda_j \lambda_k}} E\left(\mathbf{p}'_j \mathbf{x}\mathbf{x}'\, \mathbf{p}_k\right) = \frac{1}{\sqrt{\lambda_j \lambda_k}} \mathbf{p}'_j \mathbf{\Sigma} \mathbf{p}_k = 0,$$

since the eigenvectors $\mathbf{p}_j$ and $\mathbf{p}_k$ are orthogonal.

The random variables $y_k = \sqrt{\lambda_k} z_k$, $k = 1, \ldots, n$, are called the *principal components* of the vector $\mathbf{x}$. In matrix language, with $\mathbf{P} = (\mathbf{p}_1, \ldots, \mathbf{p}_n)$ as the matrix with the eigenvectors as columns,

$$\mathbf{y} = \mathbf{P}\mathbf{x} \quad \text{with inverse} \quad \mathbf{x} = \mathbf{P}'\mathbf{y}$$

is a vector of uncorrelated variables with decreasing variances λ_k.

Since the matrix $\mathbf{P}$ is orthogonal, $\mathbf{P}^{-1} = \mathbf{P}'$, the original x-variables can be expressed as a linear combination of the uncorrelated variables y_k and z_k,

$$x_k = \mathbf{p}'_k \mathbf{y} = \sum_{j=1}^{n} \sqrt{\lambda_j} p_{jk} z_j. \tag{5.8}$$

In practice, when one wants to simulate a large vector of many correlated Gaussian variables, one can use (5.8) to generate successively the most important variational modes and truncate the sum when, for example, it describes 99% of the variation of the x_k-variable.

5.3.2 *Expansion along eigenfunctions*

We shall now generalize the finite-dimensional formulation in the previous section to continuous parameter stochastic processes. In fact, every continuous process, stationary or not, can be expressed on a finite interval $[a,b]$ as a sum of deterministic functions $c_k(t)$ with random coefficients z_k,

$$x(t) = \lim_{n\to\infty} \sum_0^n c_k(t) z_k.$$

The convergence is uniform for $a \le t \le b$, i.e., the approximation error,

$$e_n(t) = x(t) - \sum_0^n c_k(t) z_k,$$

is uniformly small, $\max_{a\le t\le b} E\left(|e_n(t)|^2\right) \to 0$, as $n \to \infty$.

The solution depends on the observation interval $[a,b]$, and the random variables z_k are elements in the Hilbert space spanned by $x(t), t \in [a,b]$,

$$z_k \in \mathcal{H}(x(t), t \in [a,b]) = \mathcal{H}(x), \quad \text{for short.}$$

Suppose we have found variables z_k that form an orthonormal basis for $\mathcal{H}(x)$. What properties do the functions $c_k(t)$ need to satisfy? If we have an expansion $x(t) = \sum_k c_k(t) z_k$ with orthogonal z_k, we have

$$r(s,t) = E(x(s)\overline{x(t)}) = \sum_{j,k} c_j(s)\overline{c_k(t)}\, E(z_j \overline{z_k}) = \sum_k c_k(s)\overline{c_k(t)}.$$

Thus, we shall investigate the existence and properties of the following pair of expansions,

$$x(t) = \sum_k c_k(t) z_k, \tag{5.9}$$

$$r(s,t) = \sum_k c_k(s)\overline{c_k(t)}. \tag{5.10}$$

There are many ways to represent $x(t)$ and $r(s,t)$ in this way, and we look for a solution that possesses some optimal property. It turns out that an optimal choice, in the sense that it gives the minimum integrated squared approximation error (5.11), is to take the functions $c_k(t)$ to be *orthogonal*, i.e.,

$$\int_a^b c_j(t)\overline{c_k(t)}\,\mathrm{d}t = 0, \quad j \ne k, \quad \text{and} \quad \int_a^b |c_k(t)|^2\,\mathrm{d}t = \lambda_k \ge 0.$$

Then, by (5.9), $z_k = \frac{1}{\lambda_k}\int_a^b \overline{c_k(t)} x(t)\,\mathrm{d}t$, and as a consequence of (5.10) and the orthogonality of the functions $c_k(t)$, we also conclude that (with $\delta_{j,k} = 1$ if $j = k$ and 0 otherwise)

$$\int_a^b r(s,t) c_j(t)\,\mathrm{d}t = \int_a^b \left\{\sum_0^\infty c_k(s)\overline{c_k(t)}\right\} c_j(t)\,\mathrm{d}t$$

$$= \sum_0^\infty c_k(s) \int_a^b c_j(t)\overline{c_k(t)}\,\mathrm{d}t = \sum_0^\infty c_k(s)\,\lambda_k \delta_{j,k} = \lambda_j c_j(s).$$

Thus, we have found that the functions $c_j(t)$ have to be eigenfunctions with eigenvalues λ_j to the transformation

$$c(\cdot) \mapsto \int_a^b r(\cdot,t)\,c(t)\,\mathrm{d}t.$$

Call the normalized eigenfunctions $\phi_j(t) = \frac{1}{\sqrt{\lambda_j}} c_j(t)$ if $\lambda_j > 0$, to make $\phi_k(t)$ a family of orthonormal eigenfunctions. With this choice, the integrated approximation error is

$$\int_a^b E(e_n^2(t))\,\mathrm{d}t = \int_a^b E(x^2(t))\,\mathrm{d}t - \sum_0^n \lambda_k. \tag{5.11}$$

5.3.3 *The Karhunen-Loève theorem*

The Karhunen-Loève theorem is named after K. Karhunen, thesis 1947 [68], and M. Loève [85, Sect. 34.5].

Theorem 5.3. *Let $\{x(t), a \le t \le b\}$ be continuous in quadratic mean with mean zero and covariance function $r(s,t) = E(x(s)\overline{x(t)})$. Then there exist orthonormal eigenfunctions $\phi_k(t), k = 0, 1, \ldots, N \le \infty$, for $a \le t \le b$, with eigenvalues $\lambda_0 \ge \lambda_1 \ge \ldots$, to the equation*

$$\int_a^b r(s,t)\,\phi(t)\,\mathrm{d}t = \lambda\,\phi(s),$$

such that the random variables $z_k = \frac{1}{\sqrt{\lambda_k}} \int_a^b \overline{\phi_k(t)}\, x(t)\,\mathrm{d}t$ are uncorrelated, $V(z_k) = 1$, and can represent $x(t)$ as

$$x(t) = \sum_0^\infty \sqrt{\lambda_k}\,\phi_k(t)\,z_k. \tag{5.12}$$

The variables z_k are sometimes called *observables* and they can be used for example to make statistical inference about the distribution of the process $x(t)$. Note, that if $x(t)$ is a normal process, then the z_k are uncorrelated normal variables, and hence independent, making inference simple.

Before we sketch the proof, we present an explicit construction of the Wiener process as an example of the Karhunen-Loève theorem.

Example 5.1. *The standard Wiener process $w(t)$, observed over $[0,T]$, has covariance function $r(s,t) = \min(s,t)$, and eigenfunctions can be found explicitly: from*

$$\int_0^T \min(s,t)\,\phi(t)\,\mathrm{d}t = \lambda\phi(s),$$

it follows by differentiating twice,

$$\int_0^s t\phi(t)\,dt + \int_s^T s\phi(t)\,\mathrm{d}t = \lambda\phi(s), \tag{5.13}$$

$$s\phi(s) - s\phi(s) + \int_s^T \phi(t)\,\mathrm{d}t = \lambda\phi'(s), \tag{5.14}$$

$$-\phi(s) = \lambda\phi''(s).$$

The initial conditions, $\phi(0) = 0$, $\phi'(T) = 0$, obtained from (5.13) and (5.14), imply the solution $\phi(t) = A\sin\frac{t}{\sqrt{\lambda}}$, with $\cos\frac{T}{\sqrt{\lambda}} = 0$. *Thus, the positive eigenvalues λ_k satisfy*

$$\frac{T}{\sqrt{\lambda_k}} = \frac{\pi}{2} + k\pi, \quad k = 0,1,2,\ldots.$$

The normalized eigenfunctions are

$$\phi_k(t) = \sqrt{\frac{2}{T}}\sin\frac{(k+\frac{1}{2})\pi t}{T},$$

with eigenvalues

$$\lambda_k = T^2\frac{1}{\pi^2(k+\frac{1}{2})^2}.$$

With

$$z_k = \frac{1}{\sqrt{\lambda_k}}\int_0^T \phi_k(t)\,w(t)\,\mathrm{d}t = \frac{\pi(k+\frac{1}{2})}{T}\int_0^T\sqrt{\frac{2}{T}}\sin\frac{t\pi(k+\frac{1}{2})}{T}\,w(t)\,\mathrm{d}t,$$

we have that the Wiener process can be defined as the infinite (uniformly convergent in quadratic mean) sum

$$w(t) = \sqrt{\frac{2}{T}} \sum_{k=0}^{\infty} \frac{\sin \frac{\pi t(k+\frac{1}{2})}{T}}{\frac{\pi(k+\frac{1}{2})}{T}} z_k,$$

with independent standard normal variables z_k.

The reader should simulate a Wiener process $w(t)$, find the variables z_k for $k = 0, 1, \ldots, n < \infty$ by numerical integration, and reproduce $w(t)$ as a truncated sum.

Proof of Theorem 5.3: We only indicate the steps in the proof, following the outline in [124]. One has to show: the mathematical facts about existence and properties of eigenvalues and eigenfunctions, the convergence of the series (5.12), and finally the stochastic properties of the variables z_k. This is done in a series of steps.

Step 1: If $\int_a^b r(s,t)\,\phi(t)\,\mathrm{d}t = \lambda\phi(s)$, then λ is real and non-negative. This follows from the non-negative definiteness of covariance functions,

$$0 \le \int_a^b \int_a^b r(s,t)\,\overline{\phi(s)}\,\phi(t)\,\mathrm{d}s\,\mathrm{d}t = \lambda \int_a^b |\phi(s)|^2\,\mathrm{d}s.$$

Step 2: There is at least one non-zero eigenvalue. The largest eigenvalue is

$$\lambda_0 = \max_{\phi;\|\phi\|=1} \int_a^b \int_a^b r(s,t)\,\overline{\phi(s)}\,\phi(t)\,ds\,\mathrm{d}t,$$

where the maximum is taken over $\|\phi\|^2 = \int_a^b |\phi(t)|^2\,\mathrm{d}t = 1$. As expressed in [124], "this is not easily proved." The corresponding eigenfunction is denoted by $\phi_0(t)$, and it is continuous.

Step 3: The function $r_1(s,t) = r(s,t) - \lambda_0\phi_0(s)\overline{\phi_0(t)}$ is a continuous covariance function, namely for the process

$$x_1(t) = x(t) - \phi_0(t) \int_a^b \overline{\phi_0(s)}\,x(s)\,\mathrm{d}s,$$

and it holds that

$$\int_a^b r_1(s,t)\,\phi_0(t)\,\mathrm{d}t = 0. \tag{5.15}$$

Repeating Step 2 with $r_1(s,t)$ instead of $r(s,t)$ we get a new eigenvalue

$(0 \leq)\lambda_1 \leq \lambda_0$ with eigenfunction $\phi_1(t)$. Since $\int_a^b r_1(s,t)\,\phi_1(t)\,\mathrm{d}t = \lambda_1\phi_1(s)$, we have

$$\int_a^b \phi_1(s)\,\overline{\phi_0(t)}\,\mathrm{d}t = \frac{1}{\lambda_1}\int_a^b \overline{\phi_0(s)}\left\{\int_a^b r_1(s,t)\,\phi_1(t)\,\mathrm{d}t\right\}\mathrm{d}s$$

$$= \frac{1}{\lambda_1}\int_a^b \phi_1(t)\left\{\int_a^b r_1(s,t)\,\overline{\phi_0(s)}\,\mathrm{d}s\right\}\mathrm{d}t = 0,$$

according to (5.15) since $r_1(s,t)$ is real. Thus ϕ_0 and ϕ_1 are orthogonal.

It also follows that ϕ_1 is an eigenfunction to $r(s,t)$,

$$\int_a^b r(s,t)\,\phi(t)\,\mathrm{d}t = \int_a^b r_1(s,t)\,\phi_1(t)\,\mathrm{d}t + \lambda_0\phi_0(s)\int_a^b \overline{\phi_0(t)}\phi_1(t)\,\mathrm{d}t$$

$$= \lambda_1\phi_1(s) + 0.$$

Step 4: Repeat Steps 2 and 3 as long as there is anything remaining of

$$r_n(s,t) = r(s,t) - \sum_{k=0}^{n} \lambda_k\phi_k(s)\overline{\phi_k(t)}.$$

Then, either there is a finite n such that $r_n \equiv 0$, or there is an infinite decreasing sequence of positive eigenvalues $\lambda_k \downarrow 0$ with $\sum_k \lambda_k < \infty$. Show as an exercise that $\sum_k \lambda_k \leq \int_a^b r(s,s)\,\mathrm{d}t$.

Step 5: If there is an infinite number of non-zero eigenvalues, then

$$\sup_{a\leq s,t\leq b}\left| r(s,t) - \sum_{k=0}^{n} \lambda_k\,\phi_k(s)\overline{\phi_k(t)}\right| \to 0$$

as $n \to \infty$, i.e.,

$$r(s,t) = \sum_{k=0}^{\infty} \lambda_k\,\phi_k(s)\overline{\phi_k(t)},$$

with uniform convergence. (This is Mercer's theorem from 1909.)

Step 6: For the representation (5.12) we have

$$E\left(\left|x(t) - \sum_{0}^{n}\sqrt{\lambda_k}\phi_k(t)\,z_k\right|^2\right) = E(|x(t)|^2) - \sum_{0}^{n}\lambda_k\,|\phi_k(t)|^2$$

$$= r(t,t) - \sum_{0}^{n}\lambda_k\,|\phi_k(t)|^2 \to 0,$$

uniformly in $a \leq t \leq b$, according to Step 5, as $n \to \infty$.

Step 7: The properties of z_k follow from the orthogonality of the eigenfunctions. The proof is finished.

Remark 5.4 (Optimality of the Karhunen-Loève expansion). *It can be shown that the Karhunen-Loève expansion is optimal in integrated mean square sense, i.e., if $\psi_k(t)$, $k = 0, 1, \ldots$, is any sequence of functions and $y_k \in \mathcal{H}(x)$ are uncorrelated random variables, then, for each n,*

$$\int_a^b E\left(\left|x(t) - \sum_0^n \sqrt{\lambda_k}\phi_k(t)z_k\right|^2\right) \mathrm{d}t \leq \int_a^b E\left(\left|x(t) - \sum_0^n \psi_k(t)y_k\right|^2\right) \mathrm{d}t.$$

5.3.4 *Testing the mean function*

As a classical engineering problem about signal detection we shall illustrate the use of the Karhunen-Loève expansion to make hypothesis testing on a mean value function. Let $x(t)$ be a Gaussian process with known covariance function $r(s,t) = C(x(s), x(t))$ but unknown mean value function $m(t) = E(x(t))$. Following Grenander [56] (see also [118]), suppose we have observed $x(t)$ for $a \leq t \leq b$, and that we have two alternative hypotheses about $m(t)$,

$$H_0 : m(t) = m_0(t),$$
$$H_1 : m(t) = m_1(t),$$

wanting to ascertain which one is most likely true.

We find the independent standard normal variables z_k, according to (5.12),

$$x(t) = m(t) + \sum_0^\infty \sqrt{\lambda_k}\, \phi_k(t)\, z_k = m(t) + \widetilde{x}(t),$$

where $z_k = \frac{1}{\sqrt{\lambda_k}} \int_a^b \overline{\phi_k(t)}\, \widetilde{x}(t)\, \mathrm{d}t$. The z_k are not observable, since they require $\widetilde{x}(t) = x(t) - m(t)$, and $m(t)$ is unknown, but we can introduce observable independent Gaussian variables y_k by a similar procedure working on $x(t)$,

$$y_k = \int_a^b \overline{\phi_k(t)}\, x(t)\, \mathrm{d}t = \int_a^b \overline{\phi_k(t)}\, m(t)\, \mathrm{d}t + z_k \sqrt{\lambda_k}.$$

Write $a_k = \int_a^b \overline{\phi_k(t)}\, m(t)\, dt$, with the true, but unknown mean function, and define

$$a_{ik} = \int_a^b \overline{\phi_k(t)}\, m_i(t)\, \mathrm{d}t, \quad i = 0, 1,$$

for the two alternatives. Then $\{y_k\}$ are normal with mean a_k and variance λ_k, and the hypotheses are transformed into hypotheses about a_k:

$$H_0 : a_k = a_{0k}, \quad \text{for } k = 0, 1, \ldots,$$
$$H_1 : a_k = a_{1k}, \quad \text{for } k = 0, 1, \ldots.$$

Testing hypotheses about an infinite number of independent Gaussian variables is no more difficult than for finitely many. The *likelihood ratio* (LR) test can be used in any case. Let the parameters to be tested be $\mathbf{T}_0 = (a_{00}, a_{01}, \ldots, a_{0n}, \ldots)$ and $\mathbf{T}_1 = (a_{10}, a_{11}, \ldots, a_{1n}, \ldots)$, respectively. The LR-test based on $(y_0, \ldots, y_n)$ rejects H_0 if the likelihood ratio $p_n(\mathbf{y}) = f_1(\mathbf{y})/f_0(\mathbf{y})$ between the two alternative probability densities is large, say, greater than a constant c, that determines the significance level. In this case

$$p_n(\mathbf{y}) = \frac{f_{y_0,\ldots,y_n}(y_0, \ldots, y_n; \mathbf{T}_1)}{f_{y_0,\ldots,y_n}(y_0, \ldots, y_n; \mathbf{T}_0)} = \frac{\prod_{k=0}^n \frac{1}{\sqrt{2\pi\lambda_k}} e^{-(y_k - a_{1k})^2/2\lambda_k}}{\prod_{k=0}^n \frac{1}{\sqrt{2\pi\lambda_k}} e^{-(y_k - a_{0k})^2/2\lambda_k}}$$

$$= \exp\left\{ -\sum_{k=0}^n \frac{y_k(a_{0k} - a_{1k})}{\lambda_k} + \frac{1}{2}\sum_{k=0}^n \frac{a_{0k}^2 - a_{1k}^2}{\lambda_k} \right\} = \exp\left\{ -\sum_{k=0}^n u_k \right\},$$

say. The LR-test thus rejects H_0 if $\sum_0^n u_k < c_\alpha$.

We now let $n \to \infty$, and examine the limit of the test quantity. If

$$\sum_0^\infty \frac{a_{0k}^2}{\lambda_k} < \infty \quad \text{and} \quad \sum_0^\infty \frac{a_{1k}^2}{\lambda_k} < \infty,$$

then the sums converge; in particular $\sum_0^\infty u_k$ converges in quadratic mean to a normal random variable.

Since

$$E(u_k) = \begin{cases} +\frac{1}{2}\frac{(a_{1k} - a_{0k})^2}{\lambda_k}, & \text{if } H_0 \text{ is true,} \\ -\frac{1}{2}\frac{(a_{1k} - a_{0k})^2}{\lambda_k}, & \text{if } H_1 \text{ is true,} \end{cases}$$

and

$$V(u_k) = \frac{(a_{1k} - a_{0k})^2}{\lambda_k}$$

under both H_0 and H_1, we have that $\sum_0^\infty u_k$ is normal with mean

$$E\left(\sum_0^\infty U_k\right) = \begin{cases} m_0 = \sum_0^\infty \frac{1}{2}\frac{(a_{1k} - a_{0k})^2}{\lambda_k} & \text{if } H_0 \text{ is true} \\ m_1 = -m_0 = -\sum_0^\infty \frac{1}{2}\frac{(a_{1k} - a_{0k})^2}{\lambda_k} & \text{if } H_1 \text{ is true} \end{cases}$$

and with variance

$$V(\sum_0^\infty u_k) = (m_0 - m_1) = 2m_0.$$

Thus, H_0 is rejected if $\sum_0^\infty u_k < m_0 - \lambda_\alpha \sqrt{2m_0}$, where λ_α is the upper normal α-quantile.

If $\sum_0^\infty \frac{(a_{0k}-a_{1k})^2}{\lambda_k^2} < \infty$, the test can be expressed in a simple way by the observation that

$$\sum_0^\infty u_k = \int_a^b f(t) \left(x(t) - \frac{m_0(t)+m_1(t)}{2} \right) \mathrm{d}t,$$

where $f(t) = \sum_0^\infty \frac{a_{0k}-a_{1k}}{\lambda_k} \phi_k(t)$.

Exercises

5:1. Give the formal (very simple) arguments for the equality in (5.5).

5:2. Prove that $\sup_{s,t\in[a,b]} |\sum_0^\infty c_k(s)\overline{c_k(t)}| \le \sup_t r(t,t) < \infty$ in (5.10).

5:3. Prove that in step (4) in the proof of Theorem 5.3,

$$\sum_k \lambda_k = \sum_k \int_a^b \lambda_k |\phi_k(t)|^2 \,\mathrm{d}t \le \max_{a\le s\le b} r(s,s)\cdot(b-a).$$

5:4. Let z_k be independent standard normal variables and define

$$x(t) = \sqrt{2} \sum_{k=1}^\infty \frac{z_k}{\pi k} \sin(\pi k t), \quad 0 \le t \le 1.$$

This is the Karhunen-Loève representation of a Gaussian process with covariance function

$$r_x(s,t) = \sum_{k=1}^\infty \frac{2}{\pi^2 k^2} \sin(\pi k s)\sin(\pi k t).$$

What process is this? Hint: Compare $r_x(1-s,1-t)$ and $r_x(s,t)$.

5:5. Let $\{w(t), 0 \le t < \infty\}$ be standard Brownian motion, and let A have $E(A) = 0$, $V(A) = 1$, and be independent of $\{w(t)\}$. Suppose that $x(s) = As + w(s)$ is observed for $0 \le s \le t$. Find the best estimate of A in mean square sense of the form $\widehat{A}(t) = \int_0^t h(t,s)\mathrm{d}x(s)$; [124].

5:6. Show that

$$\int_a^b E\left(\left|x(t) - \sum_0^n \sqrt{\lambda_k}\phi_k(t) z_k\right|^2 \right) \mathrm{d}t = \sum_{n+1}^\infty \lambda_k,$$

and prove the optimality claim statement in Remark 5.4.

Chapter 6

Classical ergodic theory and mixing

The concept of ergodicity is one of the most fundamental in probability, since it links the mathematical theory of probability to what can be observed in a deterministic mechanical world. It also plays an important role in theoretical physics and, in fact, the term *ergodic* was coined by Ludwig Boltzmann in 1887 in his study of the time development of mechanical particle systems. The term itself stems from the Greek *ergos* = "work" and *hodos* = "path," possibly meaning that ergodic theory codifies how the energy in a system evolves with time. Ergodicity in itself is not a probabilistic concept and it can be studied within a purely deterministic framework, but in its stochastic setting, it helps in the interpretation of probabilities and expectations as long run relative frequencies.

The main result in this chapter is the Birkhoff ergodic theorem, which in 1931 settled the question of asymptotic properties of dynamical systems. For an account of the parallel development in statistical physics and probability theory, see the interesting historical work by von Plato [99]. The account in this chapter is based on [19] and [35].

As the title of this chapter indicates the focus is on the classical ergodic theorems, and their interpretation in probabilistic and statistical terms. We do not touch upon the modern development of ergodic theory in dynamical systems, which has flourished during the last decades, but refer the reader to specialized books, like [45], for more results on ergodic behavior of random and non-random sequences, and [73, 98] for general stochastic aspects of dynamical systems.

6.1 The basic ergodic theorem in $\mathscr{L}^2$

We met our first ergodic theorem in Section 2.5, Theorem 2.19, stating covariance conditions under which the time average $T^{-1}\int_0^T x(t)\,\mathrm{d}t$ tends, in quadratic mean and with probability one, to the expectation of the stationary process $\{x(t), t \in \mathbb{R}\}$. In the special case when $x(t)$ is stationary with covariance function $r(t) = C(x(s+t), x(s))$, the quadratic mean convergence becomes particularly simple. If $E(x(t)) = 0$,

$$\frac{1}{T}\int_0^T r(t)\,dt \to 0 \quad \text{implies} \quad \frac{1}{T}\int_0^T x(t)\,dt \overset{q.m.}{\to} 0, \tag{6.1}$$

as $T \to \infty$. This was proven by elementary calculation in Corollary 2.2 to Theorem 2.19(a).

By means of the spectral representation $x(t) = \int e^{i\omega t}\,dZ(\omega)$, we can formulate a more precise theorem and see what happens if this sufficient condition does not hold. In Section 3.3.2 we stated an explicit expression (3.21) for the spectral process $Z(\omega)$. In fact, if ω_1 and ω_2 are continuity points of the spectral distribution function $F(\omega)$, then we have the parallel expressions

$$F(\omega_2) - F(\omega_1) = \frac{1}{2\pi}\lim_{T\to\infty}\int_{-T}^{T} \frac{e^{-i\omega_2 t} - e^{-i\omega_1 t}}{-it} r(t)\,dt,$$

$$Z(\omega_2) - Z(\omega_1) = \frac{1}{2\pi}\lim_{T\to\infty}\int_{-T}^{T} \frac{e^{-i\omega_2 t} - e^{-i\omega_1 t}}{-it} x(t)\,dt.$$

We repeat the statement from Theorem 3.8, that when the spectral distribution F is a step function, then

$$\lim_{T\to\infty}\frac{1}{T}\int_0^T r(t)e^{-i\omega_k t}\,dt = \Delta F_k, \tag{6.2}$$

$$\lim_{T\to\infty}\frac{1}{T}\int_0^T |r(t)|^2\,dt = \sum_k (\Delta F_k)^2, \tag{6.3}$$

$$\lim_{T\to\infty}\frac{1}{T}\int_0^T x(t)e^{-i\omega_k t}\,dt = \Delta Z_k. \tag{6.4}$$

Note that (6.2) implies that if the spectral distribution function is continuous at $\omega = 0$, then $\frac{1}{T}\int_0^T r(t)\,dt \to 0$ as $T \to \infty$. Thus, if there is no spectral mass at the zero frequency, i.e., at the constant spectral component, then by (6.1) the process tends in average to its mean value.

Example 6.1 (Non-ergodic sum of harmonics)**.** *The discrete sum of random harmonics with frequencies* $\omega_k > 0$, $x(t) = \sum_k A_k \cos(\omega_k t + \phi_k)$, *is a simple example of a stationary process that is not ergodic. We know that* $E(x(0)^2) = \sum E(A_k^2/2)$, *which is (proportional to) the "expected energy" of the process.*

But what happens with the "average energy,"

$$E_T = \frac{1}{T}\int_0^T x(t)^2\,dt = \frac{1}{T}\Big\{\sum_k A_k^2 \int_0^T \cos^2(\omega_k t + \phi_k)\,dt$$
$$+\sum_{j\neq k} A_j A_k \int_0^T \cos(\omega_j t + \phi_j)\cos(\omega_k t + \phi_k)\,dt\Big\},$$

when $T \to \infty$? It is easily verified that E_T tends to $\sum_k A_k^2/2$, and, thus, the time average of $x(t)^2$ does not tend to the expectation $E(x(t)^2) = \sum_k E(A_k^2/2)$.

6.2 Stationarity and transformations

6.2.1 Pseudo randomness and transformation

For strictly stationary processes one can obtain limit theorems of quite different character from those valid for processes that satisfy a covariance stationarity condition. These theorems also require much deeper conditions than simple covariance conditions. Remember, however, that a Gaussian (weakly) stationary process is also strictly stationary, so the general ergodicity properties of Gaussian processes can be inferred already from its covariance function. We start by giving all results for stationary sequences $\{x_n,\, n \in \mathbb{N}\}$.

For a strictly stationary sequence the location of the origin is unessential for the stochastic properties, i.e., $P((x_1, x_2, \ldots) \in B) = P((x_{k+1}, x_{k+2}, \ldots) \in B)$ for every Borel set $B \in \mathscr{B}_\infty$; see Section 1.2.3. This also means that we can assume the sequence to be double ended, and to have started in the remote past.[1] From now on in this chapter, by a stationary process we mean a process that is strictly stationary.

How do we – or nature – construct stationary sequences? Obviously, first we need something (call it "a game") that can go on forever, and second, we need a game where the rules remain the same forever.

Example 6.2 (The modulo game)**.** *A simple game, that can go on forever and has almost all interesting properties of a stationary process, is the adding of an irrational number. Take a random number x_0 with uniform distribution between 0 and 1, and let θ be an irrational number. Define*

$$x_{k+1} = x_k + \theta \mod 1,$$

i.e., x_{k+1} is the fractional part of $x_k + \theta$. It is easy to see that x_k is a stationary

---

[1]More precisely, for every strictly stationary sequence $\{x_n, n \in \mathbb{N}\}$ there exists a strictly stationary sequence $\{\widetilde{x}_n,\, n \in \mathbb{Z}\}$ such that $\widetilde{x}_n,\, n \geq 0$, have the same finite-dimensional distributions as $x_n, n \geq 0$; prove this in Exercise 6:4.

stochastic sequence, even if the only random element is the choice of starting value. We shall soon see why this game is more interesting with an irrational θ *than with a rational one. If computers could handle irrational numbers, this type of pseudo random number generator would be even more useful in Monte Carlo simulations than it is.*

Example 6.3 (More complicated games). *One can define other stationary sequences by applying any time-invariant rule to a stationary sequence* x_n*, e.g., with* $0 \le x_0 \le 1$*, we can take* $y_n = 1$*, if* $x_{n+1} > x_n^2$ *and* $y_n = 0$ *otherwise. An even more complicated rule is* $y_n = x_n + \max_{k>0} e^{-k} x_{n+k}$.

The well-known quadratic transformation,

$$x_{k+1} = c x_k (1 - x_k),$$

is an example of an interesting transformation giving rise to a stochastic sequence when x_0 *is random; in Exercise 6:5 you are asked to find its stationary distribution.*

6.2.2 *Stationarity and measure preserving transformations*

Deterministic explicit rules, like $x_{k+1} = x_k + \theta \mod 1$, are just examples of transformations of a sample space which can produce stationary sequences for certain probability distributions. We need to define a general concept of a *measure preserving transformation*, which makes the resulting process stationary.

Definition 6.1. *a) Consider a probability space* $(\Omega, \mathcal{F}, P)$*. A measurable transformation*[2] T *on* $(\Omega, \mathcal{F}, P)$ *is called measure preserving if*

$$P(T^{-1}A) = P(A) \quad \text{for all } A \in \mathcal{F}. \tag{6.5}$$

b) Given a measurable space $(\Omega, \mathcal{F})$ *and a measurable transformation* T *on* $(\Omega, \mathcal{F})$*, a probability measure* P *is called invariant if (6.5) holds.*

In statistical physics or dynamical systems theory, a probability space is

[2] A measurable transformation T on a measurable space $(\Omega, \mathcal{F})$ is a function defined on Ω, such that the inverse images under T of all sets in $\mathcal{F}$ are again in $\mathcal{F}$; that is $T^{-1}A = \{\omega; T\omega \in A\} \in \mathcal{F}$ for all $A \in \mathcal{F}$.

regarded as a model for the "universe," with the outcomes ω representing all its different "states," e.g., the location and velocity of all its particles. The measure P defines the probabilities for the universe to be in such and such a state that certain events occur. A transformation $T\omega$ is just a law of nature that changes the state of the universe; that a transformation is measure preserving means that events occur with the same probability before and after the transformation.

From a measure preserving transformation to a stationary sequence

Every measure preserving transformation of a probability space $(\Omega, \mathscr{F}, P)$ can generate a stationary stochastic sequence. To see how, take a random variable $x(\omega)$ on $(\Omega, \mathscr{F}, P)$, i.e., a "measurement" on the state of the universe, for example its temperature. Then $x(T\omega)$ is the measurement taken after the transformation, and it probably has a different value. Define the random sequence

$$x_1(\omega) = x(\omega),\, x_2(\omega) = x(T\omega), \ldots, x_n(\omega) = x(T^{n-1}\omega), \ldots.$$

Since T is measure preserving, this sequence is strictly stationary: With $B \in \mathscr{B}_\infty$ and

$$A = \{\omega; (x_1(\omega), x_2(\omega), \ldots) \in B\} = \{\omega; (x(\omega), x(T\omega), \ldots) \in B\},$$

$P(A) = P((x_1, x_2, \ldots) \in B)$. Further,

$$\begin{aligned} T^{-1}A = \{\omega; T\omega \in A\} &= \{\omega; (x(T\omega), x(T^2\omega), \ldots) \in B\} \\ &= \{\omega; (x_2(\omega), x_3(\omega), \ldots) \in B\}, \end{aligned}$$

and thus $P(T^{-1}A) = P((x_2, x_3, \ldots) \in B)$. Since T is measure preserving, $P(A) = P(T^{-1}A)$, and hence $(x_1, x_2, \ldots)$ and $(x_2, x_3, \ldots)$ have the same distribution, that is, $\{x_n\}$ is a stationary sequence.

From a stationary process to a measure preserving transformation

We have just seen how to construct a stationary sequence from a measure preserving transformation. Conversely, every stationary sequence generates a measure preserving transformation on $\mathbb{R}^\infty$, the space of realizations for the stationary sequence.

Take the probability space $(\mathbb{R}^\infty, \mathscr{B}_\infty, P)$ with outcomes $\omega = (x_1, x_2, \ldots)$, and define the *shift transformation* T by

$$T\omega = (x_2, x_3, \ldots).$$

As an example, take the set $A = \{\omega; x_1 < x_2\}$ to be the outcomes for which the first co-ordinate is smaller than the second one. Then $T^{-1}A = \{\omega; T\omega \in A\}$ is the set $\{(x_1, x_2, \ldots); (x_2, x_3, \ldots) \in A\} = \{(x_1, x_2, \ldots); x_2 < x_3\}$, that is, the second co-ordinate is smaller than the third. The transformation just shifts the event criterion one step to the right.

Take the co-ordinate process $\mathbf{x}(\omega) = \omega = (x_1, x_2, \ldots)$, for which the nth variable in the sequence is equal to the nth co-ordinate in the outcome, and assume that P is such that $\mathbf{x}$ is a strictly stationary sequence. Then the shift transformation T is measure preserving – the shifted sequence has the same distribution as the original one.

Remark 6.1. *The property that a transformation is measure preserving of course depends on the probability measure P and not only on the transformation itself. In probability theory, where the probability measure is often given a priori, it is natural to put the request on the transformation.*

In the mathematical study of dynamical systems one often starts with a transformation T and seeks a measure under which T is measure preserving. Such a measure is called invariant. Thus, invariant measures in the dynamical systems theory are equivalent to the strictly stationary processes in probability theory.

In the next two sections we present the ergodic theorem, first with the transformation/ergodic system view in mind, and then as a stationary process property, more in line with statistical applications. Both approaches have their own merits, and the choice is partly a matter of taste (and rigor).

6.3 The ergodic theorem, transformation view

The classical ergodic theory for dynamical systems deals with what happens in the long run when one observes some characteristic $x(\omega)$ of the system. The state of the system changes by the transformation T, and our interest lies in the time average of the sequence of measurements $x(\omega), x(T\omega), x(T^2\omega), \ldots$, taken for a fixed initial outcome ω,

$$\frac{1}{n}\sum_{1}^{n} x(T^{k-1}\omega),$$

as $n \to \infty$. Think of Ω as all the possible states our universe can be in, and, to be concrete, think of $x(\omega)$ as the temperature at one specific location. The universe changes states from day to day, and if ω is the state of the universe today, and $T\omega$ its state tomorrow, then $\frac{1}{n}\sum_1^n x(T^{k-1}\omega)$ is the average temperature observed over an n-day period.

In this section we shall take the *transformation view* on ergodic theory, and prove the ergodic theorem in that vein. In the next section we shall do exactly the same thing, but take a probabilistic *strictly stationary process* view, and prove the ergodic theorem in terms of random variables and expectations. The reader may want to turn directly to Section 6.4 and return to this section later.

6.3.1 Invariant sets and invariant random variables

To motivate the introduction of invariant sets and invariant random variables, we consider the possible limits of the average of a sequence of random variables,

$$S_n/n = \frac{1}{n}\sum_1^n x_k.$$

If S_n/n converges, as $n \to \infty$, what are the possible limits? Say, if $S_n/n \to y$, a random variable, then obviously $x_n/n = S_n/n - S_{n-1}/n \to y - y = 0$, and[3]

$$y = \lim \frac{\sum_2^{n+1} x_k}{n} + \lim \frac{x_1 - x_{n+1}}{n} = \lim \frac{\sum_2^{n+1} x_k}{n} - \lim \frac{x_{n+1}}{n} = \lim \frac{\sum_2^{n+1} x_k}{n}.$$

We see that the limit of S_n/n is the same for the sequence $x_1, x_2, \ldots$ as it would be for the shifted sequence $x_2, x_3, \ldots$.

Definition 6.2. *Consider a probability space $(\Omega, \mathscr{F}, P)$ and a measure preserving transformation T of Ω onto itself.*

(a) A random variable x on $(\Omega, \mathscr{F}, P)$ is called invariant under T if $x(\omega) = x(T\omega)$, for almost every $\omega \in \Omega$.

(b) A set $A \in \mathscr{F}$ is called invariant under T if $T^{-1}A = A$.

Example 6.4. *The limit of S_n/n is invariant (when it exists) under the shift transformation $(x_1, x_2, \ldots) \mapsto (x_2, x_3, \ldots)$ of $\mathbb{R}^\infty$. The random variables $y = \limsup x_n$ and $\limsup S_n/n$ are other examples of invariant variables.*

Example 6.5 (Markov chain). *Let x_n be a Markov chain with four states,*

[3] Show as an exercise, if you have not done so already, that if x_n is a stationary sequence with $E(|x_n|) < \infty$, then $P(x_n/n \to 0, \text{ as } n \to \infty) = 1$; Exercise 6:6.

$\{1,2,3,4\}$, probability transition matrix

$$\mathbf{P} = \begin{pmatrix} 1/2 & 1/2 & 0 & 0 \\ 1/3 & 2/3 & 0 & 0 \\ 0 & 0 & 2/3 & 1/3 \\ 0 & 0 & 1/2 & 1/2 \end{pmatrix},$$

and with starting distribution $\mathbf{p}^{(0)} = (2p/5, 3p/5, 3q/5, 2q/5)$. This chain is stationary for every $p+q=1$. The empty set $\emptyset$ and the whole sample space $\{1,2,3,4\}^\infty$ are trivially invariant. The sets $A_1 = \{\omega = (x_1, x_2, \dots); x_k \in \{1,2\}\}$ and $A_2 = \{\omega = (x_1, x_2, \dots); x_k \in \{3,4\}\}$ are also invariant under the shift transformation.

The proof of the following simple theorem is left to the reader.

Theorem 6.1. *(a) The family of invariant sets,*

$$\mathscr{J} = \{\text{invariant sets } A \in \mathscr{F}\},$$

is a σ-field.

(b) A random variable y is invariant if and only if it is measurable with respect to the family $\mathscr{J}$ of invariant sets.

6.3.2 Ergodicity

Ergodicity is one of the fundamental concepts in statistical physics. Already Ludwig Boltzmann stated that a physical system over time will pass through all its possible states. The idea was later called the "ergodic hypothesis," and it played a major role in the attempts both to provide a solid mathematical meaning to probability concepts, and to formulate links between statistics and physics. For a historical account, see [99]. The study of ergodic transformations is now an important part of the dynamic systems and chaos theory; see [73].

The fundamental property of ergodic transformations $\omega \mapsto T\omega$ with an invariant (or stationary) distribution P is that the $T^k\omega$, with increasing k, visits every corner of the state space, exactly with the correct frequency as required by the probability measure P, regardless of the starting position ω. As a consequence, the asymptotic time average equals the probabilistic ensemble average. Therefore, the ergodic theorem is the general "law of large numbers" for random sequences and processes.

Another way of saying this is that the "histogram," counting the number of visits to any neighborhood of states, converges to a limiting "density," namely the density for the invariant distribution P over the state space. If we make a measurement $x(\omega)$ on the system, then the expected value $E(x)$ is the "ensemble average" with respect to the measure P,

$$E(x) = \int_{\omega \in \Omega} x(\omega)\, \mathrm{d}P(\omega),$$

and – here is the ergodicity – this is exactly the limit of the "time average"

$$\frac{1}{n}\sum_{1}^{n} x(T^{k-1}\omega).$$

In the Markov Example 6.5, if neither of p and q is equal to 0, there is no possibility for the process to visit every state the correct number of times, since either it starts in the invariant set A_1 and then it always takes the values $1,2$, or it starts in A_2, and then it stays there forever and takes only values $3,4$. This is the key to the definition of ergodicity.

Definition 6.3. *A measure preserving transformation T on a probability space $(\Omega, \mathscr{F}, P)$ is called ergodic if every invariant set $A \in \mathscr{J}$ has either $P(A) = 0$ or $P(A) = 1$, that is, all invariant sets are trivial. The term "metrically transitive" is sometimes used instead of "ergodic."*

Example 6.6 (The modulo game)**.** *The modulo game in Example 6.2, considered as a transformation on the unit interval, $([0,1], \mathscr{B}, \ell)$ (ℓ is the Lebesgue measure, i.e., the uniform distribution),*

$$Tx = x + \theta \mod 1,$$

is ergodic when θ is irrational, and non-ergodic for rational θ.

If θ is rational, $\theta = m/n$, one has $T^n x = x$ and every set of the form $A \cup \{A+1/n\} \cup \{A+2/n\} \ldots \cup \{A+(n-1)/n\}$, mod 1, *is invariant but can have any probability $\in (0,1)$. There are (at most) n possible values that $T^n x$ can take, and these repeat themselves.*

To see what happens when θ is irrational we need some further results for ergodic transformations.

Theorem 6.2. *A measure preserving transformation T of $(\Omega, \mathscr{F}, P)$ is ergodic if and only if every invariant random variable x is a.s. a constant. It is sufficient that every bounded invariant random variable is a.s. constant.*

Proof. First assume that every bounded invariant variable is constant, and take an invariant set A. Its indicator function $\chi_A(\omega) = 1$ if $\omega \in A$ is then an invariant random variable, and by assumption it is a.s. constant. This means that the sets where it is 0 and 1, respectively, have probability either 0 or 1, i.e., $P(A) = 0$ or 1. Hence T is ergodic.

Conversely, take an ergodic T, and consider an arbitrary invariant random variable x. We shall show that x is a.s. constant. Define, for real x_0,

$$A_{x_0} = \{\omega; x(\omega) \le x_0\}.$$

Then $T^{-1}A_{x_0} = \{\omega; T\omega \in A_{x_0}\} = \{\omega; x(T\omega) \le x_0\} = A_{x_0}$, since x is invariant. But then, by ergodicity, $P(A_{x_0}) = 0$ or 1, depending on x_0, and it is easy to see that then there is an $\widetilde{x}_0$ such that $P(x = \widetilde{x}_0) = 1$, hence x is constant. □

Example 6.7 (The irrational modulo game). *We can now show that the modulo transformation*

$$Tx = x + \lambda \mod 1$$

is ergodic if λ is irrational. (If $\lambda = p/q$ is rational, the transformation repeats itself after q iterations, and hence is not ergodic.)

Take any Borel-measurable function $f(x)$ on $[0,1)$ with $\int_0^1 f^2(x)\,\mathrm{d}x < \infty$. It can be expanded in a Fourier series

$$f(x) = \sum_{k=-\infty}^{\infty} c_k e^{2\pi ikx},$$

with $\sum |c_k|^2 < \infty$ and with convergence in quadratic mean for almost all x. Then $y = f(\omega)$ is a random variable. We assume it is invariant, and prove that it is a.s. constant. But that f is invariant means that

$$f(x) = f(Tx) = \sum_{k=-\infty}^{\infty} c_k e^{2\pi ikx} \cdot e^{2\pi ik\lambda},$$

which implies $c_k(1 - e^{2\pi ik\lambda}) = 0$ for all k. But $e^{2\pi ik\lambda} \ne 1$ for $k \ne 0$ when λ is irrational, and hence $c_k = 0$ for all $k \ne 0$, which means that $f(x) = c_0$, constant. By Theorem 6.2 we conclude that T is ergodic.

6.3.3 The Birkhoff ergodic theorem

In the introductory remarks, Section 6.3.1, we noted that any limit of the time average

$$\frac{1}{n}\sum_{1}^{n} x(T^{k-1}\omega)$$

is invariant under the transformation T. The limit could be a constant, and it could be a random variable, but it needs to be invariant. If it is constant, we need to find the value of the constant, and if it is a random variable, we want to find out as much as possible about its distribution. We now formalize our previous finding that the limit is the same regardless of if we start at an outcome ω or at $T\omega$.

In fact, for measure preserving transformations, the limit *always* exists, and is equal to the conditional expectation of x given the invariant sets. Since we have not dealt with this concept in detail previously in this course, we state the basic properties of conditional expectations by giving its definition. More details are found in Appendix A, Section A.3.3.

Definition 6.4. *If x is a random variable on $(\Omega, \mathscr{F}, P)$ with $E(|x|) < \infty$, and $\mathscr{A}$ a sub-σ-field of $\mathscr{F}$, then by the conditional expectation of x given $\mathscr{A}$, $E(x \mid \mathscr{A})$, is meant any $\mathscr{A}$-measurable random variable u that satisfies*

$$\int_{\omega \in A} x(\omega)\,\mathrm{d}P(\omega) = \int_{\omega \in A} u(\omega)\,\mathrm{d}P(\omega),$$

for all $A \in \mathscr{A}$. Note that the value of $E(y \mid \mathscr{A}) = u$ is defined only almost surely, and that any $\mathscr{A}$-measurable variable that has the same integral as x when integrated over $\mathscr{A}$-sets works equally well as the conditional expectation.

In particular, if $\mathscr{A}$ only contains sets that have probability 0 or 1, then the conditional expectation $E(x \mid \mathscr{A})$ is a.s. constant and equal to $E(x)$.

In particular, we shall consider the conditional expectation $E(x \mid \mathscr{J})$, given the σ-field $\mathscr{J}$ of sets which are invariant under a measure preserving transformation.

Example 6.8 (Markov chain). *The intention with this example is to show* *that the concept of ergodicity for a Markov chain agrees with ergodicity in the sense of this chapter. Consider the Markov chain from Example 6.5. The invariant sets of outcomes are* $\mathscr{J} = \{\emptyset, A_1 = \{1,2\}^\infty, A_2 = \{3,4\}^\infty, \{1,2,3,4\}^\infty\}$. *Any random variable that is measurable with respect to the* σ*-field* $\mathscr{J}$ *of invariant sets can take only two possible values and is constant over* A_1 *and over* A_2. *With* $x(\omega) = x_1$, *and the starting distribution* $\mathbf{p}^{(0)} = (2p/5, 3p/5, 3q/5, 2q/5)$, *we find*

$$E(x \mid \mathscr{J}) = \begin{cases} 1 \cdot 2/5 + 2 \cdot 3/5 = 8/5, & \text{if } \omega \in A_1, \\ 3 \cdot 3/5 + 4 \cdot 2/5 = 17/5, & \text{if } \omega \in A_2. \end{cases}$$

Depending on the starting state, the time average, $\frac{1}{n}\sum_1^n x(T^{k-1}\omega)$, *converges as* $n \to \infty$ *to* $8/5$ *or to* $17/5$. *With Markov chain terminology the chain is reducible and it is not ergodic. The limiting behavior depends on initial state.*

Time averages always converge

Theorem 6.3 (Birkhoff ergodic theorem, 1931)**.** *Let T be a measure preserving transformation on* $(\Omega, \mathscr{F}, P)$. *Then, for any random variable x with* $E(|x|) < \infty$,

$$\lim_{n\to\infty} \frac{1}{n} \sum_0^{n-1} x(T^k\omega) = E(x \mid \mathscr{J}), \text{ a.s.}$$

The proof is based on the following famous lemma, the *Maximal ergodic lemma*, with a proof given by Adriano M. Garsia in 1965; [52].

Lemma 6.1 (Maximal ergodicity lemma)**.** *Let T be a measure preserving transformation and x a random variable with* $E(|x|) < \infty$. *Define*

$$S_k(\omega) = x(\omega) + \ldots + x(T^{k-1}\omega),$$
$$M_n(\omega) = \max(0, S_1, S_2, \ldots, S_n).$$

Then

$$\int_{\omega; M_n > 0} x(\omega)\, \mathrm{d}P(\omega) \geq 0.$$

Proof of lemma: Consider $S'_k = x(T\omega)+\ldots+x(T^k\omega) = S_k - x(\omega)+x(T^k\omega)$, and $M_n(T\omega) = M'_n = \max(0, S'_1, \ldots, S'_n)$. For $k = 1, \ldots, n$, $M'_n \geq S'_k$, so

$$x + M'_n \geq x + S'_k = S_{k+1},$$

while for $k = 0$, trivially $x + M'_n \geq S_1 (= x)$. Hence

$$x \geq S_k - M'_n, \quad \text{for } k = 1, 2, \ldots, n+1,$$

which implies

$$x \geq \max(S_1, \ldots, S_n) - M'_n.$$

Thus (with $M'_n = M_n(T\omega)$),

$$\int_{M_n>0} x(\omega)\, dP(\omega) \geq \int_{M_n>0} \{\max(S_1(\omega), \ldots, S_n(\omega)) - M_n(T\omega)\}\, \mathrm{d}P(\omega).$$

But on the set $\{\omega; M_n(\omega) > 0\}$, one has that $M_n = \max(S_1, \ldots, S_n)$, and thus

$$\int_{M_n>0} x(\omega)\, dP(\omega) \geq \int_{M_n>0} \{M_n(\omega) - M_n(T\omega)\}\, \mathrm{d}P(\omega)$$
$$\geq \int \{M_n(\omega) - M_n(T\omega)\}\, \mathrm{d}P(\omega) = 0,$$

since increasing the integration area does not change the integral of $M_n(\omega)$, while it can only make the integral of $M_n(T\omega)$ larger. Further, T is measure preserving, i.e., shifting the variables one step does not change the distribution, nor the expectation. The lemma is proved. □

Proof of theorem: We first assume that $E(x \mid \mathscr{J}) = 0$ and prove that the average converges to 0, with probability one. For the general case consider $x - E(x \mid \mathscr{J})$ and use that

$$E(x \mid \mathscr{J})(T\omega) = E(x \mid \mathscr{J})(\omega),$$

since $E(x \mid \mathscr{J})$ is invariant by Theorem 6.1(b).

We show that $\overline{x} = \limsup S_n/n \leq 0$ and, similarly, $\underline{x} = \liminf S_n/n \geq 0$, giving $\lim S_n/n = 0$. Take an $\varepsilon > 0$ and denote $D = \{\omega; \overline{x} > \varepsilon\}$: we shall show that $P(D) = 0$. Since, from Example 6.4, $\overline{x}$ is an invariant random variable, also the event D is invariant. Define a new random variable,

$$x^*(\omega) = \begin{cases} x(\omega) - \varepsilon, & \text{if } \omega \in D, \\ 0, & \text{otherwise}, \end{cases}$$

and set $S_n^*(\omega) = \sum_1^n x^*(T^{k-1}\omega)$, with $M_n^* = \max(0, S_1^*, \dots, S_n^*)$. Then we know, from Lemma 6.1, that

$$\int_{M_n^*>0} x^*(\omega)\,\mathrm{d}P(\omega) \geq 0. \tag{6.6}$$

We now only replace this inequality by an inequality for a similar integral over the set D to be finished. The sets

$$F_n = \{M_n^* > 0\} = \left\{\max_{1\leq k\leq n} S_k^* > 0\right\}$$

increase towards the set

$$F = \left\{\sup_{k\geq 1} S_k^* > 0\right\} = \left\{\sup_{k\geq 1} \frac{S_k^*}{k} > 0\right\} = \left\{\sup_{k\geq 1} \frac{S_k}{k} > \varepsilon\right\} \cap D.$$

But since $\sup_{k\geq 1} S_k/k \geq \limsup S_k/k = \overline{x}$, we have that $F = D$. In order to take the limit of (6.6) we must be sure the expectations are finite, i.e., $E(|x^*|) \leq E(|x|) + \varepsilon$, and then bounded convergence gives

$$0 \leq \int_{M_n^*>0} x^*(\omega)\,\mathrm{d}P(\omega) \to \int_D x^*(\omega)\,\mathrm{d}P(\omega). \tag{6.7}$$

Here, the right hand side is

$$\begin{aligned}
\int_D x^*(\omega)\,\mathrm{d}P(\omega) &= \int_D x(\omega)\,\mathrm{d}P(\omega) - \varepsilon P(D) \\
&= \int_D E(x \mid \mathscr{J})\,\mathrm{d}P - \varepsilon P(D) = -\varepsilon P(D),
\end{aligned}$$

since $\int_D E(x \mid \mathscr{J})\,\mathrm{d}P = 0$, by assumption. Together with (6.7) this implies $P(D) = 0$, and hence $\overline{x} \leq \varepsilon$. But $\varepsilon > 0$ was arbitrary so $\overline{x} \leq 0$, a.s. The same chain of arguments leads to $\underline{x} \geq 0$, and hence $\limsup S_n/n \leq 0 \leq \liminf S_n/n$. The limit therefore exists and is 0, which was to be shown. □

Time averages of ergodic transformations go to the mean

It is now a simple corollary that time averages for ergodic transformations converge to the expected value. This corollary is the mathematical formulation of the idea that the development of an ergodic system, in the long run, will generate all its possible states and with the correct frequency.

Corollary 6.1. *If T is a measure preserving ergodic transformation on the probability space $(\Omega, \mathscr{F}, P)$, then for any random variable x with $E(|x|) < \infty$,*

$$\lim_{n\to\infty} \frac{1}{n} \sum_{0}^{n-1} x(T^k\omega) = E(x), \ a.s.$$

Proof. When T is ergodic, every invariant set has probability 0 or 1 and therefore the conditional expectation is constant, $E(x \mid \mathscr{J}) = E(x)$, a.s. □

Remark 6.2. *If x is non-negative with $E(x) = \infty$ then $S_n/n \to \infty$ if T is ergodic. Show that as Exercise 6:10.*

Ergodic non-random walks

One can regard a measure preserving transformation T as a non-random walk ω, $T\omega$, $T^2\omega$, ..., over the sample space. In the beginning of this section we interpreted the ergodic statement as a convergence result for the number of visits to any neighborhood of a fixed outcome. This can now be made precise. Take a set $A \in \mathscr{F}$ and consider its indicator function $\chi_A(\omega)$. The ergodic theorem says, that if T is ergodic,

$$\frac{1}{n} \sum_{0}^{n-1} \chi_A(T^k\omega) \overset{a.s.}{\to} P(A).$$

Example 6.9 (The irrational modulo game)**.** *In the irrational modulo game, $Tx = x + \theta \mod 1$, the relative number of points falling in an interval $[a,b)$ converges to the length of the interval if θ is irrational. Thus, the number of points becomes asymptotically equidistributed over $[0,1)$. This is the weak Weyl's equidistribution theorem.*

6.4 The ergodic theorem, process view

6.4.1 Ergodic stationary sequences

It is easy to formulate the convergence theorem for transformations in terms of time averages of stationary sequences. First we need to define invariant events and ergodicity in $(\Omega, \mathscr{F}, P)$.

Definition 6.5 (Reformulation of Definitions 6.2 and 6.3). *Let $\{x_n, n \in \mathbb{N}\}$ be a stationary sequence. A random variable z is called invariant for $\{x_n\}$, if it is a function of $x_1, x_2, \ldots$ and remains unchanged under the shift transformation, i.e., if there exists a random variable ϕ on $(\mathbb{R}^\infty, \mathscr{B}_\infty)$ such that $z = \phi(x_n, x_{n+1}, \ldots)$ for all $n \geq 1$.*

An event $A \in \mathscr{F}$ is called invariant for $\{x_n\}$, if there exists a $B \in \mathscr{B}_\infty$ such that for any $n \geq 1$,

$$A = \{(x_n, x_{n+1}, \ldots) \in B\}.$$

The sequence is called ergodic if every invariant set is trivial, i.e., has probability 0 or 1.

From the correspondence between transformations of the sample space and stationary sequences, one can formulate an ergodic theorem for a stationary sequence $\{x_n, n \in \mathbb{N}\}$.

Theorem 6.4. *(a) If $\{x_n, n \in \mathbb{N}\}$ is a stationary sequence with $E(|x_1|) < \infty$, and $\mathscr{J}$ denotes the σ-field of invariant sets, then*

$$\frac{1}{n}\sum_1^n x_k \overset{a.s.}{\to} E(x_1 \mid \mathscr{J}), \text{ a.s.}$$

(b) If $\{x_n, n \in \mathbb{N}\}$ is stationary and ergodic, then

$$\frac{1}{n}\sum_1^n x_k \overset{a.s.}{\to} E(x_1), \text{ a.s.}$$

Proof. Theorem 6.4 is a reformulation of Theorem 6.3. In order to get a slightly better understanding in probabilistic terms, we shall formulate the important steps of the proof using process arguments. However, we warn the reader of the last step in the chain, which contains an argument involving a conditional expectation that is not uniquely defined; see the end of the proof. We give a parallel proof of part (a), and leave the rest to the reader.

(a) First the counterpart of Lemma 6.1: From the sequence $\{x_n, n \in \mathbb{N}\}$, define

$S_k = \sum_{j=1}^{n} x_j$ and $M_n = \max(0, S_1, S_2, \ldots, S_n)$. We prove that

$$E(x_1 \mid M_n > 0) \geq 0. \tag{6.8}$$

To this end, write $S'_k = \sum_{j=2}^{k+1} x_j = S_k - x_1 + x_{n+1}$, and define the corresponding maximum, $M'_n = \max(0, S'_1, S'_2, \ldots, S'_n)$, and note that, for $k = 1, \ldots, n$,

$$x_1 + M'_n \geq x_1 + S'_k = S_{k+1}.$$

Since $M'_n \geq 0$ we also have $x_1 + M'_n \geq S_1 (= x_1)$, so

$$x_1 \geq \max(S_1, \ldots, S_n) - M'_n.$$

Now, when $M_n > 0$, we can replace $\max(S_1, \ldots, S_n)$ by M_n, so taking the conditional expectation, given $M_n > 0$, we get

$$\begin{aligned} E(x_1 \mid M_n > 0) &\geq E(\max(S_1, \ldots, S_n) - M'_n \mid M_n > 0) \\ &= E(M_n - M'_n \mid M_n > 0). \end{aligned}$$

Further, since $M'_n \geq 0$, one easily argues that

$$E(M_n - M'_n \mid M_n > 0) \geq E(M_n - M'_n)/P(M_n > 0),$$

which is 0, since M_n and M'_n have the same expectation. This proves (6.8).

We continue with the rest of the proof of part (a). Suppose $E(x_1 \mid \mathscr{J}) = 0$ and consider the invariant random variable $\overline{x} = \limsup S_n/n$. Take an $\varepsilon > 0$ and introduce the invariant event $D = \{\overline{x} > \varepsilon\}$. Then, by the assumption,

$$E(x_1 \mid \overline{x} > \varepsilon) = 0, \tag{6.9}$$

a fact which is basic in the proof. We intend to prove that $P(D) = 0$ for every $\varepsilon > 0$, thereby showing that $\overline{x} \leq 0$. Similarly, $\underline{x} = \liminf S_n/n$ can be shown to be non-negative, and hence $\overline{x} = \underline{x} = 0$.

However, before we prove that $P(D) = P(\overline{x} > \varepsilon) = 0$, we need to discuss the meaning of (6.9). A conditional expectation is defined as a random variable, measurable with respect to the conditioning σ-field, in this case $\mathscr{J}$. In (6.9) we conditioned on one of the events $D \in \mathscr{J}$ and that is fine if $P(D) > 0$, but if $P(D) = 0$, the claim (6.9) makes no sense. The conditional expectation given an event of probability 0 can be given any value we like, since the only requirement on the conditional expectation is that it should give a correct value when integrated over a $\mathscr{J}$-event. If that event has probability 0 the integral is 0 regardless of how the expectation is defined. We return to this at the end of the proof.

From $(x_1, x_2, \dots)$, define a new sequence of variables

$$x_k^* = \begin{cases} x_k - \varepsilon, & \text{if } \overline{x} > \varepsilon, \\ 0, & \text{otherwise,} \end{cases}$$

and define $S_k^* = \sum_{j=1}^k x_k^*$ and $M_n^* = \max(0, S_1^*, S_2^*, \dots, S_n^*)$, in analogy with S_k and M_n. The sequence $\{x_k^*\}$ is stationary, since $\overline{x}$ is an invariant random variable, so we can apply (6.8) to get

$$E(x_1^* \mid M_n^* > 0) \geq 0. \tag{6.10}$$

On the other hand, from the definition of x_k^* , we have

$$E(x_1^* \mid \overline{x} > \varepsilon) = E(x_1 \mid \overline{x} > \varepsilon) - \varepsilon = -\varepsilon < 0, \tag{6.11}$$

since $E(x_1 \mid \overline{x} > \varepsilon) = 0$ by the assumption. These two inequalities go in opposite directions, and in fact they will turn out to be in conflict, unless $P(D) = 0$, proving the assertion.

So we would like to have a relation between the events $D = \{\overline{x} > \varepsilon\}$ and $F_n = \{M_n^* > 0\} = \{\max_{1 \le k \le n} S_k^* > 0\}$, and we see that as n increases, F_n increases to

$$F = \left\{ \sup_{k \geq 1} S_k^* > 0 \right\} = \left\{ \sup_{k \geq 1} \frac{S_k^*}{k} > 0 \right\} = \left\{ \sup_{k \geq 1} \frac{S_k}{k} > \varepsilon \right\} \bigcap \{\overline{x} > \varepsilon\}.$$

But $\overline{x} = \limsup S_k/k \leq \sup_{k \geq 1} S_k/k$, so the right hand side is just $D = \{\overline{x} > \varepsilon\}$; $F_n \uparrow D$. This implies (here $E(|x_1^*|) \leq E(|x_1|) + \varepsilon < \infty$ is needed)

$$0 \leq \lim_{n \to \infty} E(x_1^* \mid F_n) = E(x_1^* \mid D) = -\varepsilon < 0,$$

which obviously is impossible.

Where did it go wrong, and where is the contradiction? By definition, for the outcomes where $\overline{x} > \varepsilon$, the variables x_k^* ARE equal to $x_k - \varepsilon$, and $S_k^*/k = S_k/k - \varepsilon$. But $F_n \uparrow D$ does not immediately imply $\lim_{n \to \infty} E(x_1^* \mid F_n) = \lim_{n \to \infty} E(x_1^* \mid D)$, since these expressions are not well defined. If $P(D) > 0$ our reasoning makes sense and leads to a contradiction: if $P(D) = 0$ we have argued with undefined quantities in (6.10) and (6.11). The reader who wants to be on safe grounds should return to the formulation and proof of Theorem 6.3, and regard these arguments as an illustration only. □

The following theorem is useful; prove it as Exercise 6:11.

Corollary 6.2. *If* $\{x_n, n \in \mathbb{N}\}$ *is stationary and ergodic, and* $\phi(x_1, x_2, \ldots)$ *is measurable on* $(\mathbb{R}^\infty, \mathscr{B}_\infty)$*, then the process* $y_n = \phi(x_n, x_{n+1}, \ldots)$ *is stationary and ergodic.*

6.4.2 *Ergodic stationary processes*

For continuous time processes $\{x(t), t \in \mathbb{R}\}$, one defines the shift transformation U_τ,

$$(U_\tau x)(t) = x(t+\tau).$$

If $x(t)$ is stationary, U_τ is measure preserving. For a set of functions B, the shifted set $U_\tau B$ is the set of functions $U_\tau x$ for $x \in B$. A Borel set $B \in \mathscr{B}_\mathbb{R}$ is called a.s. invariant if B and $U_\tau B$ differ by, at most, sets of P-measure 0. Let $\mathscr{J}$ denote the σ-field of invariant sets. The process $\{x(t), t \in \mathbb{R}\}$ is called ergodic if all invariant sets have probability 0 or 1.

Theorem 6.5. *(a) For any stationary process* $\{x(t), t \in \mathbb{R}\}$ *with* $E(|x(t)|) < \infty$ *and integrable sample paths, as* $T \to \infty$,

$$\frac{1}{T}\int_0^T x(t)\,\mathrm{d}t \overset{a.s.}{\to} E(x(0) \mid \mathscr{J}).$$

(b) If further $\{x(t), t \in \mathbb{R}\}$ *is ergodic, then*

$$\frac{1}{T}\int_0^T x(t)\,\mathrm{d}t \overset{a.s.}{\to} E(x(0)).$$

Proof. One first considers the stationary sequence $x_n = \int_{n-1}^n x(t)\,\mathrm{d}t$, and uses the ergodic theorem to get convergence for integer n,

$$\frac{1}{n}\int_0^n x(t)\,\mathrm{d}t \overset{a.s.}{\to} E(x_1 \mid \mathscr{J}) = E\left(\int_0^1 x(t)\,\mathrm{d}t \mid \mathscr{J}\right),$$

as $n \to \infty$. By invariance and stationarity this conditional expectation is equal to $E(x(0) \mid \mathscr{J})$. Finally, with $n = [T]$,

$$\frac{1}{T}\int_0^T x(t)\,\mathrm{d}t = \frac{n}{T}\frac{1}{n}\int_0^n x(t)\,\mathrm{d}t + \frac{1}{T}\int_n^T x(t)\,\mathrm{d}t.$$

The first term has the same limit as $\frac{1}{n}\int_0^n x(t)\,\mathrm{d}t$, while the second is bounded by

$$\frac{1}{T}\int_n^{n+1} |x(t)|\,\mathrm{d}t.$$

But also $|x(t)|$ is a stationary process, to which we can apply Theorem 6.4(a), getting convergence, and hence can conclude that the last term tends to 0. Thus we obtain the desired limit. □

6.5 Ergodic Gaussian sequences and processes

We shall now give simple conditions for ergodicity for Gaussian stationary processes, characterized by their covariance function $r(t)$ in continuous or discrete time.

Theorem 6.6. *Let $\{x(t), t \in \mathbb{R}\}$ be stationary and Gaussian with $E(x(t)) = 0$ and $V(x(t)) = 1$, and let its covariance function be $r(t)$.*

a) Then $x(t)$ is ergodic if and only if its spectral distribution function $F(\omega)$ is continuous everywhere.

b) If $r(t) \to 0$ as $t \to \infty$, then $x(t)$ is ergodic.

If the spectral distribution has a density $f(\omega)$, $F(\omega) = \int_{-\infty}^{\omega} f(x)\,\mathrm{d}x$, then $F(\omega)$ is obviously a continuous function and $x(t)$ is ergodic. However, it is by no means necessary that F has a density. It suffices that $F(\omega)$ is a continuous function.

Proof of "only if" part: If $x(t)$ is ergodic, so is $x^2(t)$, by Corollary 6.1, and therefore the time average of $x^2(t)$ tends to $E(x^2(0)) = 1$,

$$\frac{S_T}{T} = \frac{1}{T}\int_0^T x^2(t)\,\mathrm{d}t \overset{a.s.}{\to} 1,$$

as $T \to \infty$. Now S_T is an integral of squared normal variables, and it follows from a property of the Gaussian distribution that $E((S_T/T)^4) \leq K$, for some constant K, for large T; cf. Exercise 4.1. Therefore the almost sure convergent S_T/T also converges in quadratic mean, i.e.,

$$E((S_T/T-1)^2) \to 0; \tag{6.12}$$

see Lemma A.1(2) in Appendix A. But this expectation can be calculated. Since, for a standard Gaussian process, $E(x(s)^2x(t)^2) = 1+2r(t-s)^2$ (Exercise 4:1), one gets

$$\begin{aligned} E((S_T/T-1)^2) &= \frac{1}{T^2}E\left(\int_0^T \int_0^T x(s)^2x(t)^2\,\mathrm{d}s\,\mathrm{d}t\right) - 1 \\ &= \frac{2}{T^2}\int_0^T \int_0^T r^2(t-s)\,\mathrm{d}s\,\mathrm{d}t = \frac{4}{T^2}\int_0^T t\cdot\left\{\frac{1}{t}\int_0^t r^2(s)\,\mathrm{d}s\right\}\mathrm{d}t. \end{aligned} \tag{6.13}$$

According to Theorem 3.8, relation (3.27), $\frac{1}{t}\int_0^t r^2(s)\,\mathrm{d}s$ tends to the sum of squares, $\sum(\Delta F_k)^2$ of all jumps of the spectral distribution function $F(\omega)$. Hence, if this sum is strictly positive, the right hand side in (6.13) has a positive limit, which contradicts (6.12), proved above. We have concluded the "only if" part of the theorem.

"Proof" of "if" part: The "if" part is more difficult, and we can prove it here only under the condition that the process has a spectral density, i.e., the spectral distribution is $F(\omega) = \int_{-\infty}^{\omega} f(x)\,\mathrm{d}x$, because then

$$r(t) = \int e^{i\omega t} f(\omega)\,\mathrm{d}\omega \to 0,$$

as $t \to \infty$, by Riemann-Lebesgue's Lemma. (The full statement and proof were given by U. Grenander and G. Maruyama, and can be found in [56, 88].) Therefore we have formulated part (b) of the theorem and we now show that if $r(t) \to 0$, then $x(t)$ is ergodic.

We show that if $r(t) \to 0$, then every invariant set has probability 0 or 1. Let S be an a.s. invariant set for the $x(t)$-process, i.e., the translated event $S_\tau = U_\tau S$ differs from S by an event of probability zero. But every event in $\mathscr{F}$ can be approximated arbitrarily well by a finite-dimensional event, B, depending only on $x(t)$ for a finite number of time points $t_k, k = 1, \ldots, n$; cf. Section 1.2.4 and also Section A.1.3 in Appendix A. From stationarity, also S_τ can be approximated by the translated event $B_\tau = U_\tau B$, with the same error, and combining S with S_τ can at most double the error. Thus, we have

$$|P(S) - P(B)| < \varepsilon,$$
$$|P(S \cap S_\tau) - P(B \cap B_\tau)| < 2\varepsilon.$$

Here $P(S \cap S_\tau) = P(S)$ since S is invariant, so $P(S)$ can be approximated arbitrarily well by both $P(B)$ and by $P(B \cap B_\tau)$.

But B depends on $x(t_i), i = 1 \dots, n$, while B_τ is defined from $x(\tau + t_j)$, $j = 1, \dots, n$, and these two sets are multivariate normal with covariances

$$C(x(t_i + \tau), x(t_j)) = r(\tau + t_j - t_i) \to 0$$

as $\tau \to \infty$. Thus, the two groups of random variables become asymptotically independent,

$$P(B \cap B_\tau) - P(B) \cdot P(B_\tau) \to 0,$$

and by stationarity, $P(B \cap B_\tau) \to P(B)^2$. Hence, both $P(B)$ and $P(B)^2$ approximate $P(S)$, and we conclude that $P(S) = P(S)^2$. This is possible only if $P(S)$ is either 0 or 1, i.e., $x(t)$ is ergodic. □

6.6 Mixing and asymptotic independence

This section deals with two important questions in probability and statistics. The first question is fundamental for the interpretation of statistical and deterministic dependence: How much of the future development of a stochastic process is pre-determined from what has already happened, and how much information about the future is there in a piece of observation of a stochastic process and how closely dependent are disjoint segments of the process?

The second question, related to the first, is very relevant for statistical estimation and inference: How much independence between past and future is needed in order that an estimation functional shall be asymptotically normal with increasing observation interval?

In this section we will briefly review some criteria on a stationary process that guarantee perfect predictability and un-predictability, respectively. A differentiable process can be predicted locally by means of a Taylor expansion. The linear prediction theory gives the Cramér-Wold decomposition (Theorem 6.7) of a stochastic process into a singular component that can be predicted linearly without error and one regular component for which the predictable part tends to zero with increasing prediction horizon.

The spectral representation in Chapter 3 relates the spectrum to the number of harmonic components which are needed to build a stationary process. The ergodic theorem touches upon the problem of asymptotic independence; for a Gaussian process with (absolutely) continuous spectrum and asymptotically vanishing covariance function, values far apart are asymptotically independent and a "law of large numbers" holds. In this section we shall try to relate these scattered results to each other. Some comments are for Gaussian processes only, while others are of general nature.

For central limit theorems some new concepts are needed. *Mixing conditions* of many different kinds have been introduced that should guarantee that a certain interesting functional has an asymptotic normal distribution. We will state and study some of these classical concepts, like *uniform, strong, and weak mixing*, but also relate some newer and more flexible conditions that better adapt to special applications.

6.6.1 *Singularity and regularity*

When predicting from $x(s), s \leq t$, a question of both practical and theoretical (or perhaps philosophical) interest is *from where* does the information about $x(t+h)$ originate, and how much new information is added with increasing observation interval? Recall the notation $\mathscr{H}(x,t) = \mathscr{S}(x(s); s \leq t)$, for the Hilbert space spanned by linear combinations of $x(s)$ for $s \leq t$; see Appendix A.6.

When $t \to -\infty$, obviously

$$\mathscr{H}(x,t) \downarrow \mathscr{H}(x,-\infty) = \cap_{t \leq t_0} \mathscr{H}(x,t),$$
$$\mathscr{H}(x,-\infty) \subseteq \mathscr{H}(x,t) \subseteq \mathscr{H}(x) = \mathscr{H}(x,\infty).$$

The subspace $\mathscr{H}(x,t)$ is the space of random variables that can be obtained as limits of linear combinations of variables $x(t_k)$ with $t_k \leq t$, and $\mathscr{H}(x,-\infty)$ is what can be obtained from old variables, regardless of how old they may be. It can be called *the infinitely remote past*, or the *primordial randomness.*[4]

Definition 6.6. *The process* $\{t(t), t \in \mathbb{R}\}$ *is called*
- *purely deterministic (or singular) if* $\mathscr{H}(x,-\infty) = \mathscr{H}(x)$,
- *purely non-deterministic (or regular) if* $\mathscr{H}(x,-\infty) = \mathbf{0}$.

A process is deterministic if all information about the future that can be obtained from the past at time t, $x(s), s \leq t$, can be obtained already from $x(s), s \leq \tau < t$, arbitrarily far back. An example of this is the low-frequency white noise Gaussian process, which we studied in Chapter 3, Example 3.6. Such a process is infinitely differentiable – this follows from the general rules

[4]Nowadays called "the primordial soup."

for differentiability – and the entire sample functions can be reconstructed from the values in an arbitrarily small interval located anywhere on the time axis.

The following theorem was proved by Cramér (1962) in the general case; the discrete stationary case was given by Herman Wold (1954); [34, 123].

Theorem 6.7 (The Cramér-Wold decomposition). *Every stochastic process $\{x(t), t \in \mathbb{R}\}$, with $E(|x(t)|^2) < \infty$, is the sum of two uncorrelated processes,*

$$x(t) = y(t) + z(t),$$

where $\{y(t), t \in \mathbb{R}\}$ is regular (purely non-deterministic) and $\{z(t), t \in \mathbb{R}\}$ is singular (deterministic).

Proof. Construct $\mathscr{H}(x,t)$ and $\mathscr{H}(x,-\infty) = \lim_{t\downarrow -\infty} \mathscr{H}(x,t)$, and define, as in Theorem A.10 on page 303,

$$z(t) = P_{-\infty}(x(t)) = \text{the projection of } x(t) \text{ on } \mathscr{H}(x,-\infty),$$
$$y(t) = x(t) - z(t).$$

To prove the theorem, we have to show that

1. $\mathscr{H}(z,-\infty) = \mathscr{H}(z,t)$, and $z(t)$ is deterministic,
2. $\mathscr{H}(y,-\infty) = \mathbf{0}$, and $y(t)$ is non-deterministic,
3. $\mathscr{H}(y) \perp \mathscr{H}(z)$, and $y(s)$ and $z(t)$ are uncorrelated.

Step (3) follows from the projection properties, since the residual $y(s) = x(s) - P_{-\infty}(x(s))$ is uncorrelated with every element in $\mathscr{H}(x,-\infty)$, i.e., $y(s) \perp \mathscr{H}(x,-\infty)$. Since $z(t) \in \mathscr{H}(x,-\infty)$, (3) follows.

Further, $\mathscr{H}(y,t) \subseteq \mathscr{H}(x,t)$ and $\mathscr{H}(y,t) \perp \mathscr{H}(x,-\infty)$. Therefore we can conclude that $\mathscr{H}(y,-\infty)$ is equal to $\mathbf{0}$, because if y is an element of $\mathscr{H}(y,-\infty)$ then both $y \in \mathscr{H}(y,t) \subset \mathscr{H}(x,t)$ for all t, i.e., $y \in \mathscr{H}(x,-\infty)$, and at the same time $y \perp \mathscr{H}(x,-\infty)$. The only element that is both in $\mathscr{H}(x,-\infty)$ and is orthogonal to $\mathscr{H}(x,-\infty)$ is the zero element, showing (2).

Finally, $\mathscr{H}(z,t) = \mathscr{H}(x,-\infty)$ for every t. To see this, note that, with $\oplus$ denoting vector sum (see Appendix A.6),

$$\mathscr{H}(x,t) \subseteq \mathscr{H}(y,t) \oplus \mathscr{H}(z,t)$$

for all t, and therefore also

$$\mathscr{H}(x,-\infty) \subseteq \mathscr{H}(y,-\infty) \oplus \mathscr{H}(z,t).$$

Since $\mathscr{H}(y,-\infty) = \mathbf{0}$, it follows that

$$\mathscr{H}(x,-\infty) \subseteq \mathscr{H}(z,t) \subseteq \mathscr{H}(x,-\infty).$$

Thus $\mathscr{H}(x,-\infty) = \mathscr{H}(z,t)$, and (1) is proved. □

As the low-frequency white noise example shows, there are natural deterministic processes. Other common process models are regular. Examples of processes combining the two properties seem to be rather artificial.

Example 6.10. *An* AR(1)*-process with an added component,*

$$x(t) = ax(t-1) + e + e(t),$$

with uncorrelated e and $e(t)$-variables, can be decomposed into

$$y(t) = \sum_{k=0}^{\infty} a^k e(t-k),$$

which is regular, and

$$z(t) = \frac{1}{1-a} e,$$

which is singular. The common ARMA*-process is regular.*

6.6.2 Asymptotic independence, regularity, and singularity

As we saw in the proof of Theorem 6.6, if $r(t) \to 0$ as $t \to \infty$, then, in a Gaussian process, finitely many variables taken sufficiently far apart are almost statistically independent. But this does definitely not mean that the process observed in an entire *interval*, $x(s), s \in I$, does not influence $x(s)$, $s \in t+I$ for large t. These two segments can be completely dependent of each other in a deterministic way. For example, the realizations can be infinitely differentiable and part of an analytic function that can be reconstructed from its derivatives in an arbitrarily small interval. It was shown by Belyaev [8] that if the covariance function of a separable process is an entire function, i.e., analytic in the entire complex plane, then the sample functions are a.s. also entire functions, which can be expressed as a convergent power series,

$$x(t) = \sum_{k=0}^{\infty} x^{(k)}(0) \frac{t^k}{k!}.$$

Examples of such covariance functions are $r(t) = e^{-t^2/2}$ and $r(t) = \frac{\sin t}{t}$. Yaglom [125, Ch. 8] contains a readable account of prediction in this case; [41, 63] give more mathematical details.

Regularity, singularity, and the spectrum

The Cramér-Wold decomposition, Theorem 6.7, deals with prediction of future values by means of linear combinations of past observations. A singular (or purely deterministic) process can be perfectly predicted linearly from old values. In a regular process the predictable part tends to zero with increasing prediction horizon. Simple conditions for singularity/regularity can be formulated in terms of the spectral distribution function $F(\omega)$.

Let $f(\omega) = \frac{d}{d\omega} F(\omega)$ be the derivative of the bounded and non-decreasing function $F(\omega)$. For almost all ω this derivative exists and is non-negative [105, Ch. 5], and its integral is bounded, $\int_{-\infty}^{\infty} f(\omega)\,\mathrm{d}\omega \leq F(\infty) - F(-\infty)$. Write

$$F^{(ac)}(\omega) = \int_{-\infty}^{\omega} f(x)\,\mathrm{d}x \leq F(\omega).$$

The spectrum is absolutely continuous with spectral density $f(\omega)$ if

$$F(\omega) = F^{(ac)}(\omega).$$

In general, $F^{(ac)}(\omega)$ need not be equal to $F(\omega)$. In particular, this is of course the case when the spectrum has jumps ΔF_k at frequencies ω_k. Write, for the discrete part,

$$F^{(d)}(\omega) = \sum_{\omega_k \leq \omega} \Delta F_k,$$

so the spectrum is discrete if $F(\omega) = F^{(d)}(\omega)$ for all ω.

The part of the spectrum that is neither absolutely continuous nor discrete is called the *singular* part:

$$F^{(s)}(\omega) = F(\omega) - F^{(ac)}(\omega) - F^{(d)}(\omega).$$

Note that both $F^{(d)}(\omega)$ and $F^{(s)}(\omega)$ are bounded non-decreasing functions, differentiable almost everywhere, with zero derivative.

Conditions for stationary sequences

Since $\int_{-\pi}^{\pi} f(\omega)\,\mathrm{d}\omega < \infty$ and $-\infty \leq \log f(\omega) \leq f(\omega) = \frac{d}{d\omega} F(\omega)$ (which exists almost everywhere), the integral

$$P = \frac{1}{2\pi} \int_{-\pi}^{\pi} \log f(\omega)\,\mathrm{d}\omega \tag{6.14}$$

is either finite or equal to $-\infty$.

Theorem 6.8. *For a stationary sequence $\{x_n, n \in \mathbb{N}\}$ the following cases can occur.*

a) If $P = -\infty$, then x_t is singular.

b) If $P > -\infty$, and the spectrum is absolutely continuous with $f(\omega) > 0$ for almost all ω, then x_t is regular.

c) If $P > -\infty$, but $F(\omega)$ is either discontinuous, or is continuous with non-vanishing singular part, $F^{(s)}(\omega) \neq 0$, then x_t is neither singular nor regular.

The proof of this and the next theorem, for continuous parameter processes, can be found in [41, Ch. XII, §4] by J.L. Doob.

The theorem is quite satisfying, and it is worth making some comments on its implications. First, if the spectrum is *discrete* with a finite number of jumps, then $f(\omega) = 0$ for almost all ω and $P = -\infty$, so the process is singular. For example, in the spectral representation (3.25), the process then depends only on a finite number of random quantities which can be recovered from a finite number of observed values.

If the spectrum is *absolutely continuous* with density $f(\omega)$, singularity and regularity depends on whether $f(\omega)$ comes close to 0 or not. For example, if $f(\omega)$ vanishes in some interval, $P = -\infty$ and $x(t)$ is singular. Singularity also occurs if $f(\omega) = 0$ at a single point ω_0 and is very close to 0 nearby, such as when $f(\omega) \sim \exp(-\frac{1}{(\omega-\omega_0)^2})$ when $\omega \to \omega_0$.

If $f(\omega) \geq c > 0$ for $-\pi < \omega \leq \pi$, then the integral is finite and the process is regular.

Theorem 6.9. *A stationary sequence $x(t)$ is regular, if and only if it can be represented as a one-sided moving average*

$$x_t = \sum_{k=-\infty}^{t} h_{t-k} y_k,$$

with uncorrelated y_k; cf. Theorem 4.3, which also implies that the sequence has a spectral density.

A sequence that is neither singular nor regular can, according to the Cramér-Wold theorem, be represented as a sum of two uncorrelated sequences, $x_t = x_t^{(s)} + x_t^{(r)} = x_t^{(s)} + \sum_{k=-\infty}^{t} h_{t-k} y_k$.

The regular part $x_t^{(r)}$ has absolutely continuous spectrum,

$$F^{(ac)}(\omega) = \int_{-\pi}^{\omega} f(x)\,\mathrm{d}x,$$

while the singular part $x_t^{(s)}$ has spectral distribution $F^{(d)}(\omega) + F^{(s)}(\omega)$.

It is also possible to express the prediction error in terms of the integral P. In fact, the one step ahead prediction error is

$$\sigma_0^2 = \inf_{h_0, h_1, \ldots} E\left(\left| x_{t+1} - \sum_{k=0}^{\infty} h_k x_{t-k} \right|^2 \right) = 2\pi \exp(P).$$

Conditions for stationary processes

Conditions for regularity and singularity for stationary processes $x(t)$ with continuous parameter can be expressed in terms of the integral

$$Q = \int_{-\infty}^{\infty} \frac{\log f(\omega)}{1+\omega^2}\,\mathrm{d}\omega,$$

where as before $f(\omega) = \frac{d}{d\omega} F(\omega)$ is the a.s. existing derivative of the spectral distribution function. For the following theorem, see [41, Ch. XII, §5].

Theorem 6.10. *For a stationary process $\{x(t), t \in \mathbb{R}\}$, one has that*

a) if $Q = -\infty$, then $x(t)$ is singular,

b) if $Q > -\infty$, and the spectrum is absolutely continuous, then $x(t)$ is regular.

The decomposition of $x(t)$ into one singular component, which can be predicted, and one regular component which is a moving average, is analogous to the discrete time case,

$$x(t) = x^{(s)}(t) + \int_{u=-\infty}^{t} h(t-u)\,\mathrm{d}\zeta(u),$$

where $\{\zeta(t), t \in \mathbb{R}\}$ is a process with uncorrelated increments.

We apply the theorem to processes with covariance functions $r(t) = e^{-t^2/2}$

and $r(t) = \frac{\sin t}{t}$. They have absolutely continuous spectra with spectral densities, $f(\omega) = \frac{1}{\sqrt{2\pi}} e^{-\omega^2/2}$ and $f(\omega) = 1/2$ for $|\omega| < 1$, respectively. Obviously $Q = -\infty$ is divergent in both cases, and we have verified the statement that these processes are deterministic, although their covariance functions tend to 0 as $t \to \infty$. The Ornstein-Uhlenbeck process with spectral density $f(\omega) = \frac{\alpha}{\pi(\alpha^2+\omega^2)}$ is an example of a regular process with $Q > -\infty$.

Note that in all three examples the covariance function $r(t)$ tends to 0 as $t \to \infty$, but with quite different rates. For the regular Ornstein-Uhlenbeck process it tends to 0 exponentially fast, while for the two singular (and hence predictable) processes, the covariance falls off either much faster, as $e^{-t^2/2}$, or much slower, as $1/t$. Hence, we learn that stochastic determinism and non-determinism are complicated matters, even if they are entirely defined in terms of correlations and distribution functions.

6.6.3 *Uniform, strong, and weak mixing*

Predictability, regularity, and singularity are probabilistic concepts, defined and studied in terms of prediction error moments and correlations. When it comes to ergodicity, we have seen examples of totally deterministic sequences, which exhibit ergodic behavior, in the sense that they obey the law of large numbers. To complicate matters, we have seen that a Gaussian process, with covariance function tending to 0 at infinity, implying asymptotic independence, is always ergodic, even if the remote future may, in a deterministic sense, be determined by the arbitrarily remote past.

Ergodicity is a law of large numbers, time averages converge to a limit. In statistics, one would also like to have some idea of the asymptotic distribution; in particular, one would like a central limit theorem for normalized sums or integrals,

$$\frac{\sum_1^N x(k) - A_N}{B_N}, \qquad \frac{\int_0^N x(t)\,\mathrm{d}t - A_N}{B_N},$$

as $N \to \infty$. This asks for general concepts, called *mixing conditions*, that guarantee asymptotic independence of *functionals* of a stochastic process, i.e., random variables defined in terms of parts of a random sequence or a continuous time process, like the maximum, an average, or a spectrum estimate. These conditions are of a more abstract character than the long memory conditions we met in Section 4.4.1. Depending on the generality, one can divide such conditions into *global* conditions, which govern dependence between the entire σ-fields of past and future, and *adapted* conditions, which focus on a restricted class of events of special interest. We will first deal with some global conditions, and then briefly touch upon the more flexible and useful class of adapted conditions. For a survey, see [18], and for more details, [38, 62].

Global mixing conditions

For any stochastic process $\{x(t), t \in \mathbb{R}\}$ define the σ-field $\mathcal{M}_a^b$ as the σ-field generated by $x(t), a \leq t \leq b$. Taking a and b as $\mp\infty$, we get $\mathcal{M}_{-\infty}^b$ and $\mathcal{M}_a^\infty$ as the σ-fields generated by $\{x(t), t \leq b\}$ and by $\{x(t), t \geq a\}$, respectively. The following mixing conditions represent successively milder conditions on the asymptotic independence. The terminology, "global conditions," refers to the fact that they are required to hold for all events in the respective σ-fields, which is a very strong restriction.

Four different types of global mixing:

uniform mixing: $\{x(t)\}$ is uniformly mixing (or ϕ-mixing) if there is a function $\phi(n)$ such that $\phi(n) \to 0$ as $n \to \infty$, and for all t and events $A \in \mathcal{M}_{-\infty}^t$ and $B \in \mathcal{M}_{t+n}^\infty$,

$$|P(A \cap B) - P(A)P(B)| \leq \phi(n) P(A).$$

strong mixing: $\{x(t)\}$ is strongly mixing (or α-mixing) if there is a non-negative function $\alpha(n)$ such that $\alpha(n) \to 0$ as $n \to \infty$, and for all t and events $A \in \mathcal{M}_{-\infty}^t$ and $B \in \mathcal{M}_{t+n}^\infty$,

$$|P(A \cap B) - P(A)P(B)| \leq \alpha(n). \tag{6.15}$$

mixing in the sense of ergodicity: $\{x(t)\}$ is mixing if, for all events $A \in \mathcal{M}_{-\infty}^t$ and $B \in \mathcal{M}_t^\infty$,

$$\lim_{k \to \infty} P(A \cap U^{-k}B) = P(A)P(B).$$

Here, $U^{-k}B = \{x(\cdot); x(k + \cdot) \in B\} \in \mathcal{M}_{t+k}^\infty$.

weak mixing: $\{x(t)\}$ is weakly mixing if, for all events $A \in \mathcal{M}_{-\infty}^t$ and $B \in \mathcal{M}_t^\infty$,

$$\lim_{n \to \infty} \frac{1}{n} \sum_{k=1}^{n} |P(A \cap U^{-k}B) - P(A)P(B)| = 0.$$

It is obvious that uniform mixing is the strongest condition and that it implies strong mixing, introduced by M. Rosenblatt, as a vehicle for a central limit theorem [103], which in turn implies weak mixing. There are important relations between these mixing conditions and the notion of ergodicity. We start with the simplest of these.

Theorem 6.11. *Strong mixing implies ergodicity.*

Proof. The proof is almost a complete repetition of the proof of Theorem 6.6, part (b). Take an invariant event S and approximate it by a finite-dimensional event B with $|P(S) - P(B)| < \varepsilon$. Suppose $B \in \mathscr{M}_a^b \subset \mathscr{M}_{-\infty}^b$. Then the translated event $B_n = U_n B$ belongs to $\mathscr{M}_{a+n}^{b+n} \subset \mathscr{M}_{a+n}^{\infty}$, and hence

$$P(B \cap B_n) - P(B) \cdot P(B_n) \to 0.$$

As in the proof of Theorem 6.6, part (b), it follows that $P(B)^2 = P(B)$ and hence $P(S)^2 = P(S)$, so $P(S)$ is either 0 or 1 and S is trivial. □

We quote without proof the following equivalence theorem for processes that are infinitely divisible, such as the Gaussian process.[5] For a proof, see [104].

Theorem 6.12. *An infinitely divisible stationary process $x(t)$ is ergodic if and only if it is weakly mixing.*

6.6.4 *Global mixing conditions for Gaussian processes*

For Gaussian processes the hierarchy of mixing conditions has the following simple equivalences in terms of covariance conditions and spectral properties.

Theorem 6.13. *Let $\{x(t), t \in \mathbb{Z}\}$ be a stationary Gaussian sequence. Then*

a) $x(t)$ is uniformly mixing if and only if it is m-dependent, i.e., there is an m such that the covariance function $r(t) = 0$ for $|t| > m$.

b) $x(t)$ is strongly mixing if it has a continuous spectral density bounded away from zero, $f(\omega) \geq c > 0$ on $-\pi < \omega \leq \pi$.

c) $x(t)$ is weakly mixing if and only if it is ergodic, and, hence, if and only if the spectral distribution function is continuous.

[5] For infinite divisibility, see Definition 4.4 on page 125.

Proof. a) That m-dependence implies uniform mixing is obvious. To prove necessity, assume that $r(t)$ is not identically 0 for large t. That $x(t)$ is uniformly mixing implies that for all $A \in \mathscr{M}_{-\infty}^{0}$, with $P(A) > 0$ and all $B \in \mathscr{M}_{n}^{\infty}$,

$$|P(B \mid A) - P(B)| < \phi(n) \to 0 \tag{6.16}$$

as $n \to \infty$. This directly implies that $r(t) \to 0$ as $t \to \infty$, since otherwise there would be an infinite number of $t = t_k$ for which $r(t_k) \geq r_0 > 0$, say. Take $A = \{x(0) > 0\} \in \mathscr{M}_{0}^{0}$ and $B = \{x(t_k) > 0\} \in \mathscr{M}_{t_k}^{t_k}$. It is easy to see that, for a Gaussian sequence, $r(t_k) \geq r_0 > 0$ implies that

$$P(B \mid A) - P(B) \geq c_0 > 0,$$

hence is bounded away from 0 by a positive constant c_0, and therefore cannot tend to 0. We conclude that $r(t) \to 0$ as $t \to \infty$.

Next, we prove that $r(t)$ is equal to 0 for all large t, assuming $E(x(t)) = 0$, $E(x(t)^2) = 1$. Assume that there is an infinite number of $t = t_k$ for which $\rho_k = r(t_k) > 0$, but still $r(t_k) \to 0$. Define $A = \{x(0) > 1/\rho_k\}$ (obviously $P(A) > 0$) and let $B = \{x(t_k) > 1\}$. Since $x(0), x(t_k)$ are bivariate normal, the conditional distribution of $x(t_k)$ given $x(0) = x$ is normal with mean $\rho_k x$ and variance $1 - \rho_k^2$. As $\rho_k \to 0$, the conditional distribution of $x(0)$ given $x(0) > 1/\rho_k$ will be concentrated near $1/\rho_k$ and then $x(t_k)$ will be approximately normal with mean one and variance one. Therefore, $P(B \mid A) \to 1/2$, but $P(B) = (2\pi)^{-1/2} \int_1^\infty \exp(-y^2/2)\, dy < 0.2$. Hence (6.16) does not hold for $\phi(t_k) \to 0$.

b) For a proof of this, see [63] or [62, Thm. 17.3.3].

c) This follows from the general theorem on infinitely divisible processes; Theorem 6.12. The proof that a Gaussian process with absolutely continuous spectrum is weakly mixing follows as in the proof of Theorem 6.6. □

Remark 6.3 (Strongly mixing Gaussian sequences)**.** *The simple condition (b) in the theorem for strong mixing of a Gaussian sequence is sufficient but not necessary. An example of this is an m-dependent* MA(q)*-process with* $q = m - 1$, *with spectral density* $f(\omega) = \frac{\sigma^2}{2\pi} \left|P(e^{i\omega})\right|^2$, *where P is a polynomial. If* $P(z) = 0$ *for some* $|z| = 1$, *the spectrum violates the condition in (b), but the process is strongly mixing, since it is m-dependent.*

6.6.5 *Adapted mixing conditions*

The Rosenblatt strong (global) mixing condition works well for central limit theorems; [62, Ch. 18]. However, even in the Gaussian case, one could ask for more transparent conditions, sufficient for statistical purposes. Furthermore, there are simple auto-regressive processes that are not strongly mixing; [4].

An early example of an adapted mixing criterion is the $D(u_n)$-condition, which restricts the degree of dependence between extreme values. It was introduced by Leadbetter [74], as a condition under which a dependent stationary random sequence shares the same asymptotic extremal behavior as an independent sequence with the same marginal distribution; see also [75, Ch. 3]. We will return to the $D(u_n)$-condition in Chapter 8.

The difference between the Rosenblatt strong mixing condition and the Leadbetter mixing condition is that, while the former requires mixing and asymptotic independence for all possible past and future events, the $D(u_n)$-condition is adapted to extreme events. This special adaptation can be generalized to a formal criterion, and we present here a condition, suggested by P. Doukhan and S. Louhichi [42].

Let $\{x_n, n \in \mathbb{N}\}$ be a (strictly) stationary random sequence, and consider a family $\mathscr{F}$ of real-valued functions defined on $\{x_n\}$, such that each function depends only on a finite number of x_n-variables.

Definition 6.7 ($\mathscr{F}$-weak dependence). *The sequence $\{x_n, n \in \mathbb{N}\}$ is called $\mathscr{F}$-weak dependent, if there exists a real sequence $\{\theta_n, n \in \mathbb{N}\}$, tending to zero as $n \to \infty$, and further, for each $g, h \in \mathscr{F}$ and $u, v \in \mathbb{N}$, there exists a finite constant $\psi(g,h,u,v)$ with the following properties:*

For any u-tuple $(i_1, \dots, i_u)$ and v-tuple $(j_1, \dots, j_v)$ with $i_1 \leq \cdots \leq i_u < i_u + r \leq j_1 \leq \cdots \leq j_v$, one has

$$\left|C(g(x_{i_1}, \dots, x_{i_u}), h(x_{j_1}, \dots, x_{j_v})\right| \leq \psi(g,h,u,v)\, \theta_r. \qquad (6.17)$$

It should be noted in what sense (6.17) represents a milder condition than (6.15): If A is a finite dimensional event, and $g(x_{i_1}, \dots, x_{i_u})$ is its indicator function, and similarly for B, then

$$C(g(x_{i_1}, \dots, x_{i_u}), h(x_{j_1}, \dots, x_{j_v})) = P(A \cap B) - P(A)P(B).$$

Examples of $\mathscr{F}$-weak mixing processes and how they are used to derive asymptotic properties of density estimates, including asymptotic normality, are given in [38, 43].

Exercises

6:1. Show that the transformation $Tx = 2x \mod 1$ of $\Omega = [0,1)$, $\mathscr{F} = \mathscr{B}$, $P =$ Lebesgue measure, is measurable and measure preserving.

6:2. (Continued.) Define the random variable $x(\omega) = 0$ if $0 \le \omega < 1/2$, $x(\omega) = 1$ if $1/2 \le \omega < 1$. Show that the sequence $x_n(\omega) = x(T^{n-1}\omega)$ consists of independent zeros and ones, with probability $1/2$ each.

6:3. Show that if T is measure preserving on $(\Omega, \mathscr{F}, P)$ and x is a random variable, then $E(x(\omega)) = E(x(T\omega))$.

6:4. Show that every one-sided stationary sequence $\{x_n, n \ge 0\}$ can be extended to a two-sided sequence $\{x_n, n \in \mathbb{Z}\}$ with the same distributions.

6:5. Find a distribution for x_0 that makes $x_{k+1} = 4x_k(1-x_k)$ stationary.

6:6. Prove that if $E(|x_n|) < \infty$, then $P(x_n/n \to 0, \text{ as } n \to \infty) = 1$.

6:7. For any sequence of random variables, x_n, and event $B \in \mathscr{B}$, show that the event $\{x_n \in B, \text{infinitely often}\}$ is invariant under shift transformation.

6:8. Find an ergodic transformation T such that T^2 is not ergodic.

6:9. Show that $\{x_n\}$ is ergodic if and only if for every $A \in \mathscr{B}_k$, $k = 1, 2, \ldots$,

$$\frac{1}{n} \sum_{j=1}^{n} \chi_A(x_j, \ldots, x_{j+k}) \to P((x_1, \ldots, x_{k+1}) \in A).$$

6:10. Show that if x is non-negative with $E(x) = \infty$ and $x_n(\omega) = x(T^{n-1}\omega)$, $S_n = \sum_1^n x_n$, then $S_n/n \to \infty$ if T is ergodic.

6:11. Prove Corollary 6.2 by looking at the invariant sets of x_n and y_n.

6:12. Take two independent, stationary, and ergodic sequences, $\{x_n\}$ and $\{y_n\}$. Take one of the two sequences at random with equal probability, $z_n = x_n, n = 1, 2, \ldots$, or $z_n = y_n, n = 1, 2, \ldots$. Show that z_n is not ergodic.

6:13. Let $\{x_n\}$ and $\{y_n\}$ be two ergodic sequences, defined on $(\Omega, \mathscr{F}, P)$, and consider the bivariate sequence $z_n = (x_n, y_n)$. Construct an example that shows that $\{z_n\}$ need not be ergodic, even if $\{x_n\}$ and $\{y_n\}$ are independent.

6:14. Comfort yourself with the strange fact that an ergodic process can be exactly predicted to the infinite future from present observations.

6:15. Prove that weak mixing implies that $\{x(t)\}$ is ergodic, without the assumption of infinite divisibility; see also Exercise 6:16.

6:16. Prove that a sufficient condition for $z(n) = (x(n), y(n))$ to be ergodic, if $\{x(n)\}$ and $\{y(n)\}$ are independent ergodic sequences, is that one of $\{x(n)\}$ and $\{y(n)\}$ is weakly mixing.

Chapter 7

Vector processes and random fields

The correlation and spectral theory used to describe the internal dependence in a stationary process can be generalized to describe dependence between two stationary processes, and how it is distributed over different frequencies. A new feature is introduced, namely frequency dependent amplitude and phase information between harmonic elements. Remember that in a univariate process the phases of different components are non-informative, since they are independent and uniformly distributed over $(0, 2\pi)$. Still, there may be relations between the phase in one process and the corresponding component in another process – even if the phases have uniform marginal distributions.

A random field is a real-valued process of a multi-dimensional parameter, often a time and a set of space co-ordinates. The second half of the chapter deals with the special spectral properties of homogeneous and isotropic fields, and on some special models for space-time variability.

7.1 Spectral representation for vector processes

The internal correlation structure of a stationary process is defined by the covariance function; the spectrum distributes the correlation over different frequencies and in the spectral representation the process is actually built by individual components, with independent amplitudes and phases, like in the discrete case (3.25). The phases are all independent and uniformly distributed over $(0, 2\pi)$. The spectrum does not contain any phase information!

When dealing with joint properties of two stationary processes, in particular their covariation, one has to rely on the spectral decomposition. Covariation can then be attributed to amplitude covariation or to phase dependencies, or both. Thus, phase information becomes important.

Suppose we have two second order stationary processes $\{x(t), t \in \mathbb{R}\}$ and $\{y(t), t \in \mathbb{R}\}$ that are *jointly stationary*. Each of the processes is described by a spectral representation of the form (3.25), or the integral representation in case the spectrum is continuous. If the two processes are correlated, the individual amplitudes and phases are still, of course, independent in each process taken separately, but for every frequency the amplitudes in the two processes can be dependent, and there can exist a complicated dependence between the phases in the form of a time delay. This dependence is described by the *cross-covariance function* and the *cross-spectrum*.

A weakly stationary vector-valued process is a vector of p stationary processes, $\mathbf{x}(t) = (x_1(t),\dots,x_p(t))$, with stationary cross-covariances, for $E(x_j(t)) = 0$, $r_{jk}(t) = E(x_j(s+t)\cdot\overline{x_k(s)}) = \overline{r_{kj}(-t)}$. If the processes are real, which we usually assume,

$$r_{jk}(t) = E(x_j(s+t)\cdot x_k(s)) = r_{kj}(-t). \tag{7.1}$$

For example, in a stationary differentiable process $x(t)$ with derivative $x'(t)$, the cross-covariance function is, by Theorem 2.3, $r_{x,x'}(t) = C(x(s+t),x'(s)) = -r_x'(t) = r_x'(-t) = r_{x',x}(-t)$, in agreement with (7.1).

7.1.1 *Spectral distribution*

The covariance function $\mathbf{R}(t) = (r_{jk}(t))$ is a matrix function of covariances, where each auto-covariance function $r_{kk}(t)$ has its marginal spectral representation $r_{kk}(t) = \int e^{i\omega t}\,\mathrm{d}F_{kk}(\omega)$. We will now investigate the spectral properties of the cross-covariance functions r_{jk}.

Theorem 7.1. *(a) To every continuous covariance matrix function* $\mathbf{R}(t)$ *there exists a spectral distribution* $\mathbf{F}(\omega)$ *such that*

$$\mathbf{R}(t) = \int_{-\infty}^{\infty} e^{i\omega t}\,\mathrm{d}\mathbf{F}(\omega).$$

The function $\mathbf{F}(\omega)$ *is of non-negative type, i.e., for every complex* $\mathbf{z} = (z_1,\dots,z_p)$, *and frequency interval* $\omega_1 < \omega_2$,

$$\sum_{j,k} z_j\overline{z_k}\,(F_{jk}(\omega_2) - F_{jk}(\omega_1)) \geq 0. \tag{7.2}$$

(b) If $F_{jj}(\omega)$, $F_{kk}(\omega)$ *are absolutely continuous with spectral densities* $f_{jj}(\omega)$, $f_{kk}(\omega)$, *then* $F_{jk}(\omega)$ *is absolutely continuous, with density* $f_{jk}(\omega)$ *such that*

$$|f_{jk}(\omega)|^2 \leq f_{jj}(\omega)\,f_{kk}(\omega).$$

The relation (7.2) says that $\mathbf{\Delta F}(\boldsymbol{\omega}) = (F_{jk}(\omega_2) - F_{jk}(\omega_1))$ is a non-negative definite Hermite matrix for each choice of ω_1, ω_2.

Proof. (a) Take the complex vector $\mathbf{z}$ and define the one-dimensional process

$$y(t) = \sum_j z_j x_j(t)$$

and use Bochner's theorem on its complex continuous covariance function

$$r^{\mathbf{z}}(t) = \sum_{jk} z_j \overline{z_k} r_{jk}(t) = \int_{-\infty}^{\infty} e^{i\omega t}\, \mathrm{d}G^{\mathbf{z}}(\omega),$$

where $G^{\mathbf{z}}(\omega)$ is a real, bounded, and non-decreasing spectral distribution function. We then take, in order, $z_j = z_k = 1$, and $z_j = i$, $z_k = 1$, the rest being 0. This gives two spectral distributions, $G_1(\omega)$ and $G_2(\omega)$, say, and we have

$$r_{jj}(t) + r_{kk}(t) + r_{jk}(t) + r_{kj}(t) = \int_{-\infty}^{\infty} e^{i\omega t}\, \mathrm{d}G_1(\omega),$$

$$r_{jj}(t) + r_{kk}(t) + i r_{jk}(t) - i r_{kj}(t) = \int_{-\infty}^{\infty} e^{i\omega t}\, \mathrm{d}G_2(\omega).$$

Together with $r_{jj}(t) = \int_{-\infty}^{\infty} e^{i\omega t}\, \mathrm{d}F_{jj}(\omega)$, $r_{kk}(t) = \int_{-\infty}^{\infty} e^{i\omega t}\, \mathrm{d}F_{kk}(\omega)$, we get

$$r_{jk}(t) + r_{kj}(t) = \int_{-\infty}^{\infty} e^{i\omega t}\, (\mathrm{d}G_1(\omega) - \mathrm{d}F_{jj}(\omega) - \mathrm{d}F_{kk}(\omega)),$$

$$i r_{jk}(t) - i r_{kj}(t) = \int_{-\infty}^{\infty} e^{i\omega t}\, (\mathrm{d}G_2(\omega) - \mathrm{d}F_{jj}(\omega) - \mathrm{d}F_{kk}(\omega)),$$

which implies

$$\begin{aligned} r_{jk}(t) &= \int_{-\infty}^{\infty} e^{i\omega t} \cdot \frac{1}{2}(\mathrm{d}G_1(\omega) - i\mathrm{d}G_2(\omega) - (1-i)(\mathrm{d}F_{jj}(\omega) + \mathrm{d}F_{kk}(\omega))) \\ &= \int_{-\infty}^{\infty} e^{i\omega t}\, \mathrm{d}F_{jk}(\omega), \text{ say,} \end{aligned}$$

which is the spectral representation of $r_{jk}(t)$.

It is easy to see that $\mathbf{\Delta F}(\omega)$ has the stated properties; in particular that

$$\sum_{jk} z_j \overline{z_k} \Delta F_{jk}(\omega) \geq 0. \tag{7.3}$$

(b) From (7.3), by taking only z_j and z_k to be non-zero, it follows that

$$|z_j|^2 \Delta F_{jj} + |z_k|^2 \Delta F_{kk} + 2\Re(z_j \overline{z_k} \Delta F_{jk}) \geq 0,$$

which implies that for any ω-interval, $|\Delta F_{jk}|^2 \leq \Delta F_{jj} \cdot \Delta F_{kk}$.Thus, if F_{jj} and F_{kk} have spectral densities, so does F_{jk} and $|f_{jk}(\omega)|^2 \leq f_{jj}(\omega) f_{kk}(\omega)$. □

For real-valued vector processes, the spectral distributions may be put in real form as for one-dimensional processes. In particular, the cross-covariance function can be written

$$r_{jk}(t) = \int_0^\infty \{\cos \omega t \, \mathrm{d}G_{jk}(\omega) + \sin \omega t \, \mathrm{d}H_{jk}(t)\}, \tag{7.4}$$

where $G_{jk}(\omega)$ and $H_{jk}(\omega)$ are real functions of bounded variation.

7.1.2 Spectral representation of $\mathbf{x}(t)$

The spectral components

Each component $x_j(t)$ in a stationary vector process has its spectral representation $x_j(t) = \int e^{i\omega t} \, \mathrm{d}Z_j(\omega)$ in terms of a spectral process $Z_j(\omega)$ with orthogonal increments. The existence of these spectral processes follows from the spectral Theorem 3.7.

Further, for $j \neq k$, the increments of $Z_j(\omega)$ and $Z_k(\omega)$ over disjoint frequency intervals are orthogonal,[1] i.e.,

$$E(\mathrm{d}Z_j(\omega) \cdot \overline{\mathrm{d}Z_k(\mu)}) = 0, \quad \omega \neq \mu. \tag{7.5}$$

That the increments at different frequencies are orthogonal (uncorrelated) is in line with the general property that different frequency components do not interact.

The cross-correlation between the components of $\mathbf{x}(t)$ are determined by the correlations between the spectral components of the separate processes,

$$E(\mathrm{d}Z_j(\omega) \cdot \overline{\mathrm{d}Z_k(\omega)}) = \mathrm{d}F_{jk}(\omega), \tag{7.6}$$

where the (generally complex) spectral distribution functions satisfy

$$\mathrm{d}F_{jk}(-\omega) = \overline{\mathrm{d}F_{jk}(\omega)}.$$

Phase, amplitude, and coherence spectrum

To avoid too heavy indexing, we restrict the discussion to two processes, $x(t)$ and $y(t)$. The spectral distribution F_{xy} may be complex, and therefore it contains both amplitude information and phase information, and to separate the two types of information, one makes the following definitions.

Let the cross-covariance function be $r_{xy}(t) = \int e^{i\omega t} \, \mathrm{d}F_{xy}(\omega)$, and assume

[1] Cf. Exercise 7.4. See also [22, Sect. 11.8].

that $F_{xy}(\omega)$ has absolutely continuous real and imaginary parts. Denote the density (note the sign of q_{xy}),

$$f_{xy}(\omega) = c_{xy}(\omega) - iq_{xy}(\omega) = A_{xy}(\omega)\, e^{i\Phi_{xy}(\omega)},$$

with $A_{xy}(\omega) \geq 0$, $-\pi < \Phi_{xy}(\omega) \leq \pi$. Then the following spectral functions are defined for $-\infty < \omega < \infty$.

Definition 7.1. *When* $f_{xy}(\omega) = c_{xy}(\omega) - iq_{xy}(\omega)$, *one defines*

co-spectrum, quadrature spectrum: $c_{xy}(\omega)$ *and* $q_{xy}(\omega)$, *respectively,*

cross-amplitude spectrum: $A_{xy}(\omega) = |f_{xy}(\omega)|$,

squared coherence spectrum: $\kappa_{xy}^2(\omega) = \frac{|f_{xy}(\omega)|^2}{f_x(\omega)f_y(\omega)} = \frac{A_{xy}(\omega)^2}{f_x(\omega)f_y(\omega)}$,

phase spectrum: $\Phi_{xy}(\omega)$.

Remark 7.1. *The nomenclature on cross-spectral functions is a little vague. Note that* $f_{xy}(-\omega) = \overline{f_{xy}(\omega)}$; *therefore, the cross-amplitude spectrum can also be defined as* $2A_{xy}(\omega), \omega \geq 0$. *In some older literature, the quadrature spectrum is defined as* $-q_{xy}(\omega)$. *Our definition agrees with that in [22].*

Case with discrete spectra

To see how the spectral representation helps to understand the structure of cross-correlation, consider two processes with discrete spectrum, located at frequencies $\omega_n > 0$, for which the spectral representation are sums of the form (3.44),

$$x(t) = \sum_n \sigma_x(n)\, \{U_x(n)\cos\omega_n t + V_x(n)\sin\omega_n t\},$$
$$y(t) = \sum_n \sigma_y(n)\, \{U_y(n)\cos\omega_n t + V_y(n)\sin\omega_n t\}.$$

Here $(U_x(n), V_x(n)) = (U_x(\omega_n), V_x(\omega_n))$, $(U_y(n), V_y(n)) = (U_y(\omega_n), V_y(\omega_n))$, are pairs of uncorrelated real random variables with mean 0, variance 1, uncorrelated for different n-values. Thus, each term in the sums consists of two components, a cosine and a sine component, which are $\pi/2$ out of phase with each other. The components have random amplitudes, and these are uncorrelated for different frequencies.

The covariance between the x- and the y-process is caused by the correlation between the U's and the V's *for the same frequency* in the two representations:

$$E(U_x(n)U_y(n)) = E(V_x(n)V_y(n)) = \rho_{xy}(\omega_n),$$
$$E(U_x(n)V_y(n)) = -E(V_x(n)U_y(n)) = -\widetilde{\rho}_{xy}(\omega_n),$$
$$E(U_x(n)V_x(n)) = E(U_y(n)V_y(n)) = 0,$$

for some $\rho_{xy}(\omega_n)$ and $\widetilde{\rho}_{xy}(\omega_n)$ such that $0 \le \rho_{xy}(\omega_n)^2 + \widetilde{\rho}_{xy}(\omega_n)^2 \le 1$. (This latter restriction comes from the zero correlation between $U(n)$ and $V(n)$ in each series, which prevents too strong correlation between the different subseries.)

The function $\rho_{xy}(\omega_n)$ gives the correlation between the components that are in phase with each other, i.e., the cosine and the sine components, respectively. The function $\widetilde{\rho}_{xy}(\omega_n)$ (with proper sign) gives the correlation between the out-of-phase components, i.e., between the cosine component in one process and the sine component in the other.

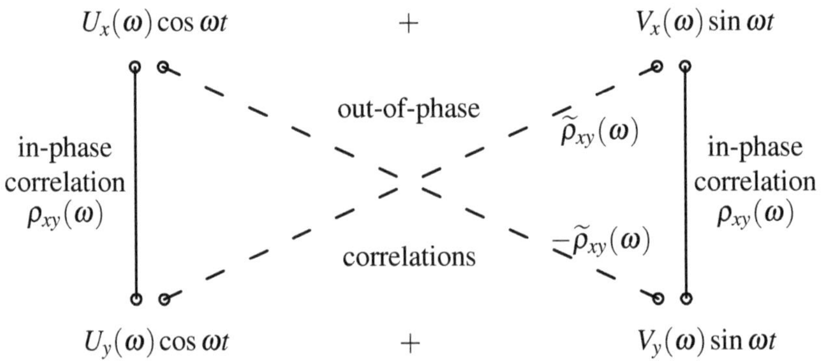

Figure 7.1: *Illustration of in-phase and out-of-phase correlations for frequency ω.*

Direct calculation of auto- and cross-covariances gives

$$r_{xy}(t) = \sum_n \sigma_x(n)\sigma_y(n)\,\{\rho_{xy}(\omega_n)\cos\omega_n t + \widetilde{\rho}_{xy}(\omega_n)\sin\omega_n t\}$$
$$= \sum_n A_{xy}(\omega_n)\,\cos(\omega_n t - \Phi_{xy}(\omega_n)), \tag{7.7}$$

$$r_x(t) = \sum_n \sigma_x(n)^2\cos\omega_n t, \quad r_y(t) = \sum_n \sigma_y(n)^2\cos\omega_n t, \tag{7.8}$$

where

$$A_{xy}(\omega_n) = \sigma_x(n)\sigma_y(n)\sqrt{\rho_{xy}(\omega_n)^2 + \widetilde{\rho}_{xy}(\omega_n)^2}$$

represents the covariances between amplitudes, while $\Phi_{xy}(\omega_n)$, with

$$\cos\Phi_{xy}(\omega_n) = \frac{\rho_{xy}(\omega_n)}{\sqrt{\rho_{xy}(\omega_n)^2+\widetilde{\rho}_{xy}(\omega_n)^2}},$$
$$\sin\Phi_{xy}(\omega_n) = \frac{\widetilde{\rho}_{xy}(\omega_n)}{\sqrt{\rho_{xy}(\omega_n)^2+\widetilde{\rho}_{xy}(\omega_n)^2}},$$

represents the phase relation.

If we write (7.7) in complex form,

$$r_{xy}(t) = \sum_n \frac{1}{2}A_{xy}(\omega_n)\,e^{-i\Phi_{xy}(\omega_n)}\,e^{i\omega_n t} + \sum_n \frac{1}{2}A_{xy}(\omega_n)\,e^{i\Phi_{xy}(\omega_n)}\,e^{-i\omega_n t},$$

we find that the cross-spectrum is concentrated to $\{\pm\omega_n\}$, with jumps of size $\Delta F_{xy}(\omega)$, equal to

$$\begin{cases} \frac{1}{2}A_{xy}(\omega_n)\,e^{-i\Phi_{xy}(\omega_n)} = \frac{1}{2}A_{xy}(\omega_n)\,\dfrac{\rho_{xy}(\omega_n)+i\widetilde{\rho}_{xy}(\omega_n)}{\sqrt{\rho_{xy}(\omega_n)^2+\widetilde{\rho}_{xy}(\omega_n)^2}}, & \omega=\omega_n, \\ \frac{1}{2}A_{xy}(\omega_n)\,e^{i\Phi_{xy}(\omega_n)} = \frac{1}{2}A_{xy}(\omega_n)\,\dfrac{\rho_{xy}(\omega_n)-i\widetilde{\rho}_{xy}(\omega_n)}{\sqrt{\rho_{xy}(\omega_n)^2+\widetilde{\rho}_{xy}(\omega_n)^2}}, & \omega=-\omega_n. \end{cases} \tag{7.9}$$

Definition 7.2. *Let $\{x(t), t\in\mathbb{R}\}$ and $\{y(t), t\in\mathbb{R}\}$ be jointly stationary processes with cross-covariance function $r_{xy}(t) = \int e^{i\omega t}\,\mathrm{d}F_{xy}(\omega)$.*

If the spectrum is discrete, with jumps $\Delta F_{xy}(\omega)$, given by (7.9), then $A_{xy}(\omega_n)$, $\omega_n \geq 0$, is called the cross-amplitude spectrum and $\Phi_{xy}(\omega_n)$ is the phase spectrum. The squared coherence spectrum is defined as

$$\frac{|\Delta F_{jk}(\omega_n)|^2}{\Delta F_{jj}(\omega_n)\Delta F_{kk}(\omega_n)} = \rho_{jk}(\omega_n)^2+\widetilde{\rho}_{jk}(\omega_n)^2.$$

Example 7.1 (Derivative, ctd.). *The cross-covariance function between a stationary process and its derivative has a cross-spectral representation,*

$$r_{xx'}(t) = -r_x'(t) = -\int i\omega e^{i\omega t}\,\mathrm{d}F_x(\omega),$$

or, for short,

$$\mathrm{d}F_{xx'}(\omega) = -i\omega\,\mathrm{d}F_x(\omega), \quad f_{xx'}(\omega) = \omega e^{-i\pi/2} f_x(\omega),$$

leading to

$$A_{xx'}(\omega) = |\omega| f_x(\omega), \quad \kappa^2_{xx'}(\omega) = \frac{\omega^2 f_x^2(\omega)}{f_x(\omega)\,\omega^2 f_x(\omega)} = 1,$$

$$\Phi_{xx'}(\omega) = -\pi/2, \quad \textit{for } \omega > 0.$$

The in-phase correlation is zero, $\rho_{xx'}(\omega) = 0$, and the out-of-phase correlation is one, $\overline{\rho}_{xx'}(\omega) = 1$, because a cosine term $A\cos(\omega t + \phi)$ in $x(t)$ becomes $-A\omega\sin(\omega t + \phi) = -A\omega\cos(\omega t + \phi - \pi/2)$ in $x'(t)$, and with a similar relation for the sine term in $x(t)$. The spectral relations can be interpreted as follows:

- *Phase spectrum constant and equal to $-\pi/2$ means that each frequency component in $x(t)$ is also present in $x'(t)$, but with a phase shift of $-\pi/2$.*
- *Coherence spectrum equal to 1 means that the relation between the amplitudes for different components in $x(t)$ and $x'(t)$ is deterministic: amplitudes in $x'(t)$ are, for each frequency, directly proportional to those in $x(t)$.*
- *Cross-amplitude spectrum is frequency dependent by a factor ω means that the proportionality factor is ω.*

7.1.3 Cross-correlation in linear filters

In a linear filter, the cross-covariance and cross-spectrum describe the relation between the input process $x(t) = \int e^{i\omega t}\,\mathrm{d}Z_x(\omega)$ and the output process $y(t) = \int_{-\infty}^{\infty} h(t-u)x(u)\,\mathrm{d}u = \int_{-\infty}^{\infty} g(\omega)e^{i\omega t}\,\mathrm{d}Z_x(\omega)$. One easily obtains

$$r_{xy}(t) = E(x(s+t)\cdot\overline{y(s)}) = E\left(\int_{-\infty}^{\infty} e^{i\omega(s+t)}\,\mathrm{d}Z_x(\omega)\cdot\overline{\int_{-\infty}^{\infty} g(\mu)e^{i\mu s}\,\mathrm{d}Z_x(\mu)}\right)$$
$$= \int_{-\infty}^{\infty} e^{i\omega t}\overline{g(\omega)}\,\mathrm{d}F_x(\omega),$$

so the cross-spectral distribution is

$$\mathrm{d}F_{xy}(\omega) = \overline{g(\omega)}\,\mathrm{d}F_x(\omega), \quad f_{xy}(\omega) = \overline{g(\omega)}\,f_x(\omega). \tag{7.10}$$

We can expand this input-output relation by adding an observation noise process, $\{z(t), t \in \mathbb{R}\}$, independent of $\{x(t), t \in \mathbb{R}\}$, so the general filter relation is

$$y(t) = \int_{-\infty}^{\infty} h(t-u)x(u)\,\mathrm{d}u + z(t) = \int_{-\infty}^{\infty} h(u)x(t-u)\,\mathrm{d}u + z(t), \tag{7.11}$$

illustrated in the following diagram.

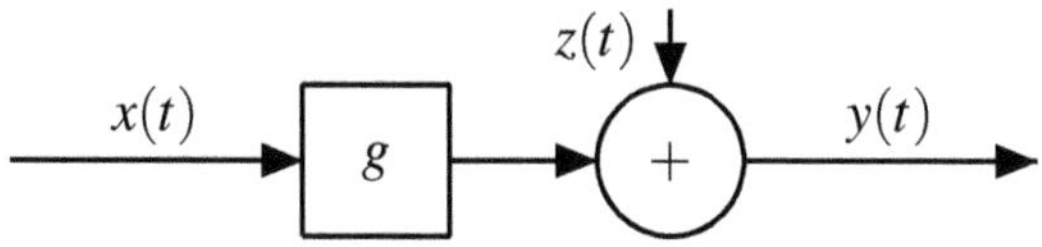

The following relations between the input and output signal characteristics are immediate consequences of (4.7) on page 121.

Theorem 7.2. *Let $\{y(t), t \in \mathbb{R}\}$ be the output of the disturbed filter (7.11) and let $f_x(\omega)$ be the spectral density of the input $\{x(t), t \in \mathbb{R}\}$. Let the impulse response function be $h(u)$ and the frequency function $g(\omega) = \int_{-\infty}^{\infty} e^{-i\omega u} h(u)\,du$. Then,*

a) The cross-covariance between input and output is

$$r_{xy}(t) = \int_{-\infty}^{\infty} h(u) r_x(t-u)\,du = \int_{-\infty}^{\infty} e^{i\omega t}\,\overline{g(\omega)} f_x(\omega)\,d\omega;$$

b) The cross-spectral density between input and output signal is

$$f_{xy}(f) = \overline{g(\omega)} f_x(\omega).$$

Remark 7.2. *The relation $f_{xy}(\omega) = \overline{g(\omega)} f_x(\omega)$ can be used to estimate the frequency function $g(\omega)$ of a linear filter. An input signal $x(t)$, with suitable frequency properties, is passed through the filter, giving an output $y(t)$. By means of frequency analysis (see, for example, [22] and [82, Chapter 9]), one then estimates the input spectral density and the cross-spectral density between input and output. The estimate of the filter frequency function is then $g^*(\omega) = \overline{f_{xy}^*(\omega)} / f_x^*(\omega) = f_{xy}^*(-\omega)/f_x^*(\omega)$, when $f_x^*(\omega) > 0$. It is important that the input contains enough power at the interesting frequencies, and that the estimates of input spectrum and cross-spectrum are reliable.*

Equation 7.11 and its schematic illustration give the impression that the variation in the y-process is directly *caused* by the x-process, except for the disturbances, of course. However, this is not a correct interpretation, since every pair of correlated processes can be written in the form of (7.11). An appealing way to illustrate a *common cause* for correlation between two processes $y_1(t)$ and $y_2(t)$ is the following diagram.

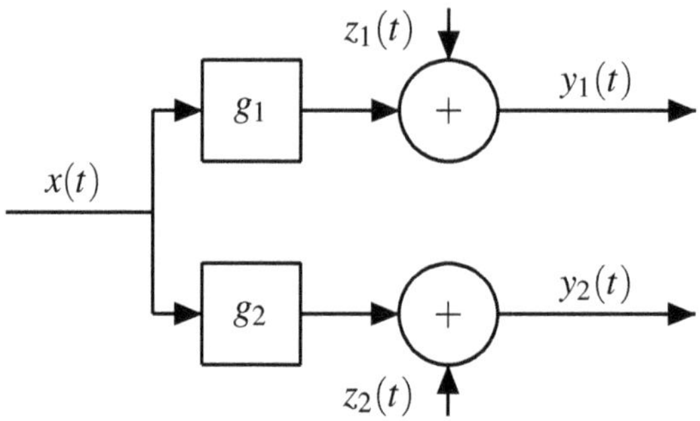

The covariance and spectral relations between $y_1(t)$ and $y_2(s)$ are found by the general formula (4.7) for covariance between integrals; for example

$$f_{y_1y_2}(\omega) = g_1(\omega)\overline{g_2(\omega)}\, f_x(\omega).$$

Of course, no conclusion about causality can be drawn from such correlation studies.

7.2 Some random field theory

A random field is a stochastic process $x(\mathbf{t})$ with multi-dimensional parameter

$$\mathbf{t} = (t_1, \dots, t_p).$$

For example, if $\mathbf{t} = (t_1, t_2)$ is two-dimensional, we can think of $(t_1, t_2, x(t_1, t_2))$ as a random surface with height $x(t_1, t_2)$ above the mean level at location with co-ordinates (t_1, t_2). A time-dependent random surface is a field $(t, s_1, s_2, x(t, s_1, s_2))$ with $\mathbf{t} = (t, s_1, s_2)$, where t is time and $(s_1, s_2) \in \mathbb{R}^2$ is location. In the general theory we use $\mathbf{t}$ as generic notation for the parameter; in special applications to random time dependent surfaces we use $(t, (s_1, s_2))$ as parameters.

Random fields have been widely used in very diverse areas, and the mathematical and statistical tools have been developed from rather different needs, often with little connection with each other. Important areas include geoscience, like geology under the name of *geostatistics*, marine science for random sea models, atmospheric science for turbulence models, radio and communication engineering, optics and image analysis. The mathematical theory for correlation and spectral properties generalizes that for one-dimensional parameter processes, and it also offers some new features of geometrical character.

Random fields are now widely used in all sorts of applications with spatial or spatio-temporal variability. Recent advances combine spectral methods (which dominate in most of this book) and spatial Markov models.

The two volume work by Yaglom [126] is almost complete on the fundamental mathematical theory, and it also contains many references. Examples on model and statistical aspects are the books by Cressie and Wikle [36], Rue and Held [107], and Stein [117], and the work by Lindgren et al. [76]. For modern accounts of mathematical and probabilistic aspects, see the books by Adler and Taylor [3], and, also for some applications, Azaïs and Wschebor [6]. The book by Adler [2] is still a standard reference.

7.2.1 Homogeneous fields

Define the mean value and covariance functions for random fields in the natural way as $m(\mathbf{t}) = E(x(\mathbf{t}))$ and $r(\mathbf{t},\mathbf{u}) = C(x(\mathbf{t}),x(\mathbf{u}))$.

The common name for the field analogue of a stationary process is a homogeneous field, even if the term "stationary" also is accepted. The field is called *homogeneous* if $m(\mathbf{t})$ is constant m and $r(\mathbf{t},\mathbf{u})$ depends only on the difference $\mathbf{t}-\mathbf{u}$, i.e., assuming a real field with $m=0$,

$$r(\mathbf{t}) = r(\mathbf{u}+\mathbf{t},\mathbf{u}) = E(x(\mathbf{u}+\mathbf{t}) \cdot x(\mathbf{u})).$$

The covariance of the process values at two locations depends on *distance* as well as on *direction* of the vector between the two points.

In spatial applications it is popular to use the *variogram*, defined by

$$2\gamma(\mathbf{u},\mathbf{v}) = E\left(|x(\mathbf{u}) - x(\mathbf{v})|^2\right),$$

or the *semi-variogram* $\gamma(\mathbf{u},\mathbf{v})$. The variogram plays the same role as the incremental variance does in a Wiener process, when $E(|w(s+t)-w(s)|^2) = t\cdot\sigma^2$ is independent of s.

A field for which the variogram only depends on the vector $\mathbf{u}-\mathbf{v}$ is *intrinsically stationary* in the sense of Definition 4.9. The semi-variogram for a homogeneous field is $\gamma(\mathbf{t}) = r(\mathbf{0}) - r(\mathbf{t})$. A homogeneous field is also intrinsically stationary, but as seen from the Wiener process the converse is not sure. The term *intrinsically stationary* has been used for random fields ever since it was introduced by Matheron in 1973 [90]. For processes with one-dimensional parameter we use a similar property under the name *stationary increments*. It is notable that in random field theory one often chooses to neglect the absolute level (average) of the field, and to concentrate on the variability, represented by the variogram. We shall discuss some aspects of this later in this section.

There are some terms which are used in many applications of random field theory, which stem from geostatistics,

nugget effect: an extra variance term (or a peaked component of the covariance function) that represents observation error or small scale variability;

sill: the limit of the variogram at large distance, usually equal to twice the variance;

range: a distance where the variogram reaches some fraction of the sill, often the distance where the covariance is a small proportion of the variance.

Examples of homogeneous spatial random fields are abundant: a segment of a forest image taken from a satellite, the internal grain structure of a piece of concrete, turbulence, local cloud and rainfall data, a homogeneous image segment without directional structure disturbed by noise, and the spatial distribution of interstellar material in a galaxy, to mention a few.

The covariance function of a homogeneous field obeys a similar spectral representation as that of a stationary process. For references on the following two theorems, see the extensive survey by Yaglom [126, Ch. 4].

Theorem 7.3 (Multi-dimensional Bochner's theorem). *A continuous function $r(\mathbf{t})$, $\mathbf{t} \in \mathbb{R}^p$, is a covariance function of a homogeneous random field if and only if there exists a p-dimensional spectral distribution function,*[2] *$F(\boldsymbol{\omega})$, depending on the frequency parameter $\boldsymbol{\omega} = (\omega_1, \dots, \omega_p)$, such that*

$$r(\mathbf{t}) = \int e^{i\boldsymbol{\omega}\cdot\mathbf{t}}\,\mathrm{d}F(\boldsymbol{\omega}), \tag{7.12}$$

where $\boldsymbol{\omega}\cdot\mathbf{t} = \omega_1 t_1 + \ldots + \omega_p t_p$. A real field has a symmetric spectral distribution, which may be absolutely continuous with spectral density $f(\boldsymbol{\omega})$.

In the same way as a stationary process can be generated as a sum or integral of cosine functions with random amplitudes and uniform random phases, a homogeneous field is a sum of spatial cosine function, like corrugated sheets, with different orientations. The total variance is divided over

[2] $F(\boldsymbol{\omega})$ is equal to a regular p-dimensional probability distribution function multiplied by a positive constant, equal to the variance of the process. Note that we here use the frequency terminology and symbol $\boldsymbol{\omega}$ also for processes in the plane and higher dimensional space, instead of the more correct terminology, *wave number*, with symbol $\mathbf{k}$ or $\boldsymbol{\kappa}$.

different frequencies, or rather wave numbers in space, and directions according to the spectral distribution $F(\boldsymbol{\omega})$. Contributions from different frequencies and directions have uncorrelated, even independent, amplitudes and phases.

Theorem 7.4. *A homogeneous stochastic process can be represented as a stochastic Fourier integral. There exists a stochastic spectral process* $Z(\boldsymbol{\omega})$ *with orthogonal increments* $\Delta Z(\boldsymbol{\omega})$ *over rectangles* $\Delta\boldsymbol{\omega} = [\omega_1, \omega_1 + \Delta_1] \times \ldots \times [\omega_p, \omega_p + \Delta_p]$, *such that* $E(\Delta Z(\boldsymbol{\omega})) = 0$ *and*

$$E(|\Delta Z(\boldsymbol{\omega})|^2) = \Delta F(\boldsymbol{\omega}), \quad \textit{and} \quad E(\Delta Z(\boldsymbol{\omega}_1) \cdot \overline{\Delta Z(\boldsymbol{\omega}_2)}) = 0,$$

for disjoint rectangles $\Delta\boldsymbol{\omega}_1$ *and* $\Delta\boldsymbol{\omega}_2$, *and*

$$x(\mathbf{t}) = \int e^{i\boldsymbol{\omega}\cdot\mathbf{t}} \, \mathrm{d}Z(\boldsymbol{\omega}). \tag{7.13}$$

Smoothness properties

Conditions for continuity, differentiability, etc., for random fields are very similar to those for one-dimensional processes. For example, a homogeneous field is continuous in quadratic mean if the covariance function is continuous at the origin, and the basic theorem, Theorem 2.4, has the following, mild twist for fields; see [6, Thm. 1.14].

Theorem 7.5. *Let* $\{x(\mathbf{t}), \mathbf{t} \in [0,1]^p\}$ *be a random field with p-dimensional parameter. If there exist two non-decreasing functions,* $g(h)$ *and* $q(h)$, $0 \le h \le 1$, *such that*

$$\sum_{n=1}^{\infty} g(2^{-n}) < \infty, \quad \sum_{n=1}^{\infty} 2^{pn} q(2^{-n}) < \infty,$$

and, for all $\mathbf{t} \in [0,1]^p$ *and* $\mathbf{t}+\mathbf{h} \in [0,1]^p$, *with* $\overline{h} = \max h_k$,

$$P(|x(\mathbf{t}+\mathbf{h}) - x(\mathbf{t})| \ge g(\overline{h})) \le q(\overline{h}), \tag{7.14}$$

then there exists an equivalent stochastic process $\{y(\mathbf{t}), \mathbf{t} \in [0,1]^p\}$ *whose sample paths are continuous on* $[0,1]^p$.

Moment conditions like those in Section 2.2.3 hold with small changes. In Corollary 2.1 the right hand side bound in (2.10) is changed according to dimension to $C|\overline{h}|^p/|\log|\overline{h}||^{1+r}$. For Gaussian fields, condition (2.16) in Theorem 2.6 is unchanged, with $\overline{h}$ instead of h; see [6] for further details on vector valued fields and on Hölder continuity. Similarly as for processes, Remark 2.1, a stationary Gaussian field defined on a compact set $\mathbf{T} \subset \mathbb{R}^p$, is either continuous or unbounded [50]. Further, with the entropy $N(\mathbf{T}, \varepsilon)$ defined as in Theorem 2.15, it is continuous if and only if

$$\int_0^\infty \sqrt{\log N(\mathbf{T}, \varepsilon)}\, \mathrm{d}\varepsilon < \infty.$$

Derivatives and spectral moments

Partial derivatives, $x_k(\mathbf{t}) = \partial x(\mathbf{t})/\partial t_k$ are obtained as derivatives along the coordinate axes and the one-dimensional rules hold for existence. Variance and covariances for the derivatives depend on the multiple spectral moments. With $\boldsymbol{\alpha} = (\alpha_1, \ldots, \alpha_p)$ and $\boldsymbol{\omega}^{\boldsymbol{\alpha}} = \prod_k \omega_k^{\alpha_k}$, the spectral moments are defined as

$$\omega_{\boldsymbol{\alpha}} = \int \boldsymbol{\omega}^{\boldsymbol{\alpha}}\, \mathrm{d}F(\boldsymbol{\omega}),$$

and the following relations are easily derived; [2, Section 2.4].

Theorem 7.6. *The partial derivatives of a homogeneous field have auto- and cross-covariance functions (with ω-index indicating appropriate integration variables),*

$$r_{x_k}(\mathbf{t}) = -\frac{\partial^2 r_x(\mathbf{t})}{\partial t_k^2}, \quad \text{with } V(x_k(\mathbf{t})) = \omega_{(0,\ldots,2,\ldots,0)},$$

$$r_{x_j,x_k}(\mathbf{t}) = -\frac{\partial^2 r_x(\mathbf{t})}{\partial t_j \partial t_k}, \quad \text{with } C(x_j(\mathbf{t}), x_k(\mathbf{t})) = \omega_{(0,\ldots,1,\ldots,1,\ldots,0)}.$$

Spatial Gaussian white noise

Gaussian white noise in space is introduced in the same way as Gaussian white noise in one dimension, namely via its correlation integral, (2.37). Let $\{w(\mathbf{t}), \mathbf{t} \in \mathbb{R}^p\}$ be a p-parametric Gaussian field such that the integrals

$$\int_{A_k} g_k(\mathbf{t}) w'(\mathbf{t})\, \mathrm{d}\mathbf{t} = \int_{A_k} g_k(\mathbf{t})\, \mathrm{d}w(\mathbf{t}), \quad k = 1, \ldots, n,$$

are zero mean jointly Gaussian random variables with covariances that satisfy

$$C\left(\int_A g(\mathbf{s})w'(\mathbf{s})\,\mathrm{d}\mathbf{s}, \int_B h(\mathbf{t})w'(\mathbf{t})\,\mathrm{d}\mathbf{t}\right) = \int_{A\cap B} g(\mathbf{t})\overline{h(\mathbf{t})}\,\mathrm{d}\mathbf{t}. \qquad (7.15)$$

Then, $w'(\mathbf{t})$ is called p-dimensional Gaussian white noise with constant spectral density. It is used only in connection with linear filters.

7.2.2 *Isotropic fields: covariance properties*

In a homogeneous isotropic field, the correlation properties are the same in all directions, i.e., the covariance function $r(\mathbf{t})$ depends only on the distance $\|\mathbf{t}\| = \sqrt{t_1^2 + \ldots + t_p^2}$. This type of process model is natural when one can not identify any special directional dependent stochastic properties in the field, which rather keeps its distributions after rotation and translation. Many phenomena in the natural world share this property while others naturally do not. Ocean waves are non-isotropic; chemical concentration in the bottom sediment in the ocean or the disturbance field in mobile phone communication might well be isotropic.

As we have seen, a spectral distribution for a homogeneous field need only satisfy the requirement that it is non-negative, integrable, and symmetric. The spectral distribution for an isotropic field needs to satisfy a special invariance condition, giving it a particular structure. To formulate the result, define the *Bessel functions of the first kind* of order $m = 0, 1/2, 1, \ldots$, as

$$J_m(z) = \frac{(z/2)^m}{\sqrt{\pi}\,\Gamma(m+1/2)} \int_0^\pi \cos(z\cos\theta)\,\sin^{2m}(\theta)\,\mathrm{d}\theta.$$

We also introduce the notation

$$\Lambda_p(z) = 2^{(p-2)/2}\,\Gamma(p/2)\,\frac{J_{(p-2)/2}(z)}{z^{(p-2)/2}}, \quad \Lambda_p(0) = \lim_{z\downarrow 0} \Lambda(z) = 1.$$

Theorem 7.7. *The covariance function $r(\mathbf{t})$ of a homogeneous isotropic field $\{x(\mathbf{t}), \mathbf{t} \in \mathbb{R}^p\}$, $p \geq 2$, is a mixture of Bessel functions, expressed as*

$$r(\mathbf{t}) = r(\|\mathbf{t}\|) = \int_0^\infty \Lambda_p(\omega \cdot \|\mathbf{t}\|)\,\mathrm{d}G(\omega), \qquad (7.16)$$

where $G(\omega)$ is a bounded, non-decreasing function, with $G(\infty) - G(0) = r(0)$.

Proof. We have $r(\mathbf{t}) = \int e^{i\boldsymbol{\omega}\cdot\mathbf{t}}\,\mathrm{d}F(\boldsymbol{\omega})$. To simplify the integral, we introduce spherical coordinates, i.e., $\boldsymbol{\omega} = \omega \cdot (\ell_1, \dots, \ell_p)$, $\omega = \|\boldsymbol{\omega}\|$, $\sum \ell_k^2 = 1$, and let $\ell_1 = \cos\theta_{p-1}$ with $0 \le \theta_{p-1} \le \pi$. For every θ_{p-1}, $(\ell_2, \dots, \ell_p)$ defines a point on a sphere with radius $\sqrt{1-\cos^2\theta_{p-1}} = \sin\theta_{p-1}$.

Since $r(\mathbf{t})$ depends only on $\|\mathbf{t}\|$, we can calculate the integral for the special point $\mathbf{t} = (t, 0, \dots, 0)$, to get

$$r(\mathbf{t}) = \int e^{i\omega_1 t}\,\mathrm{d}F(\boldsymbol{\omega}) = \int e^{i\omega t\cos\theta_{p-1}}\,\mathrm{d}F(\boldsymbol{\omega}).$$

With $G(\omega) = \int_{\|\boldsymbol{\omega}\|\le\omega} \mathrm{d}F(\boldsymbol{\omega})$, we find that

$$r(t) = \int_{\omega=0}^{\infty} \left\{ \int_{\boldsymbol{\theta}} e^{i\omega t\cos\theta_{p-1}}\,\mathrm{d}\sigma(\boldsymbol{\theta}) \right\} \mathrm{d}G(\omega),$$

where $\mathrm{d}\sigma$ is the area measure on the unit sphere in dimension p. For fixed θ_{p-1}, integrate the constant function $e^{i\omega t\cos\theta_{p-1}}$ over the $(p-1)$-dimensional sphere with radius $\sin\theta_{p-1}$, which has area $C_{p-1}\sin^{p-2}\theta_{p-1}$, to find

$$\begin{aligned} r(t) &= \int_{\omega=0}^{\infty} C_{p-1} \left\{ \int_{\theta=0}^{\pi} e^{i\omega t\cos\theta}\,\sin^{p-2}\theta\,\mathrm{d}\theta \right\} \mathrm{d}G(\omega) \\ &= \int_0^{\infty} \Lambda_p(\omega t)\,\mathrm{d}G(\omega), \end{aligned}$$

if we merge C_{p-1} with the constant in Λ_p, using $\Lambda_p(0) = 1$. □

Isotropic fields with special structure

Of course, a homogeneous field $x(t_1,t_2)$ with two-dimensional parameter and spectral density $f_x(\omega_1,\omega_2)$ and covariance function $r_x(t_1,t_2)$ gives rise to a stationary process when observed along a single straight line, for example along $t_2 = 0$. Then $x(t_1,0)$ has covariance function $r(t_1) = r_x(t_1,0)$ and spectral density $f(\omega_1) = \int_{\omega_2} f_x(\omega_1,\omega_2)\,\mathrm{d}\omega_2$. In fact, every one-dimensional covariance function/spectral density can appear as "marginal covariance function/spectral density" in a multi-dimensional homogeneous field.

It is important to realize that even if any non-negative definite function can act as covariance function for a stationary process, not every stationary process, and corresponding covariance function, can occur as marginal process/covariance function of a section in an *isotropic* field in higher dimension, only those which satisfy (7.16). The functions $\Lambda_p(\omega t)$ play a similar role in the spectral decomposition of $r(t)$ as $\cos\omega t$ in (3.12) for the one-dimensional case.

For fields in the plane, with $p = 2$, only those functions are valid covariance functions that can be expressed as

$$r(\mathbf{t}) = \int_0^\infty J_0(\omega\|\mathbf{t}\|)\,\mathrm{d}G(\omega), \tag{7.17}$$

with bounded non-decreasing $G(\omega)$.

One should also note the particularly simple form of the covariance function for the case $p = 3$,

$$r(\mathbf{t}) = \int_0^\infty \frac{\sin(\omega\|\mathbf{t}\|)}{\omega\|\mathbf{t}\|}\,\mathrm{d}G(\omega). \tag{7.18}$$

Example 7.2 (Mono-chromatic spatial wave). *Mono-chromatic waves in the plane (or any higher dimension) are waves with a single wave frequency but with random direction. In one dimension there is only one random amplitude and one phase to determine before the entire process is defined as a cosine function, and not much randomness can be observed. In the plane there is more variability, since each direction has its own amplitude and phase.*

For a mono-chromatic spatial wave with frequency ω_0, the G-function in (7.17) has one single jump $\Delta G_0 = \sigma^2$, located at $\omega = \omega_0$ and is otherwise constant. The covariance function is therefore $r(\mathbf{t}) = \sigma^2 J_0(\omega_0\|\mathbf{t}\|)$.

We will derive this covariance by direct calculation from the spectral properties of a mono-chromatic wave. Consider a mono-chromatic wave in the plane with frequency $\omega_0 = 1$, for simplicity. This means that for each direction $\theta \in [0, 2\pi)$ there is one wave component

$$A\cos(t_1\cos\theta + t_2\sin\theta + \phi),$$

with a random (infinitesimal) amplitude A and a uniform phase $\phi \in (0, 2\pi)$. When observed along the t_1-axis, the observed wavelength will be $1/|\cos\theta|$, and frequency $|\cos\theta|$; draw a figure! If the wave field is isotropic, the average energy in each direction θ is the same, and the total energy is concentrated uniformly along the unit circle ($\omega_0 = 1$) with density $\sigma^2/(2\pi)$. Now, when θ is random and uniformly distributed in $(0, 2\pi)$, the probability density of $|\cos\theta|$ is $f_{|\cos\theta|}(y) = \frac{2}{\pi\sqrt{1-y^2}}$, and the one-sided spectral density for $x(t_1, 0)$ is therefore

$$f_1(\omega) = \frac{2\sigma^2}{\pi\sqrt{1-\omega^2}}, \quad 0 < \omega < 1.$$

Now, one of the alternative characterizations of the Bessel function J_0 is

$J_0(t) = \frac{2}{\pi}\int_0^1 \frac{\cos \omega t}{\sqrt{1-\omega^2}}\,\mathrm{d}\omega$ *(see [55, 8.41(8)]), from which we get the desired result,*

$$r(t_1,0) = \int_0^1 f_1(\omega)\cos \omega t_1\,\mathrm{d}\omega = \sigma^2 J_0(t_1).$$

Lower bound on covariance

Note the lower bounds on the correlations in isotropic fields that follow from Theorem 7.7 and the explicit formulas (7.17-7.18). In two dimensions, the minimum correlation is $\inf_x J_0(x) \approx -0.403$ and in three dimensions it is $\inf_x \frac{\sin x}{x} \approx -0.218$. The lower bound increases with dimension and approaches 0 as $p \to \infty$. As a simple rule of thumb one can use $-1/p$ as a lower bound for the correlation.

7.2.3 Isotropic fields: spectral properties

The spectral distribution for an isotropic field inherits the simple rotational invariance structure of the covariance function. We will describe some properties and relations between different types of spectral densities, which exist if, for example,

$$\int_0^\infty t^{p-1}|r(t)|\,\mathrm{d}t < \infty. \tag{7.19}$$

First of all we have to make clear some definitions of "spectral density."

Three definitions of spectral density

p-dimensional spectral density: the density $f(\boldsymbol{\omega})$ in the representation (7.12),

$$r(\mathbf{t}) = \int_{\boldsymbol{\omega}} e^{i\boldsymbol{\omega}\cdot\mathbf{t}} f(\boldsymbol{\omega})\,\mathrm{d}\boldsymbol{\omega} = \int_{\boldsymbol{\omega}} \cos(\boldsymbol{\omega}\cdot\mathbf{t})\, f(\boldsymbol{\omega})\,\mathrm{d}\boldsymbol{\omega}.$$

For an isotropic field, $f(\boldsymbol{\omega})$ is a function of $\|\boldsymbol{\omega}\| = \omega$, and we denote it $f_p(\omega) = f(\boldsymbol{\omega})$;

one-dimensional (Fourier) spectral density: the density

$$f_1(\omega_1) = \int_{\omega_2}\cdots\int_{\omega_p} f(\boldsymbol{\omega})\,\mathrm{d}\omega_2\ldots\mathrm{d}\omega_p$$

in the representation

$$\begin{aligned} r((t_1,0,\ldots,0)) &= \int_{-\infty}^{\infty} e^{i\omega_1 t_1} f_1(\omega_1)\,\mathrm{d}\omega_1 \\ &= 2\int_0^\infty \cos(\omega_1 t_1)\, f_1(\omega_1)\,\mathrm{d}\omega_1; \end{aligned}$$

p-dimensional (Bessel) spectral density: the density $g(\omega) = G'(\omega)$ in (7.16),

$$r(\mathbf{t}) = \int_0^\infty \Lambda_p(\omega\,\|\mathbf{t}\|)\,g(\omega)\,\mathrm{d}\omega, \tag{7.20}$$

and equal to

$$g(\omega) = \frac{2\pi^{p/2}}{\Gamma(p/2)}\omega^{p-1}f(\omega). \tag{7.21}$$

The relation (7.21) follows from the definition $G(\omega) = \int_{\|\boldsymbol{\omega}\|\le\omega} \mathrm{d}F(\boldsymbol{\omega})$, and the formula for the surface area of a p-dimensional sphere with radius ω.

Theorem 7.8. *The spectral density $f(\boldsymbol{\omega})$ for an isotropic random field with covariance function satisfying* (7.19) *is given by (with $\omega = \|\boldsymbol{\omega}\|$)*

$$f(\boldsymbol{\omega}) = f(\omega) = \frac{1}{(2\pi)^{p/2}}\int_0^\infty \frac{J_{(p-2)/2}(\omega t)}{(\omega t)^{(p-2)/2}}\,t^{p-1}r(t)\,\mathrm{d}t, \tag{7.22}$$

with the inverse relation (with $t = \|\mathbf{t}\|$),

$$r(\mathbf{t}) = r(t) = (2\pi)^{p/2}\int_0^\infty \frac{J_{(p-2)/2}(\omega t)}{(\omega t)^{(p-2)/2}}\,\omega^{p-1}f(\omega)\,\mathrm{d}\omega. \tag{7.23}$$

Proof. Formula (7.22) follows by arguments like those for Theorem 7.7. Formula (7.23) is (7.21) rephrased. □

For easy reference, we give the explicit relations for dimensions two and three.

Corollary 7.1. *The covariance/spectrum relations for isotropic fields in two and three dimensions are*

$$f(\omega) = \frac{1}{2\pi}\int_0^\infty J_0(\omega t)\,t r(t)\,\mathrm{d}t, \quad \textit{for } p = 2,$$
$$f(\omega) = \frac{1}{2\pi^2}\int_0^\infty \frac{\sin\omega t}{\omega t}\,t^2 r(t)\,\mathrm{d}t, \quad \textit{for } p = 3.$$

The Matérn field

Theorem 7.7 gives all possible isotropic covariance functions valid in dimension p. Every isotropic random field $x(\mathbf{t})$ can generate a stationary process in lower dimension, e.g., $y(t) = x(t, 0, \ldots, 0)$, with the same covariance, $r_y(t) = r_x(t, 0, \ldots, 0)$, and hence every isotropic covariance function in $\mathbb{R}^p$ is also a covariance function in $\mathbb{R}$, but as we have just seen, the opposite is not true. In particular, a function that can be used as a covariance function in all dimensions, such as

$$r(\|\mathbf{t}\|) = \sigma^2 \exp(-\phi \|\mathbf{t}\|^{\alpha}),$$

$\alpha \in (0, 2]$, is necessarily non-negative.

Another useful class of covariance functions valid in any dimension is the *Whittle-Matérn family*, or, for short, Matérn family, after the Swedish forest statistician Bertil Matérn [89]. We met the Matérn covariance function already in Example 4.11, (4.58), and it has the explicit form

$$r(\|\mathbf{t}\|) = \frac{\sigma^2}{A} \cdot \frac{1}{2^{\nu-1}\Gamma(\nu)} (\kappa \|\mathbf{t}\|)^{\nu} K_{\nu}(\kappa \|\mathbf{t}\|), \tag{7.24}$$

where K_{ν} is a *modified Bessel function of the second kind* of order $\nu > 0$, and

$$A = \frac{\Gamma(\nu + p/2)}{\Gamma(\nu)} (4\pi)^{p/2} \kappa^{2\nu}.$$

The spectral density for the Matérn covariance can be obtained from [55, 8.432.5]:

$$f(\boldsymbol{\omega}) = \frac{\sigma^2}{(2\pi)^p} \frac{1}{(\kappa^2 + \|\boldsymbol{\omega}\|^2)^{\nu + p/2}}. \tag{7.25}$$

In this parametrization, σ^2/A is the variance of the field. The value of ν determines the smoothness of the field, and it is related to the number of existing derivatives. Since $r(t)$ in (7.24) is also a covariance function in one dimension, as in Example 4.11, and its one-dimensional spectral density is (4.57), one can use the simple differentiability conditions in Sections 2.1 and 2.3 to conclude that the field is m times (q.m.) differentiable if $\nu > m$.

The parameter κ is a spatial scale parameter, related to the range of correlation. The empirical relation $\rho = \sqrt{8\nu}/\kappa$ indicates the distance ρ where the correlation is near 0.1, for all ν. When $\nu \to \infty$, the Matérn covariance functions tend to a Gaussian shape.

The book by Stein [117, Ch. 2] contains a thorough discussion of the practical and theoretical aspects of the choice of covariance function for random fields, and it strongly argues for the use of the Matérn family.

The following theorem was proved by Whittle [120, 121]. It represents a generalization of Example 4.10 on page 148, with the Matérn field as solution to a stochastic partial differential equation (SPDE) driven by spatial Gaussian white noise $w'(\mathbf{t})$. Let

$$\Delta = \sum_{1}^{p} \frac{\partial^2}{\partial t_k^2}$$

be the Laplace differential operator working on a twice differentiable random field $x(\mathbf{t})$.

Theorem 7.9 (Matérn field as SPDE solution). *Let* $\{x(\mathbf{t}), \mathbf{t} \in \mathbb{R}^p\}$ *be a homogeneous Gaussian random field with Matérn covariance, spectral density* (7.25)*, and marginal variance*

$$\frac{\Gamma(\nu)}{\Gamma(\nu + p/2)(4\pi)^{p/2}\kappa^{2\nu}}.$$

Then the filter output $(\kappa^2 - \Delta)^{\alpha/2} x(\mathbf{t})$ *is Gaussian spatial white noise with variance* 1*, i.e.,* $x(\mathbf{t})$ *solves the stochastic differential equation*

$$(\kappa^2 - \Delta)^{\alpha/2} x(\mathbf{t}) = w'(\mathbf{t}), \tag{7.26}$$

with $\alpha = \nu + p/2$, $\kappa, \nu > 0$.

When α is an even number, the meaning of (7.26) is clear; for fractional $\alpha/2$, the solution is given in terms of a Fourier transform. It is obvious that any deterministic or random function that solves $(\kappa^2 - \Delta)^{\alpha/2} y(\mathbf{t}) = 0$ is also a solution. The family of random fields $x(\mathbf{t}) + y(\mathbf{t})$ is *intrinsically homogeneous* in the terminology of Section 4.4.2.

Markov random fields

Markov random fields, in particular those with Gaussian distribution, form a very useful and flexible model class, which lends itself to simple and natural deviations from the stationarity restriction. Their role is similar to that of the autoregressive (AR) processes in the single parameter case, with the difference that there is no natural order of past, present, and future in the multi-parameter cases. While most aspects of the AR-processes and its relatives are well described in the time-series literature, and part of common statistical knowledge, the Markov random fields are still under rapid development. First of all, they are natural generalizations of the stationary AR-processes, and second, they offer considerable numerical simplifications in model building, compared with methods that rely entirely on correlation or spectral methods. Just as the ARMA-processes can be used to model most kinds of stationary processes, the Markov-based random fields can be used for many types of homogeneous fields, by including also an MA-like member.

The Gaussian Markov random field model [106, 107] offers considerable computational advantages over full covariance models for simulation and estimation of random fields. In combination with the Matérn family and the SPDE-formulation it also allows anisotropic, non-stationary, and oscillatory models. For more details on the technique, together with history and applications, see [17, 76].

7.2.4 Space-time fields

In a space-time random field, the parameter $\mathbf{t}$ is composed of one time and one space parameter, $\mathbf{t} = (t, s_1, \dots, s_p)$. This is a very common situation in practice, and the statistical challenges are countless. The standard building blocks treated in previous chapters need to be adapted to the application at hand, and often new approaches introduced; see [36].

We start with some remarks on covariance structure. In the simplest homogeneous models, called *separable*, the time and space dependence are separated and the covariance function factorizes

$$r(t, \mathbf{s}) = r_{(t)}(t)\, r_{(s)}(\mathbf{s})$$

into time and space factors. With this restrictive structure, correlation in time is independent of location. Correlation in space remains unchanged by time. Most applications from nature involve some development also of the basic setup, from physical or other considerations. We shall treat one such model later in this section.

One important application of space-time random fields is the modeling of environmental variables, like the concentration of a hazardous pollutant. Over a reasonably short period of time the concentration variation may be regarded as statistically stationary in time, at least averaged over a 24 hour period. But it is often unlikely that the correlation structure in space is independent of the absolute location. Topography, location of cities, and pollutant sources make the process inhomogeneous in space.

One way to overcome the inhomogeneity is to make a transformation of the space map and move each observation point (s_1, s_2) to a new location $(\hat{s}_1, \hat{s}_2)$ so that the field $\hat{x}(t, \hat{s}_1, \hat{s}_2) = x(t, s_1, s_2)$ is homogeneous. This may not be exactly attainable, but the technique is often used in environmental statistics for planning of measurements.

Concerning isotropy one could hope for such only in $\mathbf{s} = (s_1, \ldots, s_p)$, in which case the spectral form becomes

$$r(t, \|\mathbf{s}\|) = \int_{\nu=-\infty}^{\infty} \int_{\omega=0}^{\infty} e^{i\nu t} \cdot H_p(\omega \|\mathbf{s}\|) \, \mathrm{d}G(\nu, \omega),$$

where

$$H_p(x) = (2/x)^{(p-2)/2} \, \Gamma(p/2) \, J_{(p-2)/2}(x).$$

A time-space field in the plane is a randomly changing surface. The spectral representation makes it easy to imagine its structure, homogeneous in time as well as in space. The spectral representation (7.13) generates a real field $x(\mathbf{t})$ as a packet of directed waves, $A_{\mathbf{t}} \cos(\boldsymbol{\omega} \cdot \mathbf{t} + \phi_{\boldsymbol{\omega}})$, with random amplitudes and phases, and constant on each plane parallel to $\boldsymbol{\omega} \cdot \mathbf{t} = 0$. For example, with $\mathbf{t} = (t, s_1, s_2)$ and t is time and (s_1, s_2) is space, and $\boldsymbol{\omega} = (\omega, \kappa_1, \kappa_2)$, the elementary waves are

$$A_{\boldsymbol{\omega}} \cos(\kappa_1 s_1 + \kappa_2 s_2 + \omega t + \phi_{\boldsymbol{\omega}}).$$

For fixed t this is a cosine-function in the plane, which is zero along lines $\kappa_1 s_1 + \kappa_2 s_2 + \omega t + \phi_{\boldsymbol{\omega}} = \pi/2 + k\pi$, k integer. For fixed (s_1, s_2) it is a cosine wave with frequency ω. The parameters κ_1 and κ_2 are the *wave numbers*.

In general, there is no particular relation between the time frequency ω and the space frequencies (κ_1, κ_2), and the frequency domain is three-dimensional.

Stochastic water waves

Stochastic models for water waves are an exceptional case of homogeneous random fields, in that the frequency representation is degenerated to only dimension two. For physical reasons, there is a unique relation between time and space frequencies (wave numbers). For a one-dimensional, time dependent Gaussian wave $x(t, s)$, where s is distance along an axis, the elementary waves have the form

$$A_{\omega} \cos(\omega t - \kappa s + \phi_{\omega}). \tag{7.27}$$

By physical considerations one can derive an explicit relation, called the *dispersion relation*, between wave number κ and frequency ω. If h is the water depth,

$$\omega^2 = \kappa g \tanh(h\kappa) = \kappa g \frac{e^{\kappa h} - e^{\kappa h}}{e^{\kappa h} + e^{\kappa h}}, \tag{7.28}$$

which for infinite depth reduces to $\omega^2 = \kappa g$. Here g is the constant of gravity.

A Gaussian random wave is a mixture of elementary waves of this form, in spectral language, with $\kappa > 0$ solving the dispersion relation,

$$\begin{aligned} x(t,s) &= \int_{\omega=-\infty}^{\infty} e^{i(\omega t - \kappa s)}\, dZ_+(\omega) + \int_{\omega=-\infty}^{\infty} e^{i(\omega t + \kappa s)}\, dZ_-(\omega) \\ &= x_+(t,s) + x_-(t,s). \end{aligned} \tag{7.29}$$

Here is a case where it is important to use both positive and negative frequencies; cf. the comments in Section 3.3.5. Waves described by $x_+(t,s)$ move to the right and waves in $x_-(t,s)$ move to the left with increasing t.

Keeping $t = t_0$ or $s = s_0$ fixed, one obtains a *space wave*, $x(t_0,s)$, and a time wave, $x(t,s_0)$, respectively. The spectral density of the time wave, $x(t,s_0) = x_+(t,s_0) + x_-(t,s_0)$, is the *wave frequency spectrum*,

$$f_x^{\text{freq}}(\omega) = f_+(\omega) + f_-(\omega),$$

and we see that it is not possible to distinguish between the two wave directions by just observing the time wave.

The space wave has a wave number spectrum given by the equation, for infinite water depth, with $\omega^2 = g\kappa > 0$,

$$f_x^{\text{time}}(\omega) = \frac{2\omega}{g} f_x^{\text{space}}(\omega^2/g), \tag{7.30}$$

$$f_x^{\text{space}}(\kappa) = \frac{1}{2}\sqrt{\frac{g}{\kappa}}\, f_x^{\text{time}}(\sqrt{g\kappa}). \tag{7.31}$$

One obvious effect of these relations is that the space process seems to have more short waves than can be inferred from the time observations. Physically this is due to the fact that short waves travel with lower speed than long waves, and they are therefore not observed as easily in the time process. Both the time wave observations and the space registrations are in a sense "biased" as representatives for the full time-space wave field.

In Chapter 1, (1.6), we introduced the *mean period* $2\pi\sqrt{\omega_0/\omega_2}$ as the average time between upcrossings of the mean level, expressed in terms of the spectral moments of a stationary time process; see more on this in Chapter 8. The corresponding quantity for the space wave is the *mean wave length*, i.e., the average distance between two upcrossings of the mean level by the space process $x(t_0, s)$. It is expressed in terms of the spectral moments of the wave number spectrum, in particular,

$$\kappa_0 = \int_\kappa f_x^{\text{space}}(\kappa)\,\mathrm{d}\kappa = \int_\omega f_x^{\text{time}}\,\mathrm{d}\omega = \omega_0,$$

$$\kappa_2 = \int_\kappa \kappa^2 f_x^{\text{space}}(\kappa)\,\mathrm{d}\kappa = \int_\omega \frac{\omega^4}{2g^2}\sqrt{\frac{g^2}{\omega^2}}\, f_x^{\text{time}}(\omega)\,\frac{2\omega}{g}\,\mathrm{d}\omega = \omega_4/g^2.$$

The average wave length is therefore $2\pi\sqrt{\kappa_0/\kappa_2} = 2\pi g\sqrt{\omega_0/\omega_4}$. We see that the average wave length is more sensitive to the tail of the spectral density than is the average wave period. Considering the practical difficulties in estimating the high frequency part of the wave spectrum, all statements that rely on high spectral moments are unreliable.

In the case of a two-dimensional time-dependent Gaussian wave $x(t, s_1, s_2)$, the elementary wave with frequency ω and direction θ, becomes

$$A_\omega \cos(\omega t - \kappa(s_1 \cos\theta + s_2 \sin\theta) + \phi_\omega), \tag{7.32}$$

where the relation between $\omega > 0$ and $\kappa > 0$ is given by the dispersion relation (7.28). With this choice of sign, θ determines the direction in which the wave moves.

The spectral density for the time-space wave field specifies the contribution to $x(t, (s_1, s_2))$ from elementary waves of the form (7.32). Summed (or rather integrated) over all directions $0 \le \theta < 2\pi$, they give the time wave $x(t, \mathbf{s}_0)$, in which one cannot identify the different directions. The spectral distribution, called the *directional spectrum*, is therefore often written in polar form, based on the spectral density $f_x^{\text{time}}(\omega)$ for the time wave, as

$$f(\omega, \theta) = f_x^{\text{time}}(\omega) g(\omega, \theta).$$

The *spreading function* $g(\omega, \theta)$, with $\int_0^{2\pi} g(\omega, \theta)\,\mathrm{d}\theta = 1$, specifies the relative contribution of waves from different directions. It may be frequency dependent.

Example 7.3 (Pierson-Moskowitz and JONSWAP spectra). *Wave spectra for the ocean under different weather conditions are important to characterize the input to (linear or non-linear) ship models. A classic reference is the book by B. Kinsman [70]. Much effort has been spent on design and estimation* *of typical wave spectra. One of the most popular is the Pierson-Moskowitz spectrum,*

$$f_{PM}^{\text{time}}(\omega) = \frac{\alpha}{\omega^5} e^{-1.25(\omega_m/\omega)^4}.$$

In a common variant, the JONSWAP[3] *spectrum, an extra factor $\gamma > 1$ is introduced to enhance the peak of the spectrum,*

$$f_J^{\text{time}}(\omega) = \frac{\alpha}{\omega^5} e^{-1.25(\omega_m/\omega)^4} \gamma^{\exp(-(1-\omega/\omega_m)^2/2\sigma_m^2)}. \tag{7.33}$$

In both spectra, α is a main parameter for the total variance, and ω_m defines the "peak frequency." The parameters γ and σ_m determine the peakedness of the spectrum. The spectrum and a realization of a Gaussian process with JONSWAP *spectrum was shown in Example 1.4 in Chapter 1 as an example of a process with moderate spectral width.*

Figure 7.2 shows, to the right, the level curves for a simulated Gaussian wave surface with the directional spectrum with frequency dependent spreading, shown on the left. Frequency spectrum is of JONSWAP *type.*

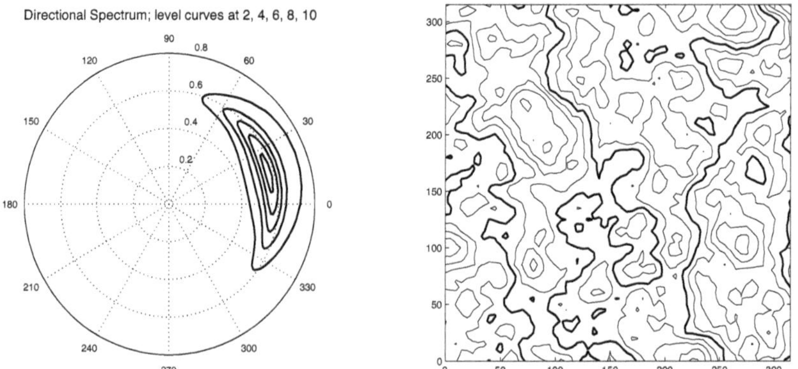

Figure 7.2 *Left: Level curves for directional spectrum with frequency dependent spreading. Right: Level curves for simulated Gaussian space sea.*

Remark 7.3 (Non-linear random water waves). *As mentioned in the historical Section 1.5.3, Gaussian waves have been used since the early 1950's, with great success in ocean engineering. However, since Gaussian processes are statistically symmetric, $x(t)$ has the same distribution as $-x(t)$ and as*

[3]The name stems from the measurement campaign, Joint North Sea Wave Project, that led to the modification.

$x(-t)$; they are not very realistic for actual water waves except in special situations with deep water and no strong wind. Much research is presently devoted to development of "non-linear" stochastic wave models, where elementary waves with different frequencies can interact. In particular, the non-linear models allow energy to be transferred between frequencies, in contrast to the "linear" Gaussian model, where energy is a characteristic of each single frequency present in the wave field, and elementary waves just add up. Markov random fields may provide a possible class of tractable models for this; see [17].

Exercises

7:1. Let $x(t)$ and $z(t)$ be uncorrelated stationary processes and define

$$y(t) = \int_0^\infty h(t-u)x(u)\,\mathrm{d}u + z(t).$$

Derive the expressions for the cross-covariance and cross-spectral density between $\{x(t), t \in \mathbb{R}\}$ and $\{y(t), t \in \mathbb{R}\}$.

7:2. Let $\{x(t), t \in \mathbb{R}\}$ and $\{y(t), t \in \mathbb{R}\}$ be jointly stationary processes, and suppose the spectrum of $x(t)$ has both a discrete component and an absolutely continuous one, while the spectrum of $y(t)$ is absolutely continuous. Examine the properties of the cross spectrum between the two processes.

7:3. Suppose the spectral processes of two stationary processes satisfy (7.6). Show that the cross-covariance function is given by $r_{ij}(t) = \int e^{i\omega t}\,\mathrm{d}F_{ij}(\omega)$.

7:4. Define two stationary processes $x(t) = A\cos(\omega_0 t + \phi)$ and $y(t) = A\cos(2\omega_0 t + \phi)$, with random A and uniform ϕ. Show that this pair does not obey relation (7.5) – the spectral component at ω_0 in $x(t)$ is correlated with the spectral component at $2\omega_0$ in $y(t)$. Explain why this does not violate the claim in (7.5).

7:5. A continuous process $\{x(t), t \in \mathbb{R}\}$ has covariance function $r_x(t) = e^{-|t|}$, and it is sampled with the sampling interval d, to give the sequence

$$z_k = x(kd), k = 0, 1, 2, \ldots.$$

Find the cross-covariance function between z_n and the exponentially smoothed sampled sequence

$$y_k = \int_0^\infty e^{-u} x(kd-u)\,\mathrm{d}u.$$

7:6. Show that the variogram function is independent of the mean value for a homogeneous field.

7:7. Construct a filter for a homogeneous field $x(m,n)$ on $\mathbb{Z} \times \mathbb{Z}$ that makes the filter output independent of any second degree polynomial mean value surface; cf. Section 4.4.2.

7:8. Consider a spatial Poisson process in $\mathbb{R}^d$, i.e., random points in $\mathbb{R}^d$ such that the number of points in disjoint regions are independent and Poisson distributed with mean proportional to the region area. Define the simple counting process $\{N(\mathbf{t}), \mathbf{t} \in \mathbb{R}^d\}$, where $N(\mathbf{t})$ is the number of points in the ball with radius one, centered at $\mathbf{t}$. Find its covariance function for $d = 2$ and $d = 3$.

7:9. Prove that $r(t_1, t_2) = e^{-|t_1|-|t_2|}$ is a covariance function for a homogeneous field in $\mathbb{R}^2$; see [117, Sect. 2.11] for some strange properties of this covariance function.

7:10. Derive, in a way similar to Example 7.2, the covariance/spectral relation for mono-chromatic waves in $\mathbb{R}^3$.

7:11. Prove that the spectral densities for water time waves and water space waves obey the relations (7.30) and (7.31), by using the spectral representation (7.29) and the dispersion relation (7.28).

7:12. Suppose a Gaussian time-space wave (7.29) is observed continuously in time at two different locations, s_1, s_2. Find expressions for the cross spectral density between the two processes $\{x(t,s_1), t \in \mathbb{R}\}$ and $\{x(t,s_2), t \in \mathbb{R}\}$.

Chapter 8

Level crossings and excursions

This chapter presents some practically important properties of stationary processes, concerning the number of times the process reaches a predetermined level. The famous Rice's formula, found by Marc Kac and Steve O. Rice [101], for the expected number of level crossings is the key tool for all crossing studies, and it is at the center of this chapter.

In many applications one encounters the question of extremes: what is the distribution of the maximum value when the process is observed over a certain time interval? With increasing observation time, the maximum will certainly grow towards the upper limit of the process distribution, at least if the process is ergodic. The exact distribution of the maximum is however not readily found, even if asymptotic extreme value theory gives reasonable answers under mild conditions; see [75]. For moderate observation times one has to rely on other methods, either exact numerical computation or approximations, often based on the number of level crossings.

In practice, level crossing counting is used as a means to describe the extremal behavior of a continuous stochastic process. For example, the maximum of the process in an interval is equal to the lowest level above which there exists no genuine level crossing, provided, of course, that the process starts below that level. Since it is often easier to investigate the statistical properties of the number of level crossings than to find the maximum distribution, crossing methods are of great practical importance.

Crossing analysis allows several consequences to be drawn about the local behavior of a continuous process in the neighborhood of crossings points. We introduce in this chapter the Slepian model, so named after David Slepian, who in 1963 introduced the type of arguments in an article about zero crossings [114]. The Slepian model is an explicit model for the conditional process behavior near level crossings. We illustrate its use on the distribution of the length and height of excursions above a fixed level, and on wave distributions.

Section 8.1 treats the basic Rice's formula, and in Section 8.2 we describe the connection between crossings and extremes. In Sections 8.3 and 8.4 we use different forms of Rice's formula to investigate the conditional behavior of a stationary process when it is observed in the neighborhood of a crossing of a predetermined level. Quantities like the height and length of the excursion above the level will be analyzed by means of explicit Slepian models. The presented distributions are *exact*, and contain no approximating assumptions.

Section 8.5, finally, treats crossings in vector processes and fields – in particular the mythical "seventh wave" is analyzed in the context of envelope crossings.

The reader may want to consult the Cramér and Leadbetter book [35] for elementary facts about crossings. The book by Azaïs and Wschebor [6] treats crossings and level sets for vector processes and random fields, with many extensions of the Rice formula.

8.1 Level crossings and Rice's formula

8.1.1 Level crossings

Before we set out to define the different crossing counting methods we shall make precise what conditions the sample functions have to satisfy. First of all, we assume that there is no interval of positive length in which the process stays constant. This is not a serious restriction; all that is required is that the process has continuous marginal distributions.[1] Secondly, we assume that the probability of a tangent of a predetermined level has probability zero; Bulinskaya's lemma, Theorem 2.14 [23], gives the simple conditions for this.

For sample functions of a continuous process $\{x(t), t \in \mathbb{R}\}$ we say that $x(t)$ has an *upcrossing* of the level u at t_0 if, for some $\varepsilon > 0$, $x(t) \leq u$ for all $t \in (t_0 - \varepsilon, t_0]$ and $x(t) \geq u$ for all $t \in [t_0, t_0 + \varepsilon)$. For any interval $I = [a, b]$, write $N_I^+(x, u)$ for the number of upcrossings by $x(t)$ in I,

$$N_I^+ = N_I^+(x, u) = \text{ the number of } u\text{-upcrossings by } x(t),\, t \in I.$$

For continuous processes that have only a finite number of u-crossing points, there must be intervals to the left and to the right of any upcrossing point such that $x(t)$ is *strictly less than* u immediately to the left and *strictly greater than* u immediately to the right of the upcrossing point. Also define

$$N_I = N_I(x, u) = \text{ the number of } t \in I \text{ such that } x(t) = u.$$

[1] Find the argument for this statement.

By the *intensity of upcrossings* we mean any function $\mu_t^+(u)$, such that

$$\int_{t\in I} \mu_t^+(u)\,\mathrm{d}t = E(N_I^+(x,u)), \quad \text{for every interval } I.$$

Similarly, we define the *intensity of crossings*, as $\mu_t(u)$, such that

$$\int_{t\in I} \mu_t(u)\,\mathrm{d}t = E(N_I(x,u)).$$

For a stationary process, $\mu_t^+(u) = \mu^+(u)$ and $\mu_t(u) = \mu(u)$ can be assumed independent of t. In general, the intensity is the mean number of events per time unit, calculated at t.

In reliability applications of stochastic processes one may want to calculate the distribution of the maximum of a continuous process $x(t)$ in an interval $I = [0,T]$. The following approximation is then often useful, and it is also sufficiently accurate for short intervals and increasing u-levels,

$$\begin{aligned}
P(\max_{0\le t\le T} x(t) > u) &= P(\{x(0)\le u\}\cap\{N_I^+(x,u)\ge 1\}) + P(x(0)>u)\\
&\le P(N_I^+(x,u)\ge 1) + P(x(0)>u) \le E(N_I^+(x,u)) + P(x(0)>u)\\
&= T\mu^+(u) + P(x(0)>u). \qquad (8.1)
\end{aligned}$$

In Section 8.2 we shall have a brief discussion of the asymptotic theory when both the observation interval and the crossed level increase in a coordinated way, $T\mu^+(u)\to\tau$. Then the term $\tau = T\mu^+(u)$ in (8.1) has to be replaced by $1-e^{-\tau}$, recognized as the probability of at least one event in a Poisson distribution with mean τ.

8.1.2 *Rice's formula for absolutely continuous processes*

The upcrossing intensity $\mu^+(u)$ was found by S.O. Rice for Gaussian processes, results which were later given strict proofs through counting methods developed by M. Kac. The classical reference is [101]. We give first a general formulation and then specialize to Gaussian processes; see [6, 75, 109].

We present the simplest version of Rice's formula, valid for processes $x(t)$ with absolutely continuous sample paths[2] and absolutely continuous distribution with density $f_{x(t)}(u) = f_{x(0)}(u)$, independent of t. For such a process, the

---

[2]A function $x(t), t\in[a,b]$ is absolutely continuous if it is equal to the integral $x(t) = \int_a^t y(s)\,\mathrm{d}s$ of an integrable function $y(s)$. This is equivalent to the requirement that for every $\varepsilon>0$ there is a $\delta>0$ such that for every collection $(a_1,b_1),(a_2,b_2),\dots,(a_n,b_n)$ of disjoint intervals in $[a,b]$ with $\sum_1^n(b_k-a_k)<\delta$ one has $\sum_1^n|x(b_k)-x(a_k)|<\varepsilon$. An absolutely continuous function is always continuous and its derivative exists almost everywhere, $x'(t) = y(t)$.

derivative $x'(t)$ exists almost everywhere, and the conditional expectations

$$E(x'(0)^+ \mid x(0) = u) \quad \text{and} \quad E(|x'(0)| \mid x(0) = u)$$

exist (with $x^+ = \max(0,x)$).

Theorem 8.1 (Rice's formula). *For any stationary process $\{x(t), t \in \mathbb{R}\}$ with density $f_{x(0)}(u)$, the crossing and upcrossing intensities are given by*

$$\mu(u) = E(N_{[0,1]}(x,u)) = f_{x(0)}(u) E(|x'(0)| \mid x(0) = u) \tag{8.2}$$

$$= \int_{-\infty}^{\infty} |z| f_{x(0),x'(0)}(u,z)\,\mathrm{d}z, \text{ if the joint density exists,}$$

$$\mu^+(u) = E(N^+_{[0,1]}(x,u)) = f_{x(0)}(u) E(x'(0)^+ \mid x(0) = u) \tag{8.3}$$

$$= \int_0^{\infty} z f_{x(0),x'(0)}(u,z)\,\mathrm{d}z, \text{ if the joint density exists.}$$

These expressions hold for almost any u.

We shall give two proofs of Rice's theorem, one short and one close to the original proof given by Rice and Kac. Before we state the short proof we shall review some facts about functions of bounded variation, proved by Banach. To formulate the proof, write for any continuous function $f(t)$, $t \in [0,1]$, and interval $I = [a,b] \subset [0,1]$,

$$N_I(f,u) = \text{ the number of } t \in I \text{ such that } f(t) = u.$$

Further, define the *total variation* of $f(t), t \in I$ as

$$V_f(I) = \sup \sum |f(t_{k+1}) - f(t_k)|,$$

where the supremum is taken over all subdivisions $a \leq t_0 < t_1 < \ldots t_n \leq b$.

Remark 8.1. *Rice's formula has been shown under successively weaker conditions both for Gaussian and for general classes of processes; for references, see [6]. Our Theorem 8.1 claims the formula to hold for "almost all levels u." This statement is somewhat unsatisfactory for practical use, since one would like it to hold for a particular u-value.*

One can prove that Rice's formula is valid for all u under somewhat stricter conditions, including the requirement that the sample functions are continuously differentiable; for a proof, see [6, Thm. 3.4]. Note that absolute continuity only implies that the derivative exists, almost everywhere.

Lemma 8.1 (Banach). *For any continuous function $f(t), t \in I$, the total variation is equal to $\int_{-\infty}^{\infty} N_I(f,u)\,\mathrm{d}u$. Further, if $f(t)$ is absolutely continuous with derivative $f'(t)$, then*

$$\int_{-\infty}^{\infty} N_I(f,u)\,\mathrm{d}u = \int_I |f'(t)|\,\mathrm{d}t.$$

Similarly, if $A \subseteq \mathbb{R}$ is a Borel measurable set, and χ_A its indicator function, then

$$\int_{-\infty}^{\infty} \chi_A(u)\,N_I(f,u)\,\mathrm{d}u = \int_I \chi_A(f(t))\,|f'(t)|\,\mathrm{d}t. \tag{8.4}$$

Proof of Rice's formula: We prove (8.2) by applying Banach's theorem on the stationary process $\{x(t), t \in \mathbb{R}\}$ with absolutely continuous and, hence, a.s. differentiable, sample paths. If $x(t)$ has a.s. absolutely continuous sample functions, then (8.4) holds for almost every realization, i.e.,

$$\int_{-\infty}^{\infty} \chi_A(u)\,N_I(x,u)\,\mathrm{d}u = \int_I \chi_A(x(t))\,|x'(t)|\,\mathrm{d}t.$$

Taking expectations and using Fubini's theorem to change the order of integration and expectation, we get, with $|I|$ for the interval length,

$$\begin{aligned}
|I| \int_{u\in A} \mu(u)\,\mathrm{d}u &= \int_{-\infty}^{\infty} \chi_A(u)\,E(N_I(x,u))\,\mathrm{d}u = E\left(\int_I \chi_A(x(t))\,|x'(t)|\,\mathrm{d}t\right) \\
&= |I|\,E\left(\chi_A(x(0))\,|x'(0)|\right) \\
&= |I| \int_{u\in A} f_{x(0)}(u)\,E(|x'(0)| \mid x(0)=u)\,\mathrm{d}u;
\end{aligned}$$

here we also used that $\{x(t), t \in \mathbb{R}\}$ is stationary.

Since A is an arbitrary measurable set, we get the desired result,

$$\mu(u) = f_{x(0)}(u)\,E(|x'(0)| \mid x(0)=u)$$

for almost all u. The proof of (8.3) is similar. □

For a proof of the following "inverse" of Rice's formula, see [109].

Theorem 8.2. *Let $\{x(t), t \in \mathbb{R}\}$ have sample paths that are continuous and of bounded variation. Let $x'(t)$ be the derivative of the absolutely continuous part of $x(t)$, and assume $E(|x'(0)|) < \infty$. Then, if Rice's formula holds, i.e.,*

$$E(N_{[0,1]}(x,u)) = f_{x(0)}(u)E(|x'(0)| \mid x(0) = u),$$

for almost all u, then $x(t)$ has absolutely continuous sample paths.

8.1.3 Alternative proof of Rice's formula

The elegant and general proof of Rice's formula just presented does not give an intuitive argument for the presence of the factor z in the integral (8.3). The following more pedestrian proof from [35] is closer to an explanation.

Theorem 8.3. *For a stationary process $\{x(t), t \in \mathbb{R}\}$ with continuous sample paths, suppose that $x(0)$ and $\zeta_n = 2^n(x(1/2^n) - x(0))$ have a joint density $g_n(u,z)$, continuous in u for all z and all sufficiently large n. Also suppose $g_n(u,z) \to p(u,z)$ uniformly in u for fixed z as $n \to \infty$ and that $g_n(u,z) \leq h(z)$ with $\int_0^\infty zh(z)\,\mathrm{d}z < \infty$. Then*

$$\mu^+(u) = E(N_{[0,1]}(x,u)) = \int_0^\infty zp(u,z)\,\mathrm{d}z. \tag{8.5}$$

It is of course tempting to think of $p(u,z)$ as the joint density of $x(0), x'(0)$, but no argument for this is involved in the following proof.

Proof. We first devise a counting technique for the upcrossings by dividing the interval $[0,1]$ into dyadic subintervals, $[(k-1)/2^n, k/2^n]$, $k = 1, \dots, 2^n$, and checking the values at the endpoints. Let N_n denote the number of points $k/2^n$ such that $x((k-1)/2^n) < u < x(k/2^n)$. Since $x(t)$ has continuous sample paths (almost surely), there is at least one u-upcrossing in every interval such that $x((k-1)/2^n) < u < x(k/2^n)$, and hence $N_n \leq N_{[0,1]}(x,u)$.

Furthermore, since $x(t)$ has a continuous distribution, we may assume that $x(k/2^n) \neq u$ for all n and $k = 1, \dots, 2^n$. When n increases to $n+1$ the number of subintervals doubles, and each interval contributing one upcrossing to N_n will contribute at least one upcrossing to N_{n+1} in at least one of the two new subintervals. Hence N_n is increasing and it is easy to see that, regardless of if $N_{[0,1]}(x,u) = \infty$ or $N_{[0,1]}(x,u) < \infty$, $N_n \uparrow N_{[0,1]}(x,u)$ as $n \to \infty$. Monotone convergence implies that $\lim_{n\to\infty} E(N_n) = E(N_{[0,1]}(x,u))$.

Now, define

$$J_n(u) = 2^n P(x(0) < u < x(1/2^n)),$$

so, by stationarity,

$$E(N_{[0,1]}(x,u)) = \lim_{n\to\infty} E(N_n) = \lim_{n\to\infty} J_n(u).$$

By writing the event $\{x(0) < u < x(1/2^n)\}$ as

$$\{x(0) < u < x(0) + \zeta_n/2^n\} = \{x(0) < u\} \cap \{\zeta_n > 2^n(u - x(0))\},$$

we have

$$\begin{aligned} J_n(u) &= 2^n P(x(0) < u, \zeta_n > 2^n(u - x(0))) \\ &= 2^n \int_{x=-\infty}^{u} \int_{y=2^n(u-x)}^{\infty} g_n(x,y)\,\mathrm{d}y\,\mathrm{d}x. \end{aligned}$$

By a change of variables, $x = u - zv/2^n$, $y = z$, $(v = 2^n(u-x)/y)$, this is equal to

$$\int_{z=0}^{\infty} z \int_{v=0}^{1} g_n(u - zv/2^n, z)\,\mathrm{d}v\,\mathrm{d}z,$$

where $g_n(u - zv/2^n, z)$ tends pointwise to $p(u,z)$ as $n \to \infty$, by the assumptions of the theorem. Since g_n is dominated by the integrable $h(z)$ it follows that the double integral tends to $\int_0^\infty zp(u,z)\,\mathrm{d}z$ as $n \to \infty$. □

Remark 8.2 (Horizontal window conditioning)**.** *The proof of Rice's formula as illustrated in Figure 8.1 shows the relation to the Kac and Slepian horizontal window conditioning: one counts the number of times the process passes through a small horizontal window $[t, t + \Delta t]$. More comments on this aspect will be given in Section 8.3.3.*

Remark 8.3 (Non-stationary process)**.** *Rice's formula can be extended to non-stationary processes, in which case the crossing intensity is time dependent. Then the density function $f_{x(t),x'(t)}(u,z)$ in the integral depends on t,*

$$E(N_{[a,b]}(x,u)) = \int_{t=a}^{b} \int_{z=-\infty}^{\infty} |z| f_{x(t),x'(t)}(u,z)\,\mathrm{d}z\,\mathrm{d}t. \tag{8.6}$$

The integral $\mu_t(u) = \int_{-\infty}^{\infty} |z| f_{x(t),x'(t)}(u,z)\,\mathrm{d}z$ is the local crossings intensity at time t, with obvious modifications for up- and downcrossings.

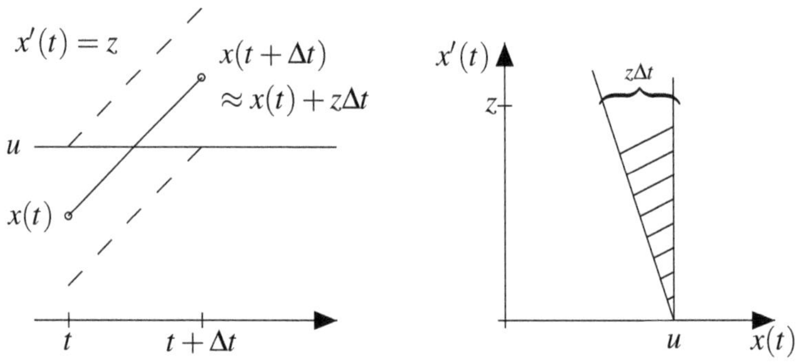

Figure 8.1 *Upcrossing occurs between t and $t+\Delta t$ when $x(t)$ is between $u - x'(t)\Delta t$ and u. The probability is the integral of the joint density $f_{x(t),x'(t)}(u,z)$ over the dashed area.*

8.1.4 *Rice's formula for differentiable Gaussian processes*

For a Gaussian stationary process, Rice's formula becomes particularly simple. We know that in a stationary differentiable Gaussian process $\{x(t), t \in \mathbb{R}\}$, the process value, $x(t)$, and the derivative, $x'(t)$, at the same time point are independent and Gaussian, with mean $E(x(t)) = m$, $E(x'(t)) = 0$, and variances given by the spectral moments, $V(x(t)) = r(0) = \omega_0$, $V(x'(t)) = -r''(0) = \omega_2$. Then

$$f_{x(0),x'(0)}(u,z) = \frac{1}{2\pi\sqrt{\omega_0\omega_2}} e^{-(u-m)^2/2\omega_0} e^{-z^2/2\omega_2}. \tag{8.7}$$

Simple integration of (8.2) and (8.3) gives that for Gaussian stationary processes,

$$\mu(u) = E(N_{[0,1]}(x,u)) = \frac{1}{\pi}\sqrt{\frac{\omega_2}{\omega_0}} e^{-(u-m)^2/2\omega_0},$$

$$\mu^+(u) = E(N^+_{[0,1]}(x,u)) = \frac{1}{2\pi}\sqrt{\frac{\omega_2}{\omega_0}} e^{-(u-m)^2/2\omega_0},$$

which are the original forms of Rice's formula. (It is easy to show by direct calculation that the conditions for Theorem 8.1 are satisfied in this special case.) The formulas hold regardless of whether ω_2 is finite or not, so $\omega_2 = \infty$ if and only if the expected number of crossings in any interval is infinite. This does not mean, however, that there necessarily are infinitely many crossings, but if there are crossings, then there may be infinitely many in the neighborhood.

Mean frequency and period: The expected number of mean-level upcrossings per time unit in a stationary Gaussian process is

$$\mu^+(m) = \frac{1}{2\pi}\sqrt{\omega_2/\omega_0} = \frac{1}{2\pi}\sqrt{\frac{\int \omega^2 f(\omega)\,\mathrm{d}\omega}{\int f(\omega)\,\mathrm{d}\omega}}, \tag{8.8}$$

and it is called the *(root) mean square frequency* of the process. The inverse is equal to the long run average time between successive mean level upcrossings,

$$T_2 = 1/\mu^+(m) = 2\pi\sqrt{\omega_0/\omega_2}; \tag{8.9}$$

this explains the name *mean period* for T_2.

Local extremes

A local extremum, minimum or maximum, of a differentiable process $\{x(t), t \in \mathbb{R}\}$ corresponds to, respectively, an upcrossing and a downcrossing of the zero level by the process derivative $\{x'(t), t \in \mathbb{R}\}$. Rice's formula applied to $x'(t)$ therefore gives the expected number of local extremes. For a Gaussian process the formulas involve the fourth spectral moment $\omega_4 = V(x''(t)) = \int \omega^4 f(\omega)\,\mathrm{d}\omega$. The general and Gaussian expressions for the expected number of local extremes per time unit, are, respectively,

$$\mu_{\min} = \int_0^\infty z f_{x',x''}(0,z)\,\mathrm{d}z = \frac{1}{2\pi}\sqrt{\frac{\omega_4}{\omega_2}},$$

$$\mu_{\max} = \int_{-\infty}^0 |z| f_{x',x''}(0,z)\,\mathrm{d}z = \frac{1}{2\pi}\sqrt{\frac{\omega_4}{\omega_2}}.$$

If we combine this with (8.8) we get the average number of local maxima per mean level upcrossing,

$$1/\alpha = \frac{\frac{1}{2\pi}\sqrt{\omega_4/\omega_2}}{\frac{1}{2\pi}\sqrt{\omega_2/\omega_0}} = \sqrt{\frac{\omega_0\omega_4}{\omega_2^2}}.$$

The parameter α is invariant under time and scale changes. It is bounded by $0 < \alpha < 1$, and it is used as an irregularity measure: an α near 1 indicates a very regular process with approximately one local maximum and minimum between mean level upcrossings. If α is near zero the process contains a high proportion of high frequency components, and then one can expect many local extremes between the upcrossings.

Remark 8.4 (Spectral width)**.** *Seen in relation to the spectrum, the parameter* α *is a measure of spectral width. A spectrum with* α *near* 1 *is narrow banded, i.e., the spectral density is concentrated to a small frequency band around a dominating center frequency. A narrow banded process has very regular sample functions, with slowly varying random amplitude; see the discussion of the envelope in Section 5.1.*

The variance of the number of level crossings

The technique to prove Rice's formula by counting crossings in a discretized version of a differentiable process can be generalized to give also higher moments of the number of level crossings. We present here the result for the second factorial moment,

$$\mu_2^+(I) = E(N_I^+(x,u)(N_I^+(x,u)-1)),$$

of the number of u-upcrossings by $x(t), t \in I = [0,T]$. In [35, Sec 10.6] the following integral expression is shown to hold for stationary Gaussian process with continuous spectrum with finite spectral moment:

$$\mu_2^+(I) = \int_0^T \int_0^T \left\{ \int_0^\infty \int_0^\infty z_1 z_2\, p_{t_1 t_2}(u,u,z_1,z_2)\, \mathrm{d}z_1\, \mathrm{d}z_2 \right\} \mathrm{d}t_1\, \mathrm{d}t_2, \qquad (8.10)$$

where $p_{t_1t_2}(u,u,z_1,z_2)$ is the joint density of $(x(t_1), x(t_2), x'(t_1), x'(t_2))$. An explicit formula in terms of the covariance function is also given in the cited work. In [6, Thm. 3.4] one can find the general formula for factorial moments, valid under general conditions.

8.2 Poisson character of high-level crossings

Rice's formula and the upcrossing intensity is an important quantity in the general asymptotic extreme value theory for continuous time stationary processes. Without going into details, we here outline the basic ideas, connecting the crossing analysis with the asymptotic extreme value distribution for a continuous time process. At the end of the section we also give precise conditions for the asymptotic Poisson character of extreme exceedances in a stationary sequence. These problems are dealt with in detail in [75].

Poisson character of crossings

Upcrossings of a high level are rare events, and under mild dependence restrictions on the process, they will occur rather independently of each other.

They are therefore approximately Poisson processes with very low intensity. Rescaling the observation interval $[0,T]$ by dividing time by T, and letting T and the level u both tend to infinity (or to the upper limit of the process distribution), in a coordinated way, so that $T\mu^+(u) \to \tau > 0$, we get a normalized point process $\widetilde{N}$ of u-upcrossings on the unit interval $[0,1]$ with asymptotic intensity τ. (Note that $T\mu^+(u)$ is the u-upcrossing intensity in $\widetilde{N}$.)

Under two conditions, one that restricts the dependence of exceedances of high levels at far distances, and one that prevents the formation of upcrossing clusters, one can prove that $\widetilde{N}$ tends in distribution to a Poisson process on $[0,1]$ with intensity τ. This means that $P(\widetilde{N}([0,1]) = k) \to e^{-\tau}\tau^k/k!$, $k = 0,1,\ldots$. In particular, $P(\widetilde{N}([0,1]) = 0) \to e^{-\tau}$.

As an example of a specific application, we quote the following theorem from [75, Thm. 8.2.7 and 9.1.2] for a continuous time Gaussian process.

Theorem 8.4 (Asymptotic maximum of Gaussian process). *Let $\{x(t), t \in \mathbb{R}\}$ be a zero mean, variance one, stationary Gaussian process with covariance function $r(t) = 1 - \omega_2 t^2/2 + o(t^2)$, as $t \to 0$, and assume $r(t)\log t \to 0$, as $t \to \infty$.*

a) If $T, u \to \infty$ so that $T\mu^+(u) \to \tau$, then the normalized point process $\widetilde{N}$ of u-upcrossings tends to a Poisson process on the interval $[0,1]$ with intensity τ.

b) The asymptotic distribution of the maximum over the interval $[0,T]$ is a Gumbel distribution with parameters a_T, b_T,

$$P(\max_{0 \le t \le T} x(t) \le x/a_T + b_T) \to e^{-e^{-x}}, \text{ as } T \to \infty, \tag{8.11}$$

$$a_T = \sqrt{2\log T}, \quad b_T = \sqrt{2\log T} + \frac{\log\sqrt{\omega_2}}{2\pi\sqrt{2\log T}}. \tag{8.12}$$

The Gumbel distribution in (8.11) is a special case of the generalized extreme value (GEV) distribution with shape parameter 0.

Of the two conditions in the theorem, the local covariance condition near $t = 0$ prevents clustering of crossings, while the condition $r(t)\log t \to 0$ at infinity is the adapted mixing condition that gives asymptotic independence.

Remark 8.5. *The condition $r(t)\log t \to 0$ is indeed very weak. If $r(t)\log t$ has a finite non-zero limit $\gamma > 0$, a type of extremal long range dependence occurs, and the typical independence property of the Poisson limit in Theo-*

rem 8.4(a) breaks down. Instead, the upcrossing process will tend to a Cox process, i.e., a Poisson process with random intensity; see Example 3.8 on page 112. Instead of constant intensity $\tau = \lim T\mu^+(u)$, the intensity in the asymptotic point process will be $\tau e^{-\gamma+\sqrt{2\gamma}\zeta}$, where ζ is a standard normal variable. As a consequence, the maximum distribution is not pure Gumbel as in (8.11) but the convolution of a Gumbel variable and an independent normal variable; this was shown for sequences in [94] and further developed in [75, Sect. 6.5]. If $r(t)\log t \to \infty$, the asymptotic distribution is normal.

Poisson exceedances in sequences

The weak dependence restriction necessary for the Poisson character of high level upcrossings is of the *adapted mixing* type, described in Section 6.6.5. The following condition, $D(u_n)$, was introduced by Leadbetter [74], for the asymptotic independence. We formulate the condition for a stationary sequence $\{x_n, n \in \mathbb{N}\}$ and a sequence $\{u_n\}$ of constants.

Definition 8.1. *The condition $D(u_n)$ is said to hold if, for any integers*

$$1 \le i_1 < \ldots < i_p < j_1 < \ldots < j_{p'} \le n,$$

for which $j_1 - i_p \ge \ell$, we have

$$|F_{i_1 \ldots i_p, j_1 \ldots j_{p'}}(u_n) - F_{i_1 \ldots i_p}(u_n) F_{j_1 \ldots j_{p'}}(u_n)| \le \alpha_{n,\ell},$$

where $\alpha_{n,\ell_n} \to 0$ as $n \to \infty$ for some sequence $\ell_n = o(n)$. Here, $F_{i_1 \ldots i_p}(u_n)$ is the p-dimensional distribution function evaluated at $(u_n, \ldots, u_n)$.

To guarantee that the resulting point process of upcrossings is simple, one needs a further condition, called $D'(u_n)$, that prevents clustering of upcrossings.

Denote by $[c]$ the integer part of c.

Definition 8.2. *The condition $D'(u_n)$ is said to hold if*

$$\limsup_{n\to\infty} n \sum_{j=2}^{[n/k]} P(x_1 > u_n, x_j > u_n) \to 0 \text{ as } k \to \infty. \qquad (8.13)$$

We quote from [75, Thm. 3.4.1].

Theorem 8.5. *Let* $\{u_n\}$ *be constants such that* $D(u_n)$, $D'(u_n)$ *hold for the stationary sequence* $\{x_n, n \in \mathbb{N}\}$. *Let* $0 \le \tau < \infty$. *Then* $P(\max_{k=1,\ldots,n} x_k > u_n) \to e^{-\tau}$ *if and only if* $nP(x_1 > u_n) \to \tau$ *as* $n \to \infty$.

8.3 Marked crossings and biased sampling

8.3.1 Marked crossings

As we shall see, there are numerous practical examples of crossing problems, where there is an interest in local properties of the process in the neighborhood of the crossings, for example, duration and height of the excursions above a level u. With point process terminology one can look at the crossings as a sequence of points in a stationary point process, and let the shape of the process around the crossing be a *mark* attached to the crossing. A simple example of a mark is the derivative, observed at the crossing, but in a more general setting, a mark can be anything that can be observed in conjunction with the crossings. In point process theory the conditional distribution of a mark is treated as a *Palm distribution*; see [35].

Let $\{x(t), t \in \mathbb{R}\}$ be a stationary differentiable process, and let $\{\mathbf{y}(t), t \in \mathbb{R}\}$ be a vector process, $\mathbf{y}(t) = (y_1(t), \ldots, y_p(t))$, jointly stationary with $x(t)$. For reasons that will be obvious later, we want the derivative $x'(t)$ to be one of the marks, and therefore define $\mathbf{z}(t) = (x'(t), \mathbf{y}(t))$ as the mark process. We will allow $\mathbf{y}(t)$ to be infinite dimensional.

We shall now formulate a generalization of Rice's formula to give the intensity of u-crossings at which the mark satisfies a certain restriction. First, some examples of natural marks and restrictions.

- **Derivative at crossing:** $\mathbf{z}(t) = x'(t)$ with restriction $x'(t) \ge z_0 > 0$ will give the intensity of upcrossings where the derivative is greater than z_0;
- **Process at a later time:** $\mathbf{z}(t) = (x'(t), x(t+s_1))$, for a fixed s_1, with restriction $x'(t) > 0, x(t+s_1) > u$, will give intensity of upcrossings of level u, such that the process is above u also a time s_1 later;
- **Length and height of excursion:** $\mathbf{z}(t) = (x'(t), x(t+s), 0 \le s \le s_0)$, with restriction $x'(t) > 0, \min_{0<s<s_0} x(t+s) < u, \max_{0<s<s_0} x(t+s) < u + h_0$,

gives intensity of upcrossings where the excursion length is less than s_0, and the excess over u is less than h_0.

We give a Rice's formula for marked crossings with finite-dimensional marks, $\mathbf{z}(t) = (x'(t), y_1(t), \dots, y_p(t))$. Write, for $\mathbf{v} = (v_1, \dots, v_p)$,

$$\mathbf{y}(t) \le \mathbf{v} \quad \text{for the event} \quad y_j(t) \le v_j, j = 1, \dots, p,$$

and define the number of restricted crossings as

$$N_I(x, u; [a,b], \mathbf{v}) = \#\{t \in I; \text{such that } x(t) = u,\, x'(t) \in [a,b],\, \mathbf{y}(t) \le \mathbf{v}\}.$$

Let I_A be the indicator function of the event A.

Theorem 8.6 (Rice's formula for marked crossings)**.** *If the mark process* $\{\mathbf{y}(t), t \in \mathbb{R}\}$ *has continuous sample paths, then*

$$\begin{aligned} E(N_{[0,1]}(x, u; [a,b], \mathbf{v})) \\ &= f_{x(0)}(u) E(|x'(0)| I_{x'(0) \in [a,b]} I_{\mathbf{y}(t) \le \mathbf{v}} \mid x(0) = u) \qquad (8.14) \\ &= \int_{z=a}^{b} |z| f_{x(0),x'(0)}(u,z) P(\mathbf{y}(0) \le \mathbf{v} \mid x(0) = u, x'(0) = z)\, \mathrm{d}z. \qquad (8.15) \end{aligned}$$

For a proof of a one-dimensional version, see [75, Lemma 7.5.2].

8.3.2 *Biased sampling and empirical distributions*

The selection of crossing points for special study is an example of *biased sampling*: if a process is observed only at time points determined by the process itself, then the observed values will most likely follow another statistical distribution than when observations are taken at a deterministic rate. Just think of the derivative of a stationary normal process, which is a zero mean normal process with variance equal to the second spectral moment, ω_2. At the upcrossings of a predetermined level u, the derivative is necessarily positive, but, as we shall see later, its sampling distribution is not a "half normal" distribution, but instead a Rayleigh distribution with parameter ω_2.

Figure 8.2 illustrates the selection procedure. In the realization of $x(t)$ in the upper diagram, one identifies the upcrossings of a level, in this case $u = 1.5$, and extracts a piece of the sample function, starting at the upcrossings. These pieces, "excursions," are then plotted in the lower diagram. With

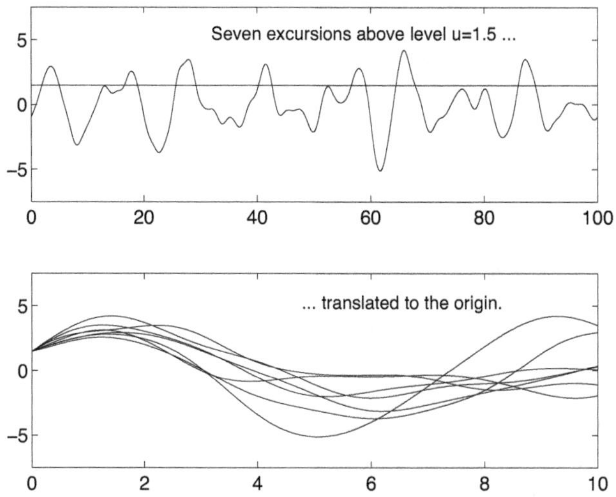

Figure 8.2 *Excursions above* $u = 1.5$*, translated to the origin, contribute to the distribution* $P^{1.5}(\cdot)$.

increasing observation interval there will be an increasing number of excursions, and if $x(t)$ is ergodic, their empirical distribution will converge to a deterministic limit that defines a probability measure. However, one should be aware that the distribution only is valid asymptotically, for increasing observation interval.

In the following definition, we let $t_k > 0, k = 1, 2, \ldots$, be the time points of upcrossings of the fixed level u, for the continuous stationary process $\{x(t), t \in \mathbb{R}\}$. We assume that the crossing intensity is finite, $\mu(u) < \infty$, which implies that there are only a finite number of t_k in any finite interval. Let A be a Borel set of continuous functions, so one can determine if the translated function $x(t_k + \cdot)$ belongs to A or not.

Definition 8.3. *For a stationary process* $\{x(t), t \in \mathbb{R}\}$ *with finite crossing intensity, the empirical (long run) conditional distribution of* $x(t_0 + \cdot)$ *after u-upcrossing is defined as*

$$P^u(A) = \lim_{T \to \infty} \frac{\#\{t_k; 0 < t_k < T \text{ and } x(t_k + \cdot) \in A)\}}{\#\{t_k; 0 < t_k < T\}}. \tag{8.16}$$

Thus, $P^u(A)$ *counts all those u-upcrossings* t_k *for which the process, taken with* t_k *as a new origin, satisfies the condition given by A.*

It is easy to see that the family of $P^u(A)$, defined for all half-open finite-dimensional rectangles, forms a consistent family of finite-dimensional distribution functions, and hence defines a probability measure on $(\mathbb{R}^{[0,\infty]}, \mathscr{B}_{[0,\infty]})$. We will now use Rice's formula for marked crossings to give an explicit expression for the finite-dimensional distribution functions.

Take $\mathbf{s} = (s_1, \dots, s_n)$, $\mathbf{v} = (v_1, \dots, v_n)$, and define

$$N_{[0,T]}(x, u; \mathbf{s}, \mathbf{v}) = \#\{t_k; 0 \le t_k \le T, \text{ and } x(t_k + \mathbf{s}) \le \mathbf{v}\},$$

as the number of u-upcrossings $t_k \in [0, T]$ that are such that the process, at each of the times s_j after the upcrossing, is less than v_j. Thus $\frac{N_{[0,T]}(x,u;\mathbf{s},\mathbf{v})}{N_{[0,T]}(x,u)}$, $\mathbf{v} \in \mathbb{R}^n$, is the empirical distribution function for the process observed at times s_j after u-upcrossings. For an ergodic process, the limit as $T \to \infty$ exists, and we get the following expression for the distribution in Definition 8.3.

In the theorem, and in the subsequent theorems,

$$A_{\mathbf{s}}(\mathbf{v})$$

denotes the subset of functions $y(t)$ in $\mathbb{R}^{[0,\infty]}$ that satisfy a condition $y(s_j) \le v_j$, for given time points $\mathbf{s} = (s_1, \dots, s_n), \mathbf{v} = (v_1, \dots, v_n)$.

Theorem 8.7. *If $\{x(t), t \in \mathbb{R}\}$ is differentiable and ergodic,*

$$P^u(A_{\mathbf{s}}(\mathbf{v})) = \frac{E\left(N_{[0,1]}(x, u; \mathbf{s}, \mathbf{v})\right)}{E\left(N_{[0,1]}(x, u)\right)} \tag{8.17}$$

$$= \frac{\int_{z=0}^{\infty} z f_{x(0),x'(0)}(u,z)\, P(x(\mathbf{s}) \le \mathbf{v} \mid x(0) = u, x'(0) = z)\, \mathrm{d}z}{\int_{z=0}^{\infty} z f_{x(0),x'(0)}(u,z)\, \mathrm{d}z}. \tag{8.18}$$

Proof. We need a result from Chapter 6, namely that for an ergodic process, the following limits exist, with probability one,

$$\frac{N_{[0,T]}(x,u)}{T} \to E\left(N_{[0,1]}(x,u)\right), \quad \frac{N_{[0,T]}(x,u;\mathbf{s},\mathbf{v})}{T} \to E\left(N_{[0,1]}(x,u;\mathbf{s},\mathbf{v})\right),$$

as $T \to \infty$. This gives (8.17). To get (8.18), use (8.15) with the mark $\mathbf{y}(t) = \mathbf{x}(t+\mathbf{s}) = (x(t+s_1), \dots, x(t+s_p))$ and the restriction to positive z, i.e., $[a,b] = [0,\infty]$. □

We can prove the claim about the Rayleigh distributed derivatives in a normal process.[3]

Theorem 8.8. *The derivatives at upcrossings of a fixed level u by an ergodic normal process are Rayleigh distributed with density*

$$p^u(z) = \frac{z}{\omega_2} e^{-z^2/2\omega_2}, \quad z \geq 0. \tag{8.19}$$

Proof. Use (8.15) with the mark $\mathbf{z}(t) = x'(t)$ and $A = \{\mathbf{y} \in \mathbb{C}; 0 < y'(0) \leq z_0\}$ to get the P^u-probability that the derivative at a u-upcrossing is less than z_0 as

$$P^u(z_0) = \frac{\int_{z=0}^{z_0} z f_{x(0),x'(0)}(u,z)\,\mathrm{d}z}{\int_{z=0}^{\infty} z f_{x(0),x'(0)}(u,z)\,\mathrm{d}z}.$$

Since $f_{x(0),x'(0)}(u,z) = \frac{1}{2\pi\sqrt{\omega_0\omega_2}} e^{-u^2/2\omega_0} e^{-z^2/2\omega_2}$, the factor $\frac{1}{\sqrt{2\pi\omega_0}} e^{-u^2/2\omega_0}$ cancels, giving the Rayleigh density. □

The representation (8.18) can be intuitively interpreted in terms of the random slope at upcrossings. If ζ is a random variable with density (8.19), then (8.18) can be formulated as

$$P^u(A) = E_\zeta\left(P(x(\mathbf{s}) \leq \mathbf{v} \mid x(0) = u, x'(0) = \zeta)\right). \tag{8.20}$$

The interpretation is that the distribution of the process values at times $\mathbf{s}$ after u-upcrossings is a mixture over the random slope ζ of the ordinary conditional distributions of $x(\mathbf{s})$, given that $x(0) = u$ and $x'(0) = \zeta$, when ζ has density $p^u(z)$.

8.3.3 *Prediction from a random crossing time*

The linear prediction theory presented in Section 1.4.3, and further elaborated on in Appendix A.6.3, can be used to predict, as accurately as possible, the unknown future value $x(t_0 + \tau)$ of a process $\{x(t), t \in \mathbb{R}\}$, from the observations that are available at time t_0. An implicit assumption is that there is no stochastic dependence between the choice of t_0 and $x(t_0 + \tau)$.

For example, given that we have a complete record of all old values, the best predictor, in the sense of smallest mean square error, is the conditional

[3] This is the solution to Exercise 8:4.

expectation $\widehat{x}(t_0+\tau) = E(x(t_0+\tau) \mid x(s), s \leq t_0)$. If the process is Gaussian, the predictor is linear in the observations. For example, if the process is stationary with mean m and we know only the value of $x(t_0)$, the optimal solution with smallest mean square error, over all outcomes of $x(t_0)$ and $x(t_0+\tau)$, is

$$\widehat{x}(t_0+\tau) = E(x(t_0+\tau) \mid x(t_0)) = m + \frac{r(\tau)}{r(0)}(x(t_0)-m). \tag{8.21}$$

There are situations where the time point from which we want to predict the future process is not a deterministic time point but a random time, determined by the process itself, for example by the presence of a level crossing. An example is an *alert predictor* of the water level in a flood protection system: when the water level reaches a certain warning level special actions are taken, such as special surveillance, more detailed prediction, etc.

We assume that the process $\{x(t), t \in \mathbb{R}\}$ has continuous sample functions. Prediction from an upcrossing time point t_0 with specified level u shall be based on the conditional distributions, given that an upcrossing of the level u has occurred at t_0,[4] in particular

$$\widehat{x}^u(t_0+\tau) = E(x(t_0+\tau) \mid x(t_0) = u, \text{ upcrossing}). \tag{8.22}$$

The conditional expectation is defined in an elementary way as $\varphi(v) = E(x \mid y = v) = \int_u u f_{x,y}(u,v)/f_y(v)\,du$. The main property is relation (A.6), $E(x) = \int_y \varphi(y) f(y)\,dy$, and its more refined version,

$$E(x \mid y \in A) = \frac{\int_{y\in A} \varphi(y) f(y)\,dy}{\int_{y \in A} f(y)\,dy},$$

for every Borel set A. For example, when the density $f(y)$ is continuous, with $A = [u-\varepsilon, u]$,

$$\varphi(u) = E(x \mid y = u) = \lim_{\varepsilon \to 0} E(x \mid u - \varepsilon \leq y \leq u). \tag{8.23}$$

The meaning of this is that $E(x \mid y = u)$ is the expected value of x for those outcomes where the value of y is close to u; for more details, see [19, Ch. 4].

To apply the technique to a stationary process $\{x(t), t \in \mathbb{R}\}$, take $y = x(t_0)$ and $x = x(t_0+\tau)$. Then

$$P(x(t_0+\tau) \leq v \mid x(t_0) = u) = \lim_{\varepsilon \to 0} P(X(t_0+\tau) \leq v \mid u - \varepsilon \leq x(t_0) \leq u), \tag{8.24}$$

$$E(x(t_0+\tau) \mid x(t_0) = u) = \lim_{\varepsilon \to 0} E(X(t_0+\tau) \mid u - \varepsilon \leq x(t_0) \leq u), \tag{8.25}$$

[4]In Markov process theory one has introduced the *strong Markov property* to handle conditioning from a time point that depends on the process.

calculated at time t_0, and (8.25) gives the best predictor of the future value $x(t_0+\tau)$ in the sense that it minimizes the squared prediction error taken as an average over all the possible outcomes of $x(t_0)$ and $x(t_0+\tau)$. By "average" we then mean expected value as well as an empirical average over many realizations observed at fixed predetermined times, chosen independently of the process.

The condition in (8.24) is the Kac and Slepian *vertical window condition*, since the process has to pass through a vertical window near u at time t_0 [66].

We now consider prediction from the times of upcrossings of a fixed level u. This differs from the previous type of conditioning in that the last observed value of the process is known, and the time points are variable, randomly determined by the process itself. The interpretation of "average future value" is then not clear at this moment but it has to be made precise. Obviously, what we should aim at is a prediction method that works well on the average, in the long run, for all the u-level upcrossings we observe in the process. Call these upcrossings time points $t_k > 0$.

The empirical long run distribution is related to Kac and Slepian's horizontal window conditioning mentioned in Remark 8.2 [66]. For a stationary process $\{x(t), t \in \mathbb{R}\}$, the (horizontal window) conditional distribution of $x(t_0+\cdot)$ after u-upcrossing at t_0 is

$$\begin{aligned} P^{hw}(A) &= P(x(t_0+\cdot) \in A \mid x(t_0) = u, \text{ "upcrossing in h.w. sense"}) \qquad (8.26)\\ &= \lim_{\varepsilon \to 0} P(x(t_0+\cdot) \in A \mid x(s) = u, \text{ upcrossing, some } s \in [t_0-\varepsilon, t_0]). \end{aligned}$$

The term, horizontal window conditioning, is natural since the process has been conditioned to pass through a horizontal window at level u somewhere near t_0; see [79] for further analysis of the difference between horizontal and vertical conditioning in prediction.

It is easy to show that the distribution just defined is identical to the distribution in Definition 8.3 and Theorem 8.7, $P^u(A) = P^{hw}(A)$, for every (ergodic) stationary process such that there are only a finite number of u-upcrossings in any finite interval; see [35].

8.4 The Slepian model

8.4.1 *Definition*

Theorem 8.7 presents the conditional distribution of a stationary process in the neighborhood of upcrossings of the fixed level u. The definition in the form of integrals in (8.18) is not very informative as it is. However, for Gaussian processes one can construct an explicit and simple process that has exactly this distribution, and lends itself to easy simulation, numerical calculation, and asymptotic expansion. The idea stems from David Slepian, who constructed a model for a normal process after crossings of the mean level [114]. A systematic account of the structure and use can be found in [83]; see also [77] and [75, Sect. 10.3].

Definition 8.4 (Slepian model). *Let $\{x(t), t \in \mathbb{R}\}$ be a stationary and ergodic stochastic process and fix a level u, such that $x(t)$ has a finite number of u-upcrossings in any finite interval. A Slepian model process for $\{x(t), t \in \mathbb{R}\}$ after u-upcrossings is any stochastic process $\{\xi_u(t), t \in \mathbb{R}\}$ with distributions given by P^u in (8.18). In particular, its finite-dimensional distributions are given by*

$$P(\xi_u(\mathbf{s}) \leq \mathbf{v}) = \int_0^\infty p^u(z) P\left(x(\mathbf{s}) \leq \mathbf{v} \mid x(0) = u, x'(0) = z\right) \mathrm{d}z, \tag{8.27}$$

where p^u is the Rayleigh density (8.19).

The Slepian model is a stochastic model for individual excursions after a level upcrossing. In the lower diagram in Figure 8.2 are realizations from that distribution. More complex Slepian models can be formulated for other crossing problems, for example the process behavior after a local maximum or minimum, since these are defined as downcrossing or upcrossing of the zero level by the derivative process $\{x'(t), t \in \mathbb{R}\}$.

Every Slepian model has two elements, which depend on the type of crossing problem: the long run distribution of the gradient at the instances of the crossings, and the conditional distribution of the process given the crossing value and the value of the gradient. Typical problems that can be analyzed by a Slepian process are

- **Prediction after crossing:** What is the best predictor of the process a time τ after one has observed a level u upcrossing?
- **Excursion shape:** How high above level u will an excursion extend, and after how long time will the process return below the level?
- **Crest shape:** What is the shape of the process near its local maxima?

We can immediately solve the first problem: the best predictor $\widehat{x}^u(t_0 + \tau)$

after u-upcrossings is the expectation of the Slepian model:

$$\widehat{x}^u(t_0+\tau) = E\left(\xi_u(\tau)\right), \tag{8.28}$$

in the sense that the average of $(x(t_k+\tau)-a)^2$, when t_k runs over all u-upcrossings, takes its minimum value when $a = E\left(\xi_u(\tau)\right)$.

8.4.2 A Slepian model in Gaussian processes

The conditional distribution of $\{x(t), t \in \mathbb{R}\}$ after u-upcrossing is particularly simple in the Gaussian case, in which the Slepian model can be expressed in a very explicit form.

First of all, Theorem 8.8 gives the density $p^u(z) = \frac{z}{\omega_2} e^{-z^2/2\omega_2}$, $z \geq 0$, for the derivative at upcrossing in the Slepian model (8.27). Next, we need the conditional distribution of $x(\mathbf{s})$ given $x(0) = u$, $x'(0) = z$ and average it over $\zeta = z$ with density $p^u(z)$. Since a conditional distribution in a multivariate normal distribution is still a normal distribution, we need only to find the conditional mean $E(x(s) \mid x(0) = u, x'(0) = z)$ and the conditional covariances $C(x(s_1), x(s_2) \mid x(0) = u, x'(0) = z)$, and these were given in Section 1.4.1, equations (1.9) and (1.10).

Take $\boldsymbol{\xi} = (x(s_1), x(s_2))$, $\boldsymbol{\eta} = (x(0), x'(0))$, and calculate the joint covariance matrix of $(\boldsymbol{\xi}, \boldsymbol{\eta})$ from the covariance function $r(t)$ for $\{x(t), t \in \mathbb{R}\}$. By Theorem 2.3, the covariance matrix is

$$\boldsymbol{\Sigma} = \left(\begin{array}{cc|cc} r(0) & r(s_2-s_1) & r(s_1) & -r'(s_1) \\ r(s_1-s_2) & r(0) & r(s_2) & -r'(s_2) \\ \hline r(s_1) & r(s_2) & r(0) & 0 \\ -r'(s_1) & -r'(s_2) & 0 & -r''(0) \end{array}\right) = \begin{pmatrix} \boldsymbol{\Sigma}_{\xi\xi} & \boldsymbol{\Sigma}_{\xi\eta} \\ \boldsymbol{\Sigma}_{\eta\xi} & \boldsymbol{\Sigma}_{\eta\eta} \end{pmatrix}.$$

With $\omega_0 = r(0)$, $\omega_2 = -r''(0)$, and $m_{\boldsymbol{\xi}} = m_{\boldsymbol{\eta}} = \mathbf{0}$, we get the conditional expectation and covariance matrix given $\boldsymbol{\eta} = \mathbf{y} = (u, z)$ as

$$E(\boldsymbol{\xi} \mid \boldsymbol{\eta} = \mathbf{y}) = \boldsymbol{\Sigma}_{\xi\eta}\boldsymbol{\Sigma}_{\eta\eta}^{-1}\mathbf{y}' = \begin{pmatrix} ur(s_1)/\omega_0 - zr'(s_1)/\omega_2 \\ ur(s_2)/\omega_0 - zr'(s_2)/\omega_2 \end{pmatrix}, \tag{8.29}$$

$$\boldsymbol{\Sigma}_{\xi\xi|\eta} = \boldsymbol{\Sigma}_{\xi\xi} - \boldsymbol{\Sigma}_{\xi\eta}\boldsymbol{\Sigma}_{\eta\eta}^{-1}\boldsymbol{\Sigma}_{\eta\xi} = \begin{pmatrix} r_\kappa(s_1,s_1) & r_\kappa(s_1,s_2) \\ r_\kappa(s_2,s_1) & r_\kappa(s_2,s_2) \end{pmatrix}, \text{ say.} \tag{8.30}$$

Here, one can evaluate $r_\kappa(s_1,s_2)$ in the covariance expression (8.30), and obtain the simple expression

$$r_\kappa(s_1,s_2) = r(s_2-s_1) - \frac{r(s_1)r(s_2)}{\omega_0} - \frac{r'(s_1)r'(s_2)}{\omega_2}. \tag{8.31}$$

This is obviously a covariance structure and it can be used as the covariance function for a non-stationary (Gaussian) process.

Note the structure of $r_\kappa(s_1, s_2)$: The first term is the unconditional covariance function of the process and the two other terms represent the reduction in covariance that is obtained by the knowledge of the uncorrelated $x(0)$ and $x'(0)$. When s_1 or s_2 tend to infinity, the reduction terms go to 0 and the influence of the conditioning vanishes.

Similarly, as for the covariance, one can use $ur(s)/\omega_0 - zr'(s)/\omega_2$ as a mean value function, with the provision that z is not a fixed number but the random value of the derivative at the upcrossing; thus, it is a random function. Combining the (random) mean value function and the covariance function (8.31) we have derived the explicit structure of the Slepian model in the Gaussian case; we formulate it as a theorem.

Theorem 8.9. *a) The Slepian model for a Gaussian process $\{x(t), t \in \mathbb{R}\}$ after u-upcrossings has the form*

$$\xi_u(t) = \frac{ur(t)}{\omega_0} - \frac{\zeta r'(t)}{\omega_2} + \kappa(t), \tag{8.32}$$

where ζ has the Rayleigh density $p^u(z) = (z/\omega_2)e^{-z^2/2\omega_2}$, $z \geq 0$, and the residual process $\{\kappa(t), t \in \mathbb{R}\}$ is a non-stationary Gaussian process, independent of ζ, with mean zero and covariance function $r_\kappa(s_1, s_2)$ given by (8.31).

b) In particular, the best prediction of $x(t_k + \tau)$ taken over all u-upcrossings t_k is obtained by replacing ζ by its expected value, $E(\zeta) = \sqrt{\pi\omega_2/2}$ and $\widehat{\kappa}(\tau) = 0$ in (8.32), to get

$$\widehat{x}^u(t_0 + \tau) = \frac{ur(\tau)}{\omega_0} - \frac{E(\zeta)r'(\tau)}{\omega_2} = \frac{u}{\omega_0}r(\tau) - \sqrt{\frac{\pi}{2\omega_2}}r'(\tau). \tag{8.33}$$

We have now found the correct way of taking the positive slope at a u-upcrossing into account in predicting the near future. Note that the simple formula (8.21),

$$\widehat{x}(t_0 + \tau) = E(x(t_0 + \tau) \mid x(t_0) = u) = \frac{u}{\omega_0}r(\tau),$$

in the Gaussian case, lacks the slope term in (8.33). The best prediction of the slope is the expectation of the Rayleigh variable ζ. If the slope at the u-upcrossing is observed and used in the prediction, the difference between the two approaches disappears; see [79].

The first two terms in (8.32) represent the *regression* on the process value at the crossing ($= u$) and the Rayleigh-distributed slope ($= \zeta$). The third term, $\kappa(t)$, is the Gaussian *residual* that describes the variation around the regression curve; it becomes the dominating term as $t \to \infty$, but is negligible near the crossing.

Example 8.1 (Excursion over a high level). *To illustrate the efficiency of the Slepian model, we shall analyze the shape of an excursion above a very high level u in a Gaussian process, by expanding the Slepian model $\xi_u(t)$ in a Taylor series as $u \to \infty$. It will turn out that the length and height of the excursion will both be of the order u^{-1}, so we normalize the scales of $\xi_u(t)$ by that factor. First, we have simply*

$$r(t/u) = \omega_0 - \omega_2 \frac{t^2}{2u^2}(1+o(1)), \quad r'(t/u) = -\omega_2 \frac{t}{u}(1+o(1)),$$

as $t/u \to 0$. Further, it is easy to see from the covariance function $r_\kappa(s_1,s_2)$ that $\kappa(t/u) = o(t/u)$, and we get, omitting all o-terms, and with ζ as the Rayleigh slope variable,

$$u\{\xi_u(t/u) - u\} = u\left\{u\left(\frac{r(t/u)}{\omega_0} - 1\right) - \zeta \frac{r'(t/u)}{\omega_2} + \kappa(t/u)\right\} \approx \zeta t - \frac{\omega_2 t^2}{2\omega_0}.$$

Thus, the excursion above a high level u takes the approximate form of a parabola with height $\frac{\zeta^2 \omega_0}{2u\omega_2}$ and length $\frac{2\omega_0 \zeta}{u\omega_2}$. It is easy to check that the normalized height of the excursion above u has an exponential distribution.

8.4.3 Local maxima in a Gaussian process

The distribution of the height of local maxima in a stationary process is of interest in reliability applications in engineering, and in general analysis of oscillatory phenomena. For a normal process, Rice, in [101], showed that the height of local maxima is distributed as a sum of a normal and a Rayleigh distributed random variable in proportions determined by the spectral width.

A local maximum of $x(t)$ is a zero-downcrossing point for the derivative $x'(t)$, and the second derivative $x''(t)$ at these local maxima has a negative Rayleigh distribution. A Slepian model for the derivative after local maxima, therefore, has the same structure as the level crossing model, with $r_x(t)$ replaced by $r_{x'}(t) = -r_x''(t)$.

If we want the distribution of the *height difference* between the maximum and the following minimum, we need a more elaborate Slepian model, since now also the height of the maximum is random, not only the curvature. The reader is encouraged to prove the following theorem, copying Theorem 8.7 with analogous notation, now with t'_k for the times of local maxima.

Theorem 8.10. *If $\{x(t), t \in \mathbb{R}\}$ is twice differentiable and ergodic, the long run empirical distribution of $x(t'_k + \mathbf{s})$ around local maxima at t'_k is equal to*

$$P_1^{\max}(A_{\mathbf{s}}(\mathbf{v})) = \frac{\int_{z=-\infty}^{0} |z| f_{x'(0),x''(0)}(0,z) P(x(\mathbf{s}) \leq \mathbf{v} \mid 0,z)\,\mathrm{d}z}{\int_{-\infty}^{0} |z| f_{x'(0),x''(0)}(0,z)\,\mathrm{d}z},$$

where

$$P(x(\mathbf{s}) \leq \mathbf{v} \mid 0,z) = P(x(\mathbf{s}) \leq \mathbf{v} \mid x'(0) = 0, x''(0) = z).$$

Similarly as for the upcrossing models, the Slepian models in Gaussian processes can be represented in explicit form as in the following theorem.

Theorem 8.11. *The Slepian model for a Gaussian process around local maxima has the explicit form*

$$\xi_1^{\max}(t) = \zeta_1^{\max} \frac{r''(t)}{\omega_4} + \Delta_1(t), \tag{8.34}$$

where the curvature $\zeta_1^{\max}$ has a negative Rayleigh distribution with density $p_1^{\max}(z) = |z|/\omega_4\, e^{-z^2/2\omega_4}$, for $z \leq 0$, and the non-stationary Gaussian process $\Delta_1(t)$ is independent of ζ_1^{max}, and has mean 0 and covariance function,

$$r_{\Delta_1}(s_1,s_2) = r(s_1 - s_2) - \frac{r'(s_1)r'(s_2)}{\omega_2} - \frac{r''(s_1)r''(s_2)}{\omega_4}. \tag{8.35}$$

Since $\Delta_1(0)$ is normal with mean 0 and variance $r_{\Delta_1}(0,0) = \omega_0 - \omega_2^2/\omega_4$, we can identify the distribution of the height of a local maximum as the distribution of

$$\xi_1^{\max}(0) = -\zeta_1^{\max}\frac{\omega_2}{\omega_4} + \Delta_1(0) = \sqrt{\omega_0}\left\{\sqrt{1-\varepsilon^2}\cdot R + \varepsilon\cdot N\right\}, \tag{8.36}$$

with R, N as standard Rayleigh and normal variables, illustrating the relevance of the spectral width parameter $\alpha = \sqrt{1-\varepsilon^2} = \sqrt{\omega_2^2/(\omega_0\omega_4)}$. The probability density function for the sum $\sqrt{1-\varepsilon^2}\cdot R + \varepsilon\cdot U$ is easily found:

$$f_{\max}(u) = \varepsilon\phi(u/\varepsilon) + \sqrt{1-\varepsilon^2}ue^{-u^2/2}\Phi(\sqrt{1-\varepsilon^2}u/\varepsilon), \tag{8.37}$$

where ϕ and Φ are the standard normal density and distribution functions. Figure 8.3 shows the density $f_{\max}(u)$ for $\varepsilon = 0$, representing the pure Rayleigh distribution, in steps of 0.1, to $\varepsilon = 1$, which gives the normal density.

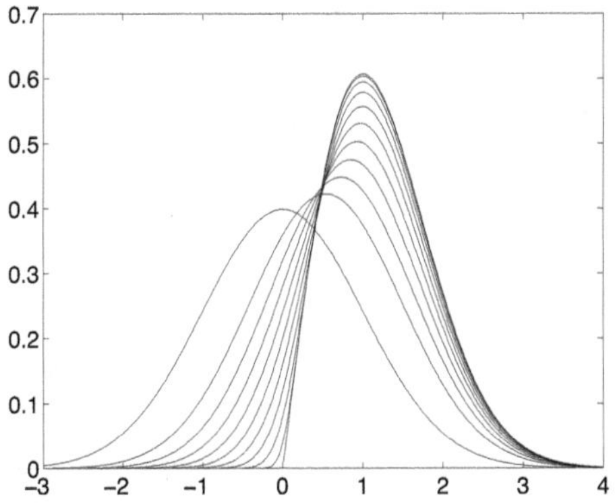

Figure 8.3 *Density* (8.37) *of height of local maxima in a standard normal process; spectral parameter* $\varepsilon = 0$ *(Rayleigh) to* $\varepsilon = 1$ *(normal).*

Extended model around local maximum

The model (8.34) contains the random curvature, and it is the simplest form of the Slepian model after maximum. There is nothing that prevents us to include also the random height of the local maximum in the model. We have seen in (8.36) how the height and the curvature depend on each other, so we can build an alternative Slepian model after maximum that explicitly includes both the height of the maximum and the curvature.

To formulate the extended model we define three regression functions,

$A(t), B(t), C(t)$, from the conditional expectation,

$$E(x(t) \mid x(0) = u, x'(0) = y, x''(0) = z) = uA(t) + yB(t) + zC(t)$$

$$= u\,\frac{\omega_4 r(t) + \omega_0\omega_2 r''(t)}{\omega_0\omega_4 - \omega_2^2} - y\,\frac{r'(t)}{\omega_2} + z\,\frac{\omega_2 r(t) + \omega_0 r''(t)}{\omega_0\omega_4 - \omega_2^2}.$$

The conditional covariance between $x(s_1)$ and $x(s_2)$ is found as previously, and the explicit expression is given in the following Theorem 8.13. To formulate the effect of observing a local maximum we will first introduce the crest height, $x(0)$, and then find the conditional properties of $x''(0)$, given $x(0)$ and $x'(0)$. We use Theorem 2.3 and define the function

$$b(t) = \frac{Cov(x(t), x''(t) \mid x(0), x'(0))}{\sqrt{V(x''(0) \mid x(0), x'(0))}} = \frac{r''(t) + (\omega_2/\omega_0)r(t)}{\sqrt{\omega_4 - \omega_2^2/\omega_0}}. \tag{8.38}$$

Theorem 8.12. *If $\{x(t), t \in \mathbb{R}\}$ is twice differentiable and ergodic, the empirical (long run) distribution of $x(t'_k + \mathbf{s})$ around local maxima at t'_k is equal to*

$$P_2^{\max}(A)$$

$$= \frac{\int_{u=-\infty}^{\infty}\int_{z=-\infty}^{0} |z| f_{x(0),x'(0),x''(0)}(u,0,z) P(x(\mathbf{s}) \le \mathbf{v} \mid u,0,z)\,\mathrm{d}z\,\mathrm{d}u}{\int_{u=-\infty}^{\infty}\int_{-\infty}^{0} |z| f_{x(0),x'(0),x''(0)}(u,0,z)\,\mathrm{d}z\,\mathrm{d}u},$$

where

$$P(x(\mathbf{s}) \le \mathbf{v} \mid u,0,z) = P(x(\mathbf{s}) \le \mathbf{v} \mid x(0) = u, x'(0) = 0, x''(0) = z).$$

To obtain the explicit Slepian model for Gaussian processes one just has to add the regression terms, depending on the maximum height, to get the following explicit representation; see [77].

Let $\Delta_2(t)$ be non-stationary Gaussian with mean zero, independent of $(\eta_2^{\max}, \zeta_2^{\max})$, and with covariance function, using (8.38),

$$r_{\Delta_2}(s,t) = r(s-t) - \frac{r(s)r(t)}{\omega_0} - \frac{r'(s)r'(t)}{\omega_2} - b(s)b(t).$$

Theorem 8.13. *An explicit representation of the Slepian models for a Gaussian process around a local maximum is*

$$\xi_2^{\max}(t) = \eta_2^{\max} A(t) + \zeta_2^{\max} C(t) + \Delta_2(t), \tag{8.39}$$

where the random $(\eta_2^{\max}, \zeta_2^{\max})$ *has the two-dimensional density (with normalizing constant* c*),*

$$p_2^{\max}(u,z) = c|z| \exp\left\{ -\frac{\omega_0 z^2 + 2\omega_2 uz + \omega_4 u^2}{2(\omega_0\omega_4 - \omega_2^2)} \right\},$$

for $-\infty < u < \infty, z < 0$*. The variables* $\eta_2^{\max}$ *and* $\zeta_2^{\max}$ *represent the random height and the curvature at the maximum, respectively, and they catch the main structure of the crest shape.*

8.4.4 Excursion length and related distributions

One of the most intriguing problems in stationary process theory is that of the distribution of the length of excursions above a critical fixed level. Even for Gaussian processes, no explicit solution is known, except in a few cases. At the same time, applications require a practical answer, given a correlation structure, and therefore many approximate solutions have been suggested, mostly based on some ad hoc assumption of independence between successive intervals, or by means of upper and lower bound including higher moments of the number of crossings, so called "Rice series"; cf. Exercise 1.10.

In this section we present a method to numerically calculate the *exact* distributions of excursion length and other crossing related quantities. The technique is a combination of the Slepian model, and a general formula on the first-passage density, given by J. Durbin, for Gaussian processes [44], and generalized by I. Rychlik [108, Thm. 2] to certain non-Gaussian processes.

Lemma 8.2. *Let* $\{y(t), 0 \le t < \infty\}$ *be a continuous process and define* $T = \inf\{t > 0; y(t) > 0\} \le \infty$*. If, with probability one,* $y(s)$ *has no tangent of the level* 0*, then*

$$P(T \ge t) = P(y(s) < 0, \text{ for all } s \in (0,t)). \tag{8.40}$$

Theorem 8.14 (Durbin's formula). *If the conditional distribution of $y(s), 0 \le s \le t$, given $y(0)$, is non-degenerate, in a sense made precise in [108], then T has density function*

$$f_T(t) = f_{y(t)}(0)\, E\left(I\{y(s) < 0, \text{ for all } s \in (0,t)\} \cdot y'(t)^+ \mid y(t) = 0\right) \tag{8.41}$$

with $x^+ = \max(0,x)$, and $I\{A\}$ denoting the indicator function of the event A.

Proof. The lemma is clear; Bulinskaya's lemma, Theorem 2.14, gives sufficient conditions for no tangents. The proof of Durbin's formula is similar to the alternative proof of Rice's formula, Theorem 8.1.3 [108]. □

Remark 8.6 (Numerical aspects). *Durbin's formula is exact. The only catch is that it contains an infinite dimensional expectation that has to be evaluated numerically. That used to be regarded as a definite obstacle in the past, but advances in numerical methods and computer software have made it quite realistic to evaluate the expectation with very high accuracy, at least for Gaussian problems.*

The simple idea in the evaluation of the conditional expectation in (8.41) *is to replace the infinite dimensional event $\{y(s) < 0, 0 < s < t\}$ by a finite-dimensional event based on a finite set of time points $s_k \in (0,t)$, and then evaluate the probability $P(y(s_k) < 0, k = 1, \dots, n)$. For smooth processes the accuracy in this approximation is very high if n is large and the points chosen in an optimal way. There exists several numerical integration algorithms for normal distributions that can do the job with $n = 100$ and even higher.*

The calculations in this section have been made by the routines in the MATLAB *package* WAFO *[119]. The central integration routine in* WAFO *is partly based on numerical routines developed by A. Genz [53, 54]. Another software package for multi-variate normal computations, based on the same integration routine, is the package* MAGP *[91].*

Slepian models in Durbin's formula

One of the advantages of a Slepian model in the Gaussian case is that it lends itself to efficient numerical calculations of important quantities related to crossings and maxima. The structure of the model is such that the random variables in the regression terms that represent slope, height, and curvature in the crossings and crest models have simple explicit density functions, which are easy to handle numerically. The regression terms act as a skeleton in the Slepian models.

The Gaussian residual processes describe the variation around the skeleton, and they are determined by their non-stationary covariance functions. To find the distribution of crossing and wave characteristics in the total model one can use Durbin's formula combined with a numerical integration algorithm. One can then use a successive conditioning technique to reduce the uncertainty. For example, consider the model (8.32) for the process after a u-upcrossing. If we introduce the value of the normal residual at an extra single point, s_1 say, and include $\kappa(s_1)$ as a separate regression term, we get an extended model. The residual process in the extended model will have correspondingly reduced variability, and if the procedure is repeated, the remaining residual can, after a few steps, be disregarded completely. This approximation technique is called the *regression approximation* in crossing theory; the technique is described in detail in [83].

Length of excursions above a level

Theorem 8.9 is the basis for numerical calculations of crossing characteristic distributions like the excursion length. The model (8.32),

$$\xi_u(t) = \frac{ur(t)}{\omega_0} - \frac{\zeta r'(t)}{\omega_2} + \kappa(t) = m_\zeta(t) + \kappa(t),$$

contains a smooth regression term, $m_\zeta(t)$, and a Gaussian residual term, $\kappa(t)$. We know that $\xi_u(0) = u$, upcrossing, and we now want to compute the distribution of T, the time of first downcrossing of u by $\xi_u(t)$, so T is the length of an excursion above u. Since $T > t$ if and only if $\xi_u(s)$ stays above the level u in the entire interval $0 < s < t$, we can use Durbin's formula with $y(s) = u - m_\zeta(s) - \kappa(s)$, and express the probability $P(T > t)$ by means of the indicator function

$$I_z(\kappa, t) = \begin{cases} 1, & \text{if } m_z(s) + \kappa(s) > u, \text{ for all } s \in (0,t), \\ 0, & \text{otherwise.} \end{cases}$$

The result is a simple weighted average,

$$P(T > t) = \int_{z=0}^{\infty} p^u(z) \cdot E(I_z(\kappa, t))\, \mathrm{d}z, \tag{8.42}$$

where $p^u(z)$ is the Rayleigh density for the derivative at u-upcrossing, and the expectation

$$E(I_z(\kappa,t)) = P(m_z(s) + \kappa(s) > u, \text{for all } s \in (0,t))$$

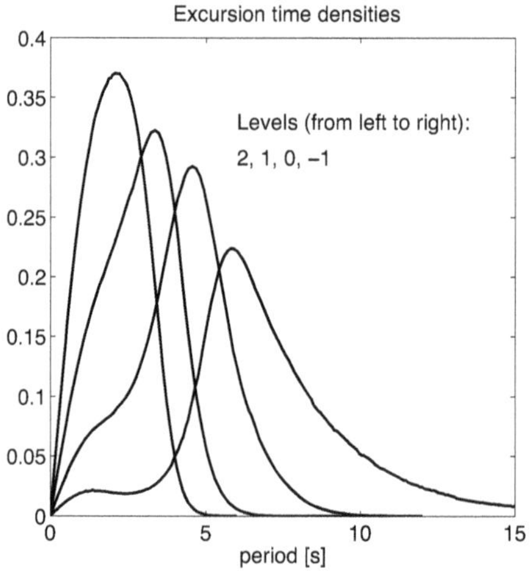

Figure 8.4 *Probability densities for excursions above* $u = -1, 0, 1, 2$ *for process with North Sea wave spectrum* JONSWAP.

is an infinite dimensional normal probability. That probability is calculated numerically, for example by means of routines in the MATLAB package WAFO [119] and integrated as in (8.42).

Figure 8.4 shows the excursion length densities for a realistic water wave process with a common JONSWAP North Sea wave spectrum. (For definition of the JONSWAP spectrum, see (7.33).)

The probability density function for the excursion time T is of course minus the t-derivative of $P(T > t)$. Expressed as in Durbin's formula, the expression reads

$$f_T(t) = f_{\xi_u(t)}(u) E\left(I\{\xi_u(s) > u, 0 < s < t\} \cdot (-\xi_u'(t)^-) \mid \xi_u(t) = u\right), \quad (8.43)$$

where $\xi_u'(t)^- = \min(0, \xi_u'(t))$ is the negative part of the derivative.

Wave shape

The distribution of wave characteristics such as drop in height and time difference between a local maximum and the next local minimum can be derived from the Slepian models in Theorem 8.11 or Theorem 8.13.

First consider model (8.34),

$$\xi_1^{\max}(t) = \zeta_1^{\max} \frac{r''(t)}{\omega_4} + \Delta_1(t),$$

which completely describes the stochastic properties of the shape around maximum. The simplest, *zero order*, approximation is to delete the residual process $\Delta_1(t)$ completely, only keeping the curvature dependent term $\zeta_1^{\max}\frac{r''(t)}{\omega_4}$. By replacing $\zeta_1^{\max}$ by its average $-\sqrt{\omega_4\pi/2}$, we can, for example, get the average shape, as

$$\widehat{\xi^{\max}}(t) = -\sqrt{\omega_0\pi/2}\,\alpha\frac{r''(t)}{\omega_2}.$$

The zero order approximation is usually too crude to be of any use. A better approximation is obtained from the model (8.39), which also includes the (random) height at the maximum point,

$$\xi_2^{\max}(t) = \eta_2^{\max}A(t) + \zeta_2^{\max}C(t) + \Delta_2(t).$$

Define the random variable T as the time of the first local minimum of $\xi_2^{\max}(t), t > 0$. The height drop is then $H = \xi_2^{\max}(0) - \xi_2^{\max}(T)$ and we ask for the joint distribution of T and H.

Using the fact that $A(0) = 1, C(0) = 0$ and $\xi_2^{\max}(T)' = 0$, we get the following relations that need to be satisfied,

$$\begin{aligned} \eta_2^{\max}A'(T) + \zeta_2^{\max}C'(T) + \Delta_2'(T) &= 0, \\ \eta_2^{\max} + \Delta_2(0) - (\eta_2^{\max}A(T) + \zeta_2^{\max}C(T) + \Delta_2(T)) &= H. \end{aligned}$$

We now describe the regression approximation of order 1, which is obtained by deleting all of the residual process terms. The relations will then be

$$\eta_2^{\max}A'(T^r) + \zeta_2^{\max}C'(T^r) = 0, \tag{8.44}$$

$$\eta_2^{\max} - (\eta_2^{\max}A(T^r) + \zeta_2^{\max}C(T^r)) = H^r, \tag{8.45}$$

where we write T^r, H^r for the approximative time and height variables.

To write the solution in a form that can be generalized to more complicated problems, define

$$G(t) = \begin{pmatrix} 1 - A(t) & C(t) \\ A'(t) & C'(t) \end{pmatrix},$$

and write equations (8.44) and (8.45) as (with ^{T} for matrix transpose)

$$G(T^r)(\eta_2^{\max}\ \zeta_2^{\max})^T = (H^r\ T^r)^T.$$

If $\det G(T^r) \neq 0$ we get from $(\eta_2^{\max}\, \zeta_2^{\max})^T = G(T^r)^{-1}(H^r\, 0)^T$ that the variables with known distribution ($\eta_2^{\max}$ and $\zeta_2^{\max}$) are simple functions of the variables with unknown distribution,

$$\eta_2^{\max} = H^r p(T^r) q(T^r), \quad \zeta_2^{\max} = H^r q(T^r),$$

where

$$p(t) = \frac{-C'(t)}{A'(t)}, \quad q(t) = \frac{-A'(t)}{(1-A(t))C'(t) - A'(t)C(t)}.$$

We want the density at the point $T^r = t, H^r = h$; let $\eta(t,h), \zeta(t,h)$ be the corresponding solution and define the indicator function $I(t,h)$ to be 1 if the approximating process $\eta(t,h)A(s) + \zeta(t,h)C(s)$ is strictly decreasing for $0 < s < t$.

The Jacobian for the transformation is $J(t,h) = hp'(t)q(t)^2$, and therefore the density of T^r, H^r is

$$\begin{aligned} f_{T^r,H^r}(t,h) &= f_{\eta_{\max},\zeta_{\max}}(hp(t)q(t), hq(t)) \cdot |J(t,h)|\, I(t,h) \\ &= \text{const} \times I(t,h)\, h^2 |q^2(t)^3 p'(t)| \\ &\qquad \times \exp\{-\frac{1}{2\varepsilon^2} h^2 q(t)^2 (T_m/\pi)^4 (((\pi/T_m)^2 p(t) + 1)^2 + \frac{\varepsilon^2}{1-\varepsilon^2})\}. \end{aligned}$$

This form of the (T,H)-distribution is common in the technical literature, where $T_m = \pi\sqrt{\omega_2/\omega_4}$ is called the *mean half wave period.* Note that this distribution depends on the spectrum only through the spectral width parameter

$$\varepsilon = \sqrt{1 - \frac{\omega_2^2}{\omega_0\omega_4}} = \sqrt{1-\alpha^2}. \tag{8.46}$$

This first order approximation of the T,H-density is not very accurate, but it illustrates the basic principle of the regression approximation. The WAFO toolbox [119] contains algorithms for very accurate higher order approximations, in which the values of the residual processes $\Delta_j(s)$ at several selected s-values are taken into account in the conditioning.

Joint distribution of excursion length and height

As a final example we show the joint density, computed in WAFO, of the duration and height of an excursion above the mean level. The computations include an extension of (8.43) to also include an upper restriction on $\xi_u(s)$ for $0 < s < t$. The process is Gaussian with a North Sea JONSWAP spectrum, defined by (7.33). In ocean engineering, excursion period and height are important characteristics, called *crest period and amplitude.*

Figure 8.5 shows the result together with a simulated section of the process. The process standard deviation is 1.75[m] and the mean period is 8.4[s]. The simulation represents more than one hour of wave observations, and there were 481 observed wave crests, shown as dots in the right diagram. The observed mean crest period and amplitude were 4.1[s] and 2[m], respectively.

From the figure, one can draw several practically important conclusions: very high waves tend to have a period near the mean period, the majority of waves last less than 2 seconds and are less than 1 meter high, the probability of short but high waves is very small.

8.5 Crossing problems for vector processes and fields

8.5.1 Crossings by vector processes

In reliability applications it is often the combined effects of a set of random load processes that represent the critical event that causes a construction to break down, for example the simultaneous occurrence of extreme wind and high waves on a marine construction. If the load processes are modeled as a stationary vector process, $\{\mathbf{x}(t), t \in \mathbb{R}\}$, $\mathbf{x}(t) = (x_1(t), \dots, x_n(t))$, one can formulate the safety problem in terms of the *exit rate* from a safe region $S \subset \mathbb{R}^n$: the system is safe as long as $\mathbf{x}(t) \in S$, and each excursion outside S represents

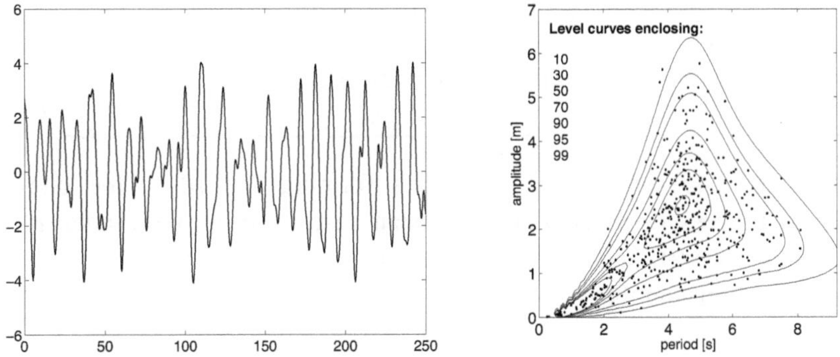

Figure 8.5 *Probability density for crest period T_c and crest height A_c for excursions above the mean level for a process with North Sea* JONSWAP *spectrum together with 481 observed cycles.*

a danger to the system. Then a critical quantity is the average number of exits per time unit by the vector process $\mathbf{x}(t)$.

A Rice formula for vector processes

Let $\{\mathbf{x}(t), t \in \mathbb{R}\}$ be a stationary vector process with p differentiable components, and assume the safe region S is bounded by a smooth surface ∂S in $\mathbb{R}^p$. We can assume that ∂S is defined as a level set $\partial S_u = \{\mathbf{x}; g(\mathbf{x}) = u\}$ of a smooth real-valued function $g(\mathbf{x})$. Denote by $\boldsymbol{\nu}_\mathbf{x} = (\nu_\mathbf{x}^1, \dots, \nu_\mathbf{x}^p)$ the normal, of unit length, pointing out from the safe region $S_u = \{\mathbf{x}; g(\mathbf{x}) \leq u\}$, at the point $\mathbf{x}$ on ∂S_u, and let $\mathbf{x}'(t)$ be the derivative vector process of the $\mathbf{x}$-process. We denote by $\mathrm{d}s(\mathbf{x})$ the surface area measure on S.

The Rice formula for vector process gives the expected number of times per time unit that the process $\mathbf{x}(t)$ crosses the boundary ∂S_u in the outward direction. Just as in the one-dimensional case, there is a slope bias, and in this case the bias factor is the positive part of the scalar product $< \boldsymbol{\nu}_\mathbf{x}, \mathbf{x}(t)' >= \sum_j \nu_\mathbf{x}^j x_j'(t)$, between the derivative vector and the unit normal. The following theorem was proved by Yu.K. Belyaev (1968) [11] and generalized in [80].

Theorem 8.15. *The expected number (per time unit) of outgoing crossings by a stationary differentiable vector process $\{\mathbf{x}(t), t \in \mathbb{R}\}$ across the smooth boundary ∂S_u is given by*

$$\mu(S_u) = \int_{\partial S_u} E\left(< \boldsymbol{\nu}_\mathbf{x}, \mathbf{x}(t)' >^+ \mid \mathbf{x}(0) = \mathbf{x}\right) f_{\mathbf{x}(0)}(\mathbf{x})\, \mathrm{d}s(\mathbf{x}) \quad (8.47)$$

$$= f_{g(\mathbf{x}(0))}(u)\, E\left(\left\{\frac{\mathrm{d}}{\mathrm{d}t} g(\mathbf{x}(t))\right\}_{t=0}^+ \mid g(\mathbf{x}(0)) = u\right), \quad (8.48)$$

where $f_{g(\mathbf{x}(0))}(u)$ is the probability density of $g(\mathbf{x}(0))$.

Note that (8.48) is the general Rice's formula applied to the one-dimensional process $\{g(\mathbf{x}(t)\}$. For a Gaussian vector process with independent components, $\mathbf{x}(0)$ and $\mathbf{x}(0)'$ are independent, so (8.47) can be slightly simplified by observing that

$$E\left(< \boldsymbol{\nu}_\mathbf{x}, \mathbf{x}(t)' >^+ \mid \mathbf{x}(0) = \mathbf{x}\right) = E\left(< \boldsymbol{\nu}_\mathbf{x}, \mathbf{x}(t)' >^+\right)$$
$$= E\left(\{\sum_j \nu_\mathbf{x}^j x_j'(t)\}^+\right). \quad (8.49)$$

Crossings by the χ^2-process

Even for a Gaussian process the integral in Theorem 8.15 may be hard to evaluate, except in special cases. As an example of such a special case we give the explicit Rice's formula for the χ^2-process, $\chi^2(t) = \sum_j x_j(t)^2$, with independent normal components, with mean zero, variance one, and common covariance function. If we denote the common variance of the derivatives by $\omega_2 = V(x_j'(t))$, the sum in (8.49) is normal and its variance is also ω_2, since $\mathbf{v_x}$ has unit length. Obviously, crossings of level u^2 by $\chi^2(t)$ are equivalent to crossings of level u by $\sqrt{\chi^2(t)}$, and we chose to formulate the upcrossing intensity in terms of the square root process. Thereby we also get the expected number of exits from a sphere with radius u by the vector process $\mathbf{x}(t)$.

Theorem 8.16. *The expected number of upcrossings of the level u by a $\sqrt{\chi^2}$-process with p degrees of freedom in which all components have the same time-scale, $V(x_j'(t)) = \omega_2$, is equal to*

$$\mu^+_{\sqrt{\chi^2}}(u) = \frac{(\omega_2/\pi)^{1/2}}{2^{(p-1)/2}\Gamma(p/2)}\, u^{p-1} e^{-u^2/2}. \tag{8.50}$$

Proof. The expectation in (8.49) is $\sqrt{\omega_2/(2\pi)}$, and the density in (8.47) is constant, $f_{\mathbf{x}(0)}(\mathbf{x}) = (2\pi)^{-p/2} e^{-\sum_j x_j^2/2} = (2\pi)^{-p/2} e^{-u^2/2}$, on the sphere $\sqrt{\chi^2(t)} = u$. Thus, the intensity $\mu_{\chi^2}(u)$ is equal to $\sqrt{\omega_2/(2\pi)}(2\pi)^{-p/2}e^{-u^2/2}$ times the area of the p-dimensional sphere with radius u, which is equal to $2\pi^{p/2}u^{p-1}/\Gamma(p/2)$. Simplifying the product we get the result. □

Envelope crossings

The χ^2-process is the sum of squares of independent Gaussian processes. The envelope process $R(t) = \sqrt{x(t)^2 + \widehat{x}(t)^2}$ to a stationary process $\{x(t), t \in \mathbb{R}\}$ is the square root of the sum of squares of two *dependent* Gaussian processes, namely $x(t)$ and its Hilbert transform $\widehat{x}(t)$; see Definition 5.2. Even if the marginal distribution of $R^2(t)$ is the same as that of the χ^2-process with two degrees of freedom, namely exponential, the level crossing properties of the two processes are quite different. The reason for this is the fact that the derivative of the process is strongly correlated with the Hilbert transform.

First, recall the properties of the Hilbert transform from Theorem 5.1. Assume $x(t)$ has no spectral mass at $\omega = 0$, so $x(t)$ and $\widehat{x}(t)$ have the same covariance function $r(t) = \int_0^\infty \cos \omega t \, dG(\omega)$, with spectral moments

$$\omega_0 = r(0) = V(x(0)) = V(\widehat{x}(0)),$$
$$\omega_1 = -C(x'(0), \widehat{x}(0)) = C(x(0), \widehat{x}'(0)) = \int_0^\infty \omega \, dG(\omega),$$
$$\omega_2 = V(x'(0)) = V(\widehat{x}'(0)) = \int_0^\infty \omega^2 \, dG(\omega).$$

The mean period is then equal to $T_2 = 2\pi\sqrt{\omega_0/\omega_2}$. Define $\rho^2 = \omega_1^2/(\omega_0\omega_2)$ as the squared correlation between Hilbert transform and derivative.

Theorem 8.17 (Envelope upcrossings). *The expected number of u-level upcrossings per time unit by the envelope $R(t)$ to a stationary Gaussian process $\{x(t), t \in \mathbb{R}\}$ is*

$$\mu_R^+(u) = \sqrt{\frac{\omega_2(1-\rho^2)}{2\pi\omega_0}} \frac{u}{\sqrt{\omega_0}} e^{-u^2/(2\omega_0)}, \tag{8.51}$$

with average number of envelope u-upcrossings per mean period

$$T_2 \, \mu_R^+(u) = \sqrt{2\pi(1-\rho^2)} \frac{u}{\sqrt{\omega_0}} e^{-u^2/(2\omega_0)}. \tag{8.52}$$

Proof. Since an envelope u-upcrossing is equivalent to a circle-outcrossing by the vector process $(x(t), \widehat{x}(t))$, one can apply Theorem 8.15 to get the result. One can also use the general Rice's formula (8.3) for upcrossings applied to the squared envelope $U(t) = R^2(t)$,

$$\mu_R^+(u) = E(U'(0)^+ \mid U(0) = u^2) \, f_{U(0)}(u^2), \tag{8.53}$$

where the density $f_{U(0)}$ is exponential with mean $2\omega_0$. Further, the conditional distribution of $U'(0) = 2x'(0)\widehat{x}(0) + 2x(0)\widehat{x}'(0)$, given that $U(0) = u^2$, is normal with mean zero and variance

$$V(2x'(0)\widehat{x}(0) + 2x(0)\widehat{x}'(0) \mid x(0)^2 + \widehat{x}(0)^2 = u^2) = 4u^2\,\omega_2\,(1-\rho^2).$$

(First condition on the individual values of $x(0)$ and $\widehat{x}(0)$.) Thus, the conditional expectation in (8.53) is $\sqrt{4u^2\,\omega_2\,(1-\rho^2)/(2\pi)}$, which, combined with

$$f_{U(0)}(u^2) = \frac{1}{2\omega_0} e^{-u^2/(2\omega_0)},$$

gives the result. □

Example 8.2 (The seventh wave). *As an example of the theorem, we give an illustration to the seventh wave phenomenon, the observation that waves on a shore or on the ocean seem to have a typical regularity of one big wave followed by six smaller ones. For a reasonably narrow wave spectrum the envelope is a smooth process with slowly varying amplitude, and the observed wave heights follow the envelope quite well. For Gaussian waves, formula* (8.52) *gives approximately the inverted number of waves per observed envelope upcrossing, as a function of the height.*

Figure 8.6 shows a JONSWAP spectrum with cut-off frequency $1.2[rad/s]$ *and the average number of waves per high wave, according to* (8.52)*. As seen, for waves with a crest height of twice the standard deviation of the sea surface, seven is the magic number, explained as a fading phenomenon between interfering waves. Of course, the regularity is much dependent on the narrowness of the spectrum as measured by the quantity* ρ^2.

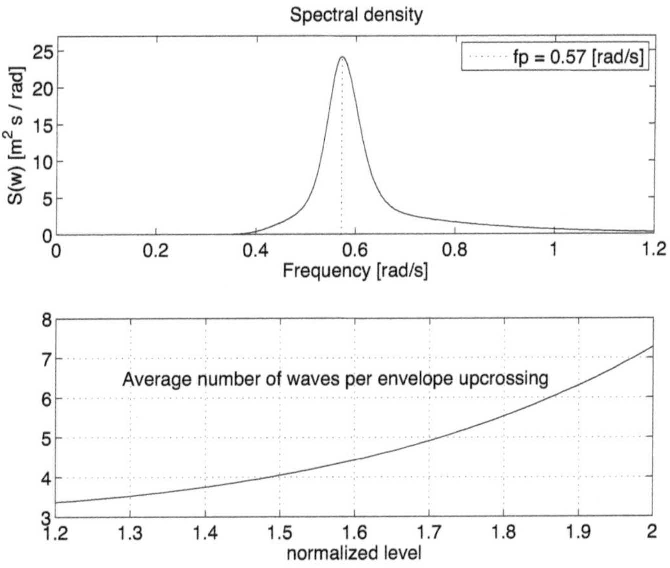

Figure 8.6 *Moderately narrow JONSWAP spectrum, and the average number of waves per envelope upcrossing as function of level.*

8.5.2 *Level sets in fields*

The theory and applications of contour curves, level sets, and other characteristics of random fields have a history almost as old as that of Rice's formula for processes. Longuet-Higgins, in the substantial paper [86] from 1957, defined and analyzed the statistical distributions of a large number of geometrical characteristics of moving random surfaces, of particular interest to marine applications; cf. also [116]. Other areas where statistics of contour curves have stimulated the theoretical development of random field theory are astronomy, forestry, meteorology, mining, optics, and turbulence; see the book by Azaïs and Wschebor [6].

Here we shall just present a very useful generalization of Rice's formula to multivariate random fields, i.e., random functions

$$\mathbb{R}^p \ni \mathbf{t} = (t_, \ldots, t_p) \mapsto (X_1(\mathbf{t}), \ldots, X_p(\mathbf{t})) = \mathbf{X}(\mathbf{t}) \in \mathbb{R}^p.$$

More specifically, we seek the expected number of solutions of the equation $\mathbf{X}(\mathbf{t}) = \mathbf{u}$, i.e.,

$$N_S(\mathbf{X}, \mathbf{u}) = \#\{\mathbf{t} \in S; \mathbf{X}(\mathbf{t}) = \mathbf{u}\},$$

for a fixed $\mathbf{u} = (u_1, \ldots, u_p)$ and S any Borel subset of $\mathbb{R}^p$.

As an example of this type of problem, we may look at a common model for laser light. After having been diffused in an opaque medium the light that falls on a plane screen can be modeled by a complex Gaussian field, $x_1(\mathbf{s}) + i x_2(\mathbf{s})$, $\mathbf{s} \in \mathbb{R}^2$, with independent zero mean Gaussian real and imaginary components. The zero-level curves for the two fields are random curves in the plane, and when these curves cross, the light intensity $\sqrt{x_1(\mathbf{s})^2 + x_2(\mathbf{s})^2}$ is exactly zero, and a black spot appears on the screen. The expected number of black spots per unit area is equal to the expected number of solutions to $(x_1(\mathbf{s}), x_2(\mathbf{s})) = (0,0)$, $\mathbf{s} \in [0,1] \times [0,1]$.

The multidimensional Rice's formula is based on the following generalization of Banach's formula in Lemma 8.1. If $f(\mathbf{s})$ is a continuously differentiable function from $\mathbb{R}^p$ to $\mathbb{R}^p$, and $g(\mathbf{u})$ is real valued, continuous, and bounded, then the following formula, called the *area formula*, holds:

$$\int_{\mathbb{R}^p} g(\mathbf{u}) N_S(\mathbf{X}, \mathbf{u})\, d\mathbf{u} = \int_S |\det \mathbf{X}'(\mathbf{t})|\, g(\mathbf{X}(\mathbf{t}))\, d\mathbf{t}.$$

Here, $\mathbf{X}'$ is the Jacobian, the matrix of partial derivatives of $\mathbf{X}$.

We quote the following theorem [6, Thm. 6.2], based on the area formula.

Theorem 8.18 (Multidimensional Rice formula). *If $\mathbf{X}(\mathbf{t})$ is a p-dimensional Gaussian random field with non-degenerate distribution, such that, with probability zero, there is no critical point such that $\mathbf{X}(\mathbf{t}) = \mathbf{u}$ and $\det \mathbf{X}'(\mathbf{t}) = 0$, then*

$$E(N_S(\mathbf{X}, \mathbf{u})) = \int_S E\left(\left|\det \mathbf{X}'(\mathbf{t})\right| \mid \mathbf{X}(\mathbf{t}) = \mathbf{u}\right) f_{\mathbf{X}(\mathbf{t})}(\mathbf{u})\,\mathrm{d}\mathbf{t}, \quad (8.54)$$

where $f_{\mathbf{X}(\mathbf{t})}(\mathbf{u})$ is the density of $\mathbf{X}(\mathbf{t})$.

The reader should compare (8.54) with the corresponding formula (8.6) for a non-stationary process. For applications of the theorem, see [1, 78, 81] and [6, Ch. 11]. In [6, Ch. 6] one can also find a discussion of the absence of critical points; cf. Bulinskaya's lemma, Theorem 2.14.

Exercises

8:1. Let (x_n, y_n) have a bivariate Gaussian distribution with mean 0, variance 1, and correlation coefficient ρ_n.

a) Show that $P(x_n < 0 < y_n) = \frac{1}{2\pi} \arccos \rho_n$.

b) Calculate the conditional density functions for

$$(x_n + y_n) \mid x_n < 0 < y_n, \quad \text{and} \quad (y_n - x_n) \mid x_n < 0 < y_n.$$

c) Let z_n and u_n be distributed with the density functions derived in (b) and assume that $\rho_n \to 1$ as $n \to \infty$. Take $c_n = 1/\sqrt{2(1-\rho_n)}$, and show that the density functions for $c_n z_n$ and $c_n u_n$ converge to density functions f_1 and f_2, respectively.

Hint: $f_2(u) = u \exp(-u^2/2), u \geq 0$ is the Rayleigh density.

8:2. Give an example of a stationary, differentiable process $\{x(t), t \in \mathbb{R}\}$, such that $x(t), x'(t)$ do not have a joint density, and calculate the crossing intensity as in (8.2).

8:3. Show that with a_T, b_T defined by (8.12) and $u_T = x/a_T + b_T$, one gets $T\mu^+(u_T) \to e^{-x}$ for a standardized Gaussian process. (This is part of Theorem 8.4.)

8:4. Let $\{x(t), t \in \mathbb{R}\}$ be a stationary Gaussian process with mean 0, and with a covariance function that satisfies

$$-r''(t) = -r''(0) + o(|t|^a), \quad t \to 0,$$

for some $a > 0$. Define $x_n = x(0)$, $y_n = x(1/n)$, $\rho_n = r(1/n)$, and use the previous exercise to derive the asymptotic distribution of

$$\left.\frac{x(1/n) - x(0)}{1/n}\ \right|\ x(0) < 0 < x(1/n)$$

as $n \to \infty$. What conclusion do you draw about the derivative at a point with an upcrossing of the zero level? What is the result if you condition on the event $x(0) < u < x(1/n)$?

8:5. Consider the Slepian model $\xi_u(t) = \frac{ur(t)}{\omega_0} - \frac{\zeta r'(t)}{\omega_2} + \kappa(t)$ in Theorem 8.9 for a stationary Gaussian process after an upcrossing of the fixed level u. Show that $\kappa(0) = \kappa'(0) = 0$ from its covariance function.

8:6. (Continuation) Formulate conditions on the covariance functions $r_x(t)$ that guarantee that the residual process $\kappa(t)$ has differentiable sample paths.

8:7. Prove Theorem 8.10.

8:8. Prove, or at least make plausible, the formula for the variance (8.10) using the idea in Figure 8.1.

8:9. Generalize Theorem 8.16 and find an expression for the upcrossing rate for a χ^2-process, when the components have different time-scales, i.e., $V(x_j'(t))$ depends on j.

8:10. Find the argument for the statement in Section 8.1.1 that if a process has continuous marginal distributions then there is, with probability one, no interval of positive length in which the process stays constant.

Appendix A

Some probability theory

In this appendix we summarize the concepts of event, σ-algebra, and random variable, the probability axioms, expectation and conditional expectation, stochastic convergence, characteristics functions, and some facts about Hilbert spaces; for more reading, see [15, 19, 25, 57, 58, 69].

A.1 Events, probabilities, and random variables

A.1.1 Events and families of events

A probability measure P assigns probabilities to certain events, i.e., subsets, in the sample space Ω, in such a way that Kolmogorov's probability axioms are satisfied. If a subset A has a probability, then also its complement A^* has a probability, and $P(A^*) = 1 - P(A)$, and further, if A is disjoint with B and B has probability $P(B)$, then also $A \cup B$ has a probability, and $P(A \cup B) = P(A) + P(B)$. These requirements lead to the conclusion that probabilities have to be defined at least on a certain minimal family of subsets of Ω.

Definition A.1. *(a) A family of subsets $\mathscr{F}_0$ to an arbitrary space Ω is called a field (or algebra) if it contains Ω and is closed under the set operations complement, A^*, union, $A \cup B$, and intersection, $A \cap B$, i.e., if A and B are sets in $\mathscr{F}_0$, then also the complement A^* and the union $A \cup B$ belong to $\mathscr{F}_0$, etc. It then also contains all unions of finitely many sets $A_1, \ldots, A_n$ in $\mathscr{F}_0$.*

(b) A field $\mathscr{F}$ of subsets is called a σ-field (or σ-algebra) if it contains all countable unions and intersections of its sets, i.e., if it is a field and furthermore

$$A_1, A_2, \ldots \in \mathscr{F} \quad \textit{implies} \quad \cup_1^\infty A_n \in \mathscr{F}.$$

Definition A.2. *A measurable space* $(\Omega, \mathscr{F})$ *is a sample space* Ω *together with a* σ*-field* $\mathscr{F}$ *of subsets.*

To every collection $\mathscr{A}$ of subsets of Ω there is always a unique *smallest* field $\mathscr{F}_0$ that contains all the sets in $\mathscr{A}$. Similarly, there always exists a (unique) smallest σ-field $\mathscr{F}$ that contains all $\mathscr{A}$-sets. That σ-field $\mathscr{F}$ is said to be *generated by* $\mathscr{A}$, and it is denoted $\mathscr{F} = \sigma(\mathscr{A})$. The term Borel field is sometimes used to mean the same as σ-field or σ-algebra. We shall reserve the term for σ-fields generated by certain simple families of sets, as in the following two examples.

Example A.1 (Fields and σ-fields in $\mathbb{R}$). *The simplest useful field* $\mathscr{F}_0$ *of subsets of the real line* $\mathbb{R}$ *consists of all finite half-open intervals,* $a < x \leq b$, *together with unions of a finite number of such intervals. In order for* $\mathscr{F}_0$ *to be a field, it is required that it also contains the complement of such unions. To anticipate the introduction of probabilities and random variables, we can remark here that* $\mathscr{F}_0$ *is a natural family of sets, since distribution functions can be used to assign probabilities to intervals.*

The *smallest* σ-field that contains all half-open intervals is called the *Borel field in* $\mathbb{R}$. It is denoted $\mathscr{B}$, and its sets are called *Borel sets.*

Example A.2 (Fields and σ-fields in $\mathbb{R}^n$). *An interval in* $\mathbb{R}^n$ *is an* n*-dimensional rectangle, and the smallest interesting field* $\mathscr{F}_0$ *in* $\mathbb{R}^n$ *consists of unions of finitely many half-open rectangles, with sides* $(a_i, b_i]$, $i = 1, \ldots, n$, *and the complements of such unions. As in* $\mathbb{R}$, *the* σ*-field generated by* $\mathscr{F}_0$ *is called the Borel field in* $\mathbb{R}^n$. *It is denoted by* $\mathscr{B}_n$ *and its sets are called the Borel sets in* $\mathbb{R}^n$.

One could note here that it is possible to start with more general "rectangles," where the "sides" are real Borel sets instead of intervals, i.e., sets $B_1 \times B_2 \times \ldots \times B_n$, *where the* B_j *are one-dimensional Borel sets. However, even if these generalized rectangles form a richer class than the simple rectangles, the smallest* σ*-field that contains all such generalized rectangles is exactly equal to* $\mathscr{B}_n$.

Example A.3 (Fields and σ-fields in $\mathbb{R}^\infty$ and $\mathbb{R}^T$). *An interval in* $\mathbb{R}^\infty$ *is a set* *of the form*

$$(a_1,b_1]\times(a_2,b_2]\times\ldots\times(a_n,b_n]\times\mathbb{R}^\infty,$$

and the family of sets that are unions of a finite number of intervals, or their complements, forms a field $\mathscr{F}_0$. It generates the Borel σ-field $\mathscr{B}_\infty$.

In $\mathbb{R}^T$, an interval is a set defined by a finite-dimensional condition of the form $(x(t_1),\ldots,x(t_n))\in(a_1,b_1]\times(a_2,b_2]\times\ldots\times(a_n,b_n]$, for a finite number of time points. The field $\mathscr{F}_0$ of finite unions, with their complements, generates the Borel σ-field $\mathscr{B}_T$.

A.1.2 Probabilities

We have now defined the sample space Ω and an interesting σ-field $\mathscr{F}$ of events in Ω. By this we have constructed a measurable space, $(\Omega,\mathscr{F})$, i.e., a space that is ready to be measured. We now discuss how to measure the events, i.e., assign probabilities to the sets in $\mathscr{F}$. We start with some properties that we feel natural for a probability measure.

Probabilities are defined for events, i.e., subsets of a sample space Ω. A probability measure should be a function P, defined for every event in a field $\mathscr{F}_0$, such that

$$0\le P(A)\le 1,\quad P(\emptyset)=0,\quad P(\Omega)=1,$$

and such that, first of all, for any finite number of disjoint events $A_1,\ldots,A_n$ in $\mathscr{F}_0$, one has

$$P(A_1\cup\ldots\cup A_n)=P(A_1)+\ldots+P(A_n). \tag{A.1}$$

That is, probabilities should be *finitely additive*. As remarked, it is easy to assign probabilities to intervals, and unions of intervals, simply by taking

$$P((a,b])=F(b)-F(a),$$

for some distribution function F.[1] By additivity and the property of fields, one then also assigns probability to elements in the field $\mathscr{F}_0$ of finite unions of intervals.

It turns out that finite additivity is a too weak property to be useful for much of probability theory.[2] In order that one could deal with limiting events

[1]That is, F is non-decreasing, right-continuous, with $0\le F(x)\le 1$, and $\lim_{x\to-\infty}F(x)=0$ and $\lim_{x\to\infty}F(x)=1$.

[2]Finite additive probabilities are popular, for example in Bayesian statistics, as models for a “uniform distribution” on the natural numbers; see [67].

and the infinity, probabilities are also required to be *countably additive*, i.e., equation (A.1) is required to hold for infinitely many disjoint events,

$$P(\cup_{k=1}^{\infty} A_k) = \sum_{k=1}^{\infty} P(A_k),$$

for all disjoint events $A_k \in \mathscr{F}_0$ such that $\cup_1^\infty A_k \in \mathscr{F}_0$. An equivalent requirement is the continuity property, that $\lim_{n\to\infty} P(A_n) = 0$ for any shrinking sequence of $A_n \in \mathscr{F}_0$, such that $A_1 \supseteq A_2 \supseteq \ldots$ with $\cap_1^\infty A_k = \emptyset$.

A natural question to ask is whether this also produces probabilities to the events in the σ-field $\mathscr{F}$ generated by $\mathscr{F}_0$. In fact, it does, and that in a unique way, as stated in Theorem A.1 in Appendix A.2.

Extension of probability measures: Every probability P, defined and countably additive on a field $\mathscr{F}_0$, can be extended to be defined for every event in the σ-field $\mathscr{F}$ generated by $\mathscr{F}_0$. This can be done in one way only. This means that a probability measure on the real Borel sets is uniquely determined by its values on the half-open intervals, i.e., it depends only on the values of the function

$$F(b) = P((-\infty, b]), \quad F(b) - F(a) = P((a, b]).$$

Probabilities on the Borel sets in $\mathbb{R}^n$ are similarly uniquely determined by the n-dimensional distribution function

$$F(b_1, \ldots, b_n) = P((-\infty, b_1] \times \ldots \times (-\infty, b_n]). \tag{A.2}$$

For example, for $n = 2$,

$$P((a_1, b_1] \times (a_2, b_2]) = F(b_1, b_2) - F(a_1, b_2) - F(b_1, a_2) + F(a_1, a_2)).$$

A probability measure P is defined on the *measurable space* $(\Omega, \mathscr{F})$, and sets in the σ-field $\mathscr{F}$ are called the *measurable sets.* The triple $(\Omega, \mathscr{F}, P)$ is called a *probability space.*

Remark A.1 (Completion of probability measure)**.** *The completion of a probability measure P is obtained as follows. Suppose P is defined on* $(\Omega, \mathscr{F})$*, i.e., it assigns a probability* $P(A)$ *to every A in* $\mathscr{F}$*. Now, if there is an event B with* $P(B) = 0$*, then it seems natural to assign probability 0 to any smaller set*

$B' \subset B$. Unfortunately, subsets of measurable sets are not necessarily measurable, so one can not immediately conclude that $P(B') = 0$. However, no other choice is possible, and it is also easy to create the σ-field that also contains all subsets of $\mathscr{F}$-sets B with $P(B) = 0$. The extended probability measure is called a complete probability measure.

A.1.3 Sets with countable basis

One may wonder how far the Borel sets in $\mathbb{R}^T$ are from the finite-dimensional intervals. The intervals were characterized by some restriction on function values at a finite number of times. A set $C \subseteq \mathbb{R}^T$ that is characterized by function values at a countable set of times, $T' = (t_1, t_2, \ldots)$, is said to have a *countable basis*. More precisely, $C \subseteq \mathbb{R}^T$ has a countable basis T' if there is a Borel set $B \subset \mathbb{R}^\infty$ (with $B \in \mathscr{B}_\infty$), such that

$$x \in C \quad \text{if and only if} \quad (x(t_1), x(t_2), \ldots) \in B.$$

The Borel sets in $\mathbb{R}^T$ are exactly those sets that have a countable basis, i.e.,

$$\mathscr{B}_T = \{C \subset \mathbb{R}^T; C \text{ has a countable basis}\}.$$

We show this, as an example of a typical σ-field argument.

First, it is clear that if B is a Borel set in $\mathbb{R}^\infty$, then

$$C = \{x \in \mathbb{R}^T; (x(t_1), x(t_2), \ldots) \in B\}$$

is a Borel set in $\mathbb{R}^T$, since $\mathscr{B}_T$ contains all intervals with base in T', and hence all sets in the σ-field generated by those intervals. This shows that

$$\{C \subset \mathbb{R}^T; C \text{ has a countable basis}\} \subseteq \mathscr{B}_T.$$

To show the other inclusion we show that the family of sets with countable basis is a σ-field that contains the intervals, and then it must be at least as large as the smallest σ-field that contains all intervals, namely $\mathscr{B}_T$. First, we note that taking complements still gives a set with countable basis. Then, take a sequence $C_1, C_2, \ldots,$ of sets, all with countable basis, and let $T_1, T_2, \ldots, T_j = \{t_1^1, t_2^1, \ldots\}, \ldots, \{t_1^j, t_2^j, \ldots\}$ be the corresponding countable sets of time points, so that

$$C_j = \{x \in \mathbb{R}^T; (x(t_1^{(j)}), x(t_2^{(j)}), \ldots) \in B_j\}, \quad \text{with } B_j \in \mathscr{B}_\infty.$$

Then $T' = \cup_j T_j$ is a countable set, $T' = (t_1', t_2', \ldots)$, and $\cup_{j=1}^\infty C_j$ is characterized by its values on T'.

A.1.4 Random variables

A (real) random variable[3] x is just a real-valued function $x(\omega), \omega \in \Omega$, on a probability space $(\Omega, \mathscr{F}, P)$, such that it is possible to talk about its distribution, i.e., the probability

$$P(x \leq a) = P(\{\omega; x(\omega) \leq a\})$$

is defined for all real a. This means that the set (event)

$$A_a = x^{-1}((-\infty, a]) = \{\omega; x(\omega) \leq a\}$$

is a member of the family $\mathscr{F}$, for all $a \in \mathbb{R}$. This is equivalent to the seemingly more general statement that

$$x^{-1}(B) \in \mathscr{F} \quad \text{for all Borel sets } B \in \mathscr{B}, \tag{A.3}$$

and of course[4] it holds that

$$P(x^{-1}(B)) = \text{Prob}(x \in B).$$

The requirement (A.3) is the formal definition of a random variable: a random variable is a *Borel measurable function.*

If x is a random variable on $(\Omega, \mathscr{F}, P)$, then we write P_x for the probability measure on $(\mathbb{R}, \mathscr{B})$ that is defined by

$$P_x(B) = P(x^{-1}(B)).$$

Figure A.1 summarizes the core properties of a random variable: For every random variable x, the set $x^{-1}((-\infty, a]) = \{\omega \in \Omega; x(\omega) \leq a\}$ must be one of the sets in $\mathscr{F}$ that has gotten probability by the probability measure P.

An n-dimensional (real) random variable $\mathbf{x} = (x_1, \ldots, x_n)$ is a family of n random variables, defined on the same probability space $(\Omega, \mathscr{F}, P)$. Each x_k is a function of the outcome ω and the sample value is $(x_1(\omega), \ldots, x_n(\omega)) \in \mathbb{R}^n$. If $B \in \mathscr{B}^n$ is a Borel set in $\mathbb{R}^n$, then

$$\mathbf{x}^{-1}(B) \in \mathscr{F},$$

and it has probability $P(\mathbf{x}^{-1}(B)) = P_{\mathbf{x}}(B) = \text{Prob}(\mathbf{x} \in B)$.

---

[3] We have chosen to denote random variables and random functions by lowercase letters instead of the customary uppercase. This notation is chosen in order to emphasize that a random function can be seen as a "point" in a space of functions; cf. Figure 1.1.

[4] We will use the notation Prob(event A happens) to be taken literally as the *probability* that the event A happens. The numerical value is given by the value of the *function* P on the corresponding set A.

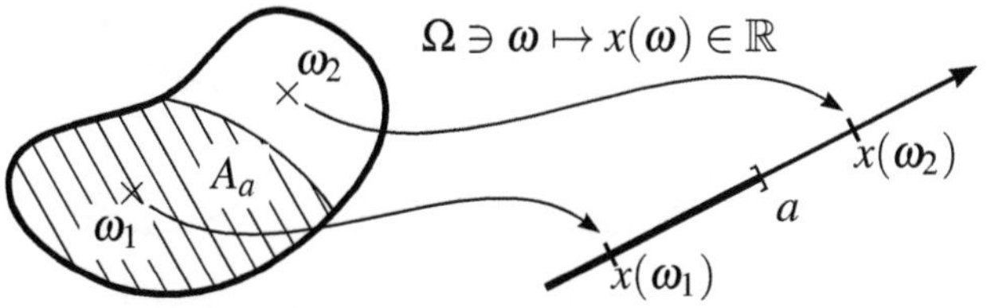

Figure A.1 *The random variable x maps outcomes ω in the sample space Ω on to the real line $\mathbb{R}$; $A_a = \{\omega; x(\omega) \leq a\}$.*

A.1.5 The σ-field generated by random variables

When x is a random variable on $(\Omega, \mathscr{F}, P)$, the set $\{\omega \in \Omega; x(\omega) \leq a\}$ belongs to $\mathscr{F}$ and hence it has a probability $\text{Prob}(x \leq a)$. Furthermore, all sets of the type $x^{-1}(B)$, where B is a Borel set, belong to $\mathscr{F}$. In fact, the family of such Ω-sets is a σ-field, and it is denoted $\mathscr{F}(x)$ or $\sigma(x)$. It is obvious that $\mathscr{F}(x) \subset \mathscr{F}$, and we already know that P assigns a probability to these sets. If x were the only random variable of interest to us, we could have worked on the probability space $(\Omega, \mathscr{F}(x), P)$. The reason for using a general, usually larger σ-field $\mathscr{F}$ is that it allows us perfect freedom to include any further random variable without changing either the σ-field or the probability measure.

A characterization of $\mathscr{F}(x)$ is that it is the smallest σ-field on Ω that makes the function x measurable, i.e., a random variable. The σ-field $\mathscr{F}(x)$ is called the *σ-field generated by the random variable x*. When there are several random variables, $x_1, \dots, x_n$, $n < \infty$, there will be a *smallest σ-field*, denoted $\mathscr{F}(x_1, \dots, x_n)$, that contains all the sub-σ-fields $\mathscr{F}(x_j)$. It is the smallest σ-field that makes all x_j random variables. Figure A.2 illustrates the sub-fields.

The figure shows three ways to partition the sample space Ω. The left partition is generated by a random variable x that takes distinct constant values on each of the NW-SE stripes. Every function that is constant on each of these stripes is a random variable on $(\Omega, \mathscr{F}(x))$. Similarly, each function that is constant on every NE-SW stripe in the middle partition is a random variable on $(\Omega, \mathscr{F}(y))$. The right partition allows many more random variables, namely all those which are constant on each small square.

Remark A.2. *When we have the σ-field generated by a random variable x we can construct a probability measure, giving values of $P(A)$ for certain $A \in \mathscr{F}$. If F is a distribution function and x is a random variable, then $P(A_a) = P(\{\omega \in \Omega; x(\omega) \leq a\}) = F(a)$ defines probabilities on the sub-class of events A_a that can be extended to a probability measure on the σ-field $\mathscr{F}(x)$.*

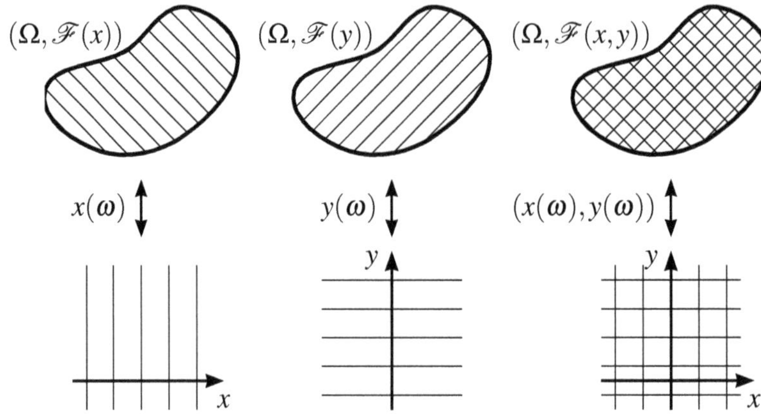

Figure A.2: *Sub-σ-fields generated by random variables x and y on Ω.*

A.2 The axioms of probability

A.2.1 The axioms

A probability P is a countably additive measure on a probability space, as defined here.

A typical field $\mathscr{F}_0$ in $\mathbb{R}$ is the family of sets which are unions of a finite number of intervals. In $\mathbb{R}^n$ the family of sets which are unions of finitely many rectangles form a field. To be specific, we let the intervals defining the rectangles be open to the left and closed to the right, i.e., a typical rectangle in $\mathbb{R}^n$ is $(a_1,b_1] \times (a_2,b_2] \times \ldots \times (a_n,b_n]$. The smallest σ-field that contains all sets in $\mathscr{F}_0$ is the family $\mathscr{B}$ of Borel sets.

Definition A.3. *A probability measure P on a sample space Ω with a σ-field $\mathscr{F}$ of events is a function defined for every $A \in \mathscr{F}$, with the following properties:*

(1) $0 \leq P(A) \leq 1$ *for all* $A \in \mathscr{F}$*;*

(2) $P(\Omega) = 1$*;*

(3) For disjoint sets $A_k \in \mathscr{F}$, $k = 1,2,\ldots$ *one has* $P(\cup_1^\infty A_k) = \sum_1^\infty P(A_k)$.

A.2.2 Extension of a probability from field to σ-field

How do we define probabilities? For real events, of course, via a statistical distribution function $F(x)$. Given a distribution function F, we can define a probability P for finite unions of real half-open disjoint intervals,

$$P(\cup_1^n (a_k, b_k]) = \sum_1^n (F(b_k) - F(a_k)). \tag{A.4}$$

In higher dimensions, half-open intervals can be assigned probabilities from a multi-dimensional distribution function, which has to obey certain rules in order to be a probability distribution. For example, in $\mathbb{R}^2$ a function $F(x,y)$ is a two-dimensional distribution function if, for $h > 0, k > 0$,

- $0 \leq F(x,y) \leq 1$,
- $\lim_{x \to -\infty} F(x,y) = \lim_{y \to -\infty} F(x,y) = 0$, $\lim_{x \to \infty, y \to \infty} F(x,y) = 1$,
- $F(x+h, y+k) - F(x+h, y) - F(x, y+k) + F(x,y) \geq 0$.

This means that it can assign a positive probability to rectangles. The conditions in higher dimensions are only notationally more complicated.

Sets that are unions of a finite number real intervals, $(a,b]$, (a,b), $[a,b)$, $[a,b]$, with $-\infty \leq a < b \leq \infty$, form a field on $\mathbb{R}$, and they can be given probabilities via a distribution function. The question is does this also give probabilities to the more complicated events (Borel sets) in the σ-field $\mathscr{F}$ generated by the intervals? The answer is yes, as stated in the following extension theorem, *Carathéodory's extension theorem*, which is valid not only for intervals and Borel sets, but for any field $\mathscr{F}_0$ and generated σ-field $\mathscr{F}$. For a proof, the reader is referred to any textbook in probability or measure theory, e.g., [15] or [122].

Theorem A.1 (Extension theorem)**.** *Suppose P is a function which is defined for all sets in a field $\mathscr{F}_0$, there satisfying the probability axioms, i.e.,*

(1) $0 \leq P(A) \leq 1$ *for all* $A \in \mathscr{F}_0$.

(2) $P(\Omega) = 1$.

(3) If $A_1, A_2 \in \mathscr{F}_0$ *are disjoint, then* $P(A_1 \cup A_2) = P(A_1) + P(A_2)$.

(4) If $A_1 \supseteq A_2 \supseteq \ldots \in \mathscr{F}_0$ *with* $\cap_1^\infty A_k = \emptyset$*, then* $\lim_{k \to \infty} P(A_k) = 0$.

Then one can extend P to be defined, in one and only one way, for all sets in the σ-field $\mathscr{F}$ generated by $\mathscr{F}_0$, so that it still satisfies the probability axioms.

Condition (4) is equivalent to any of the following conditions:

(4b) If $A_1, A_2, \ldots \in \mathscr{F}_0$ are disjoint and $\cup_1^\infty A_k \in \mathscr{F}_0$, then

$$P(\cup_1^\infty A_k) = \sum_1^\infty P(A_k).$$

(4c) If $A_1, A_2, \ldots \in \mathscr{F}_0$ are disjoint, $\cup_1^\infty A_k = \Omega$, then

$$\sum_1^\infty P(A_k) = 1.$$

We can now state and prove the existence of probability measures on the real line with a given distribution function.

Theorem A.2. *Let $F(x)$ be a cumulative distribution function on $\mathbb{R}$, i.e., non-decreasing, right-continuous, with $F(-\infty) = 0$, $F(\infty) = 1$. Then there exists exactly one probability measure P on $(\mathbb{R}, \mathscr{B})$, such that $P((a,b]) = F(b) - F(a)$.*

Proof. We shall use the extension Theorem A.1. The Borel sets $\mathscr{B}$ equal the σ-field generated by the field $\mathscr{F}_0$ of unions of finitely many intervals. Equation (A.4) extended by singletons defines P for each set in $\mathscr{F}_0$, and it is easily checked that properties (1), (2), and (3) hold. The only difficult part is (4), which we prove by contradiction.

The idea is to use Cantor's theorem that every decreasing sequence of compact, non-empty sets has a non-empty intersection. Assume, for a decreasing sequence of sets $A_n \in \mathscr{F}_0$, that $P(A_n) \downarrow h > 0$. We show that then the intersection of the A_n-sets is not empty. Each A_n consists of finitely many half-open intervals. It is then possible to remove from A_n a short piece from the left end, to make it closed and bounded, i.e., there exists a compact, non-empty $K_n \subset A_n$, such that

$$P(A_n - K_n) \leq \varepsilon/2^n.$$

(Convince yourself that $P(A_n - K_n)$ is defined.) Then

$$L_m = \cap_1^m K_n \subseteq K_m \subseteq A_m$$

form a decreasing sequence,

$$L_1 \supseteq L_2 \supseteq \ldots .$$

If we can prove that the L_m can be taken nonempty, we can use Cantor's theorem, and conclude that they have a nonempty intersection, i.e., there exists at least one point $x \in \cap_1^\infty L_m$, which also implies $x \in \cap_1^\infty A_m$, so the A_k do not decrease to the empty set. The proof would be finished.

It remains to prove that we can choose each L_m nonempty. Take $\varepsilon < h$. Then

$$\begin{aligned} P(A_m - L_m) &= P(A_m - \cap_1^m K_n) = P(\cup_1^m (A_m - K_n)) \\ &\leq \sum_1^m P(A_m - K_n) \leq \sum_1^m P(A_n - K_n) \leq \sum_1^m \varepsilon/2^n \leq \varepsilon, \end{aligned}$$

which implies

$$P(L_m) = P(A_m) - P(A_m - L_m) \geq h - \varepsilon > 0,$$

and so L_m is non-empty. □

A.2.3 Kolmogorov's extension to $\mathbb{R}^\infty$ and $\mathbb{R}^T$

Kolmogorov's existence, or extension, theorem from 1933 allows us to define a stochastic process through its family of finite-dimensional distribution functions. Kolmogorov's book [71] appeared after a period of about 30 years of attempts to give probability theory a solid mathematical foundation; in fact, Hilbert's sixth problem (1900) asked for a logical investigation of the axioms of probability.

Theorem A.3 (Existence of stochastic process)**.** *To every consistent family of finite-dimensional distribution functions,* $\mathbf{F} = \{F_{\mathbf{t}^n}\}_{n=1}^\infty$, $t_k \in T$, *there exists one and only one probability measure* P *on* $(\mathbb{R}^T, \mathscr{B}_T)$ *with*

$$P(x(t_1) \leq b_1, \ldots, x(t_n) \leq b_n) = F_{t_1 \cdots t_n}(b_1, \ldots, b_n), \textit{ for } n = 1, 2, \ldots .$$

Proof. We indicate how one can construct a canonical, or co-ordinate, process on $(R^T, \mathscr{B}_T)$, i.e., the sample space will be the set of real functions defined on T, and each outcome $\omega \in \mathbb{R}^T$ is a sample function by itself, $x(t,\omega) = \omega(t)$.

Consider the intervals in $\mathbb{R}^T$, i.e., the sets of the form

$$I = \{x \in \mathbb{R}^T; x(t_i) \in (a_i, b_i], i = 1, \dots, n\},$$

for some n, and some base, $t_i \in T$. The finite-dimensional distribution functions define a probability, $P(I)$, for each such interval.

Let I_1 and I_2 be two disjoint intervals. They may have different dimensions ($n_1 \neq n_2$), and be based on different sets of t_i-values, but setting suitable a_i or b_i equal to $\pm\infty$, we may assume that they have the same dimension, and are based on the same set of t_i. (The consistency of the family $\{P_n\}$ guarantees that this does not change their probabilities.) Furthermore, the additivity property holds, that is, if also $I_1 \cup I_2$ is an interval, then $P(I_1 \cup I_2) = P(I_1) + P(I_2)$.

It is easy to extend P with additivity to all finite unions of intervals. By this we have defined P on the field $\mathscr{F}_0$ of finite unions of intervals, and checked that properties (1), (2), and (3) of Theorem A.1 hold.

Now check property (4) in the same way as for Theorem A.2, for a decreasing sequence of non-empty intervals with empty intersection,

$$I_1 \supseteq I_2 \supseteq \dots, \quad \text{with} \quad \cap_1^\infty I_n = \emptyset,$$

and suppose $P(I_n) \downarrow h > 0$.[5] We can always assume I_n to have dimension n,

$$I_n = \{x \in \mathbb{R}^T; a_i^{(n)} < x(t_i) \le b_i^{(n)}, i = 1, \dots, n\},$$

and we can always assume the a_i and b_i to be bounded. As in the proof of Theorem A.2, remove a small piece of the lower side of each interval to get a compact K_n, and define $L_m = \cap_1^m K_n$. By removing a small enough piece one can obtain that $P(L_m) \ge h/2 > 0$ so L_m is non-empty.

If we write

$$\begin{array}{llll}
L_1 : \alpha_1^{(1)} \le x(t_1) \le \beta_1^{(1)} & & & \\
L_2 : \alpha_1^{(2)} \le x(t_1) \le \beta_1^{(2)} & , \alpha_2^{(2)} \le x(t_2) \le \beta_2^{(2)} & & \\
L_3 : \alpha_1^{(3)} \le x(t_1) \le \beta_1^{(3)} & , \alpha_2^{(3)} \le x(t_2) \le \beta_2^{(3)} & , \alpha_3^{(3)} \le x(t_3) \le \beta_3^{(3)} & \\
\vdots \qquad\quad \vdots & \qquad \vdots & \qquad \vdots &
\end{array}$$

[5]Property (4) deals with a decreasing sequence of finite unions of intervals. It is easy to convince oneself that it suffices to show that (4) holds for a decreasing sequence of intervals.

we get, for each i, $[\alpha_i^{(n)}, \beta_i^{(n)}]$, $n = i, i+1, \ldots$, that is a decreasing sequence of non-empty, closed, and bounded intervals, and by Cantor's theorem they have at least one common point, $\bar{x}(t_i) \in \cap_{n=i}^{\infty}[\alpha_i^{(n)}, \beta_i^{(n)}]$. Then, $\bar{x} = (\bar{x}(t_1), \bar{x}(t_2), \ldots) \in \overline{L}_n$ for all n. Hence $\bar{x} \in L_n \subseteq I_n$ for all n and the intersection $\cap_1^{\infty} I_n$ is not empty. This contradiction shows that $P(I_n) \downarrow 0$, and (4) is thus shown to hold.

The conditions (1), (2), (3), and (4) of Theorem A.1 are thus all satisfied, and hence P can be extended uniquely to the σ-field $\mathscr{F}$ generated by the intervals. □

A.3 Expectations

A.3.1 Expectation of a random variable

Expectation for a simple random variable

If x takes only a finite number of values, $a_1, a_2, \ldots, a_n$, it is called a *simple* random variable, and its expectation is defined as

$$E(x) = \sum a_k P(x = a_k).$$

Expectation for a non-negative random variable

For non-negative random variable x, construct an increasing sequence of simple random variables x_n, such that $x_n \uparrow x$; take, for example,

$$x_n(\omega) = \begin{cases} k/2^n, & \text{when } k/2^n < x(\omega) \leq (k+1)/2^n, k = 0, 1, \ldots, 2^{2n} - 1, \\ 0, & \text{when } x(\omega) > 2^n. \end{cases}$$

Define $E(x) = \lim_{n \to \infty} E(x_n) \leq \infty$. It is easy to see that $E(x_n)$ is increasing so the limit exists. The limit is the same for all increasing sequences of simple random variables that converge to x. The expectation is written as an integral

$$E(x) = \int_{\Omega} x(\omega)\, dP(\omega).$$

Expectation of a real-valued random variable

Define the positive and negative parts of x as

$$x^+ = \max(0, x), \quad x^- = -\min(0, x),$$

and define

$$E(x) = E(x^+) - E(x^-) = \int_\Omega x(\omega)\,\mathrm{d}P(\omega),$$

provided at most one of the expectations is infinite. Then, also $E(|x|) = E(x^+) + E(x^-)$. If both expectations are infinite, the expectation $E(x)$ remains undefined.

A.3.2 *Expectations and the space $\mathscr{L}^p(\Omega, \mathscr{F}, P)$*

In everyday terms, the *expectation* of a random variable x, defined on a probability space $(\Omega, \mathscr{F}, P)$, is the average of its possible values, weighted by their probabilities,

$$E(x) = \int_\Omega x(\omega)\,\mathrm{d}P(\omega).$$

One should be aware that the expectation of x is defined only if the integral $\int_\Omega |x(\omega)|\,\mathrm{d}P(\omega) < \infty$, or, if not, at least one of $\int_\Omega x^+(\omega)\,\mathrm{d}P(\omega)$ and $\int_\Omega x^-(\omega)\,\mathrm{d}P(\omega)$ is finite.

The set of random variables on $(\Omega, \mathscr{F}, P)$ such that $E(|x|^p) < \infty$, is denoted by $\mathscr{L}^p(\Omega, \mathscr{F}, P)$, for $1 \leq p < \infty$. In this book, we are specifically using $p = 1$ and $p = 2$, and for short write $\mathscr{L}^1$ and $\mathscr{L}^2$, when the probability space is clear from the context.

For a real random variable $x \in \mathscr{L}^2$, the *variance* is defined as $V(x) = E((x - E(x))^2) = E(x^2) - E(x)^2 > 0$, as usual.

Remark A.3 (Complex random variables). *Complex-valued random variables appear frequently in this book. They are of the form $z = x + iy$, where the real and imaginary part are defined on the same probability space and have a joint distribution. The expectation of a complex random variable is simply $E(z) = E(x) + iE(y)$. The notation $\mathscr{L}^2$ is used to denote also the set of complex random variables with $E(|z|^2) < \infty$.*

A.3.3 *Conditional distribution and conditional expectation*

One of the most important concepts in probability theory is the idea of "conditioning" – the "if-then" reasoning is an extremely effective way of successively reducing uncertainty in any random system. Conditional expectation is the mathematical concept that sets the rules for how to handle such arguments.

Elementary definition

Here we will first give an elementary definition of *conditional distribution* and *conditional expectation.* If x, y are two random variables, where y may be multivariate, with joint density $f_{xy}(u, v)$, and with marginal y-density $f_y(v) = \int_u f_{xy}(u, y)\,du$, the conditional distribution of x given $y = v$ is given by the density

$$f_{x|y=v}(u) = \begin{cases} \frac{f_{x,y}(u,v)}{f_y(v)}, & \text{when } f_y(v) > 0, \\ 0, & \text{otherwise.} \end{cases}$$

Obviously, $f_{x|y=v}(\cdot)$ is a probability density function, since it is non-negative and integrates to one. Its expectation $\varphi(v)$ defines the conditional expectation for those v for which $f_y(v) > 0$,

$$\varphi(v) = E(x \mid y = v) = \int_u u \frac{f_{xy}(u, v)}{f_y(v)}\,du. \tag{A.5}$$

For outcomes v with $f_y(v) = 0$, $\varphi(v)$ can be defined arbitrarily.

We write the conditional expectation as a function of the random variable y, $\varphi(y) = E(x \mid y)$, to indicate that it itself is a random variable. It satisfies

$$E(x) = E(\varphi(y)) = E(E(x \mid y)) = \int_y \varphi(y) f(y)\,dy, \tag{A.6}$$

$$V(x) = E(V(x \mid y)) + V(E(x \mid y)), \tag{A.7}$$

where $V(x \mid y) = \int_x (x - \varphi(y))^2 f(x \mid y)\,dx$.

The expectation $m = E(x)$ is the constant c that minimizes $E\left((x-c)^2\right)$. This, together with (A.6), shows the following theorem.

Theorem A.4. *The best predictor of x given y in least squares sense is given by $\varphi(y)$, i.e.,*

$$E\left((x - \varphi(y))^2\right) \le E\left((x - \psi(y))^2\right)$$

for every function $\psi(y)$.

Proof. With $m = E(x)$ and $E((x-a)^2) = E((x-m)^2) + (m-a)^2$, one gets

$$\begin{aligned} E\left((x-\psi(y))^2\right) &= \int_u \int_v (u-\psi(v))^2 f_{xy}(u,v)\,\mathrm{d}u\,\mathrm{d}v \\ &= \int_v f_y(v)\left[\int_u (u-\psi(v))^2 f_{x|y=v}(u)\,\mathrm{d}u\right]\mathrm{d}v, \end{aligned}$$

where the bracket expression is minimized when $\psi(v) = E(x \mid y = v)$. □

General properties of conditional expectation

The full definition of conditional expectation aims to generalize (A.6).

Definition A.4. *If x is a random variable on $(\Omega, \mathscr{F}, P)$ with $E(|x|) < \infty$, and $\mathscr{A}$ a sub-σ-field of $\mathscr{F}$, then by the conditional expectation of x given $\mathscr{A}$, $E(x \mid \mathscr{A})$, is meant any $\mathscr{A}$-measurable random variable u that satisfies*

$$\int_{\omega \in A} x(\omega)\,\mathrm{d}P(\omega) = \int_{\omega \in A} u(\omega)\,\mathrm{d}P(\omega), \qquad \text{(A.8)}$$

for all $A \in \mathscr{A}$. Note that the value of $E(x \mid \mathscr{A}) = u$ is defined only "almost surely," and that any $\mathscr{A}$-measurable variable which has the same integral as x when integrated over $\mathscr{A}$-sets works equally well as the conditional expectation.

In particular, if $\mathscr{A}$ only contains sets that have probability 0 *or* 1, *then $E(x \mid \mathscr{A})$ is a.s. constant and equal to $E(x)$.*

To understand the meaning of this general definition of conditional expectation, consider again Figure A.2. Suppose that the sub-σ-field in the definition is generated by a random variable y, i.e., $\mathscr{A} = \mathscr{F}(y)$, and that, furthermore, y takes distinct values, $y_1, \ldots, y_n$, on the NE-SW stripes $A_1, \ldots, A_n$ in the middle partition. Then, any random variable u that is measurable with respect to $\mathscr{A}$ is constant over each stripe, $u(\omega) = u_j$ if $\omega \in A_j$, say. Then, if u fulfills the requirements for a conditional expectation, from (A.8),

$$\int_{\omega \in A_j} x(\omega)\,\mathrm{d}P(\omega) = \int_{\omega \in A_j} u(\omega)\,\mathrm{d}P(\omega) = u_j P(A_j),$$

which implies, provided $P(A_j) > 0$, that

$$u_j = \frac{1}{P(A_j)} \int_{\omega \in A_j} x(\omega)\,\mathrm{d}P(\omega).$$

We see that $E(x \mid \mathscr{F}(y)) = u_j = E(x \mid y = y_j)$ when $\omega \in A_j$, so $E(x \mid \mathscr{A})$ is simply the average value of $x(\omega)$ over each A_j.

The *existence* of a conditional expectation needs to be shown; [122, Ch. 9] gives a simple proof, based on the completeness of the space $\mathscr{L}^2(\Omega, \mathscr{A}, P)$.

A.4 Convergence

A.4.1 Stochastic convergence

Here we summarize the basic types of stochastic convergence and the ways we have to check the convergence of a random sequence with specified distributions. For further reading on stochastic convergence, the classic book by K.L. Chung [26] is to be recommended.

Definition A.5. *Let $\{x_n\}_{n=1}^{\infty}$ be a sequence of random variables $x_1(\omega), x_2(\omega), \ldots$ defined on the same probability space, and let $x = x(\omega)$ be a random variable, defined on the same probability space. Then, the convergence $x_n \to x$ as $n \to \infty$ can be defined in three ways:*

- **almost surely, with probability one** ($x_n \stackrel{a.s.}{\to} x$): $P(\{\omega; x_n \to x\}) = 1$;
- **in quadratic mean** ($x_n \stackrel{q.m.}{\to} x$): $E\left(|x_n - x|^2\right) \to 0$;
- **in probability** ($x_n \stackrel{P}{\to} x$): *for every* $\varepsilon > 0$, $P(|x_n - x| > \varepsilon) \to 0$.

Furthermore, x_n tends **in distribution** *to x (in symbols $x_n \stackrel{\mathscr{L}}{\to} x$) if*

$$P(x_n \le a) \to P(x \le a)$$

for all a such that $P(x \le u)$ is a continuous function of u at $u = a$.

Lemma A.1. *1.* $x_n \overset{a.s.}{\to} x \Rightarrow x_n \overset{P}{\to} x$ *and* $x_n \overset{\mathscr{L}}{\to} x$ *;*

2. $x_n \overset{a.s.}{\to} x$ *and* $E(|x_n|^a) \leq K$*, some* $a > 2, K < \infty \Rightarrow x_n \overset{q.m.}{\to} x$*;*

3. $x_n \overset{q.m.}{\to} x \Rightarrow x_n \overset{P}{\to} x$ *and* $x_n \overset{\mathscr{L}}{\to} x$*.*

A.4.2 The Borel-Cantelli lemmas

The two Borel-Cantelli lemmas are simple and extremely useful when studying almost sure convergence of random variables. Let $\{A_k\}$ be a sequence of events in a probability space $(\Omega, \mathscr{F}, P)$, and define the two events

$$\limsup_k A_k = \cap_{n=1}^{\infty} \cup_{k=n}^{\infty} A_k, \qquad \liminf_k A_k = \cup_{n=1}^{\infty} \cap_{k=n}^{\infty} A_k.$$

The first is interpreted as "infinitely many of the events A_k occur," or, for short, "A_k, infinitely often." The second event simply says that "there is an index n such that from that point on, all A_k occur." Then the following statements hold.

Theorem A.5 (Borel-Cantelli lemmas)**.**

a) For any sequence of events A_k*,* $\sum_k P(A_k) < \infty$ *implies that* $P(A_k$*, infinitely often*$) = 0$*.*

b) If the events A_k *are independent, then* $\sum_k P(A_k) = \infty$ *implies that* $P(A_k$*, infinitely often*$) = 1$*.*

A.4.3 Criteria for convergence with probability one

In order for a random sequence x_n to converge almost surely (i.e., with probability one), to the random variable x, it is necessary and sufficient that

$$\lim_{m \to \infty} P(|x_n - x| > \delta \text{ for at least one } n \geq m) = 0 \tag{A.9}$$

for every $\delta > 0$.

To prove this, note that if ω is an outcome such that the real sequence $x_n(\omega)$ does not converge to $x(\omega)$, then

$$\omega \in \left\{ \bigcup_{q=1}^{\infty} \bigcap_{m=1}^{\infty} \bigcup_{n=m}^{\infty} |x_n(\omega) - x(\omega)| > 1/q \right\}.$$

Here, the inner union, $B_{m,q} = \cup_{n=m}^{\infty} \{|x_n(\omega) - x(\omega)| > 1/q\}$, has probability

$$P(B_{m,q}) = P\left(|x_n - x| > 1/q \text{ for at least one } n \geq m\right).$$

Thus, condition (A.9) is equivalent to the condition that, for all q, $P(B_{m,q}) \to 0$ as $m \to \infty$. Now, $B_1 \supseteq B_2 \supseteq \ldots$ is a non-increasing sequence of events. Hence, with $A_q = \cap_m B_{m,q}$, it follows that $P(A_q) = \lim_{m\to\infty} P(B_{m,q}) = 0$. This proves the statement.

Now,

$$P(|x_n - x| > \delta \text{ for at least one } n \geq m) \leq \sum_{n=m}^{\infty} P(|x_n - x| > \delta),$$

and hence a simple *sufficient* condition for (A.9) and a sufficient condition for almost sure convergence is that for all $\delta > 0$,

$$\sum_{n=1}^{\infty} P(|x_n - x| > \delta) < \infty. \tag{A.10}$$

(In fact, the first Borel-Cantelli lemma directly shows that (A.10) is sufficient for almost sure convergence.)

A simple moment condition is obtained from Markov's inequality $P(|x_n - x| > \delta) \leq E(|x_n - x|^h)/\delta^h$, giving that a sufficient condition for almost sure convergence is

$$\sum_{n=1}^{\infty} E(|x_n - x|^h) < \infty, \tag{A.11}$$

for some $h > 0$.

Theorem A.6. *From* (A.11) *follows that if* $x_n \overset{q.m.}{\to} x$, *then there is a subsequence* x_{n_k} *that converges almost surely to* x.

A Cauchy convergence type condition is the following *sufficient condition for almost sure convergence*: if there exist two sequences of positive numbers δ_n and ε_n such that $\sum_{n=1}^{\infty} \delta_n < \infty$ and $\sum_{n=1}^{\infty} \varepsilon_n < \infty$, and such that

$$P(|x_{n+1} - x_n| > \delta_n) < \varepsilon_n, \tag{A.12}$$

then there exists a random variable x such that $x_n \overset{a.s.}{\to} x$.

To see this, use the Borel-Cantelli lemma to conclude that

$$P(|x_{n+1} - x_n| > \delta_n \text{ for infinitely many } n) = 0.$$

Thus, for almost all ω, there is a number N, depending on the outcome ω, such that

$$|x_{n+1} - x_n| < \delta_n \quad \text{for all } n \geq N.$$

Since $\sum \delta_n < \infty$, the sequence $x_n(\omega)$ converges to a limit $x(\omega)$ for these outcomes. For ω where the limit does not exist, set $x(\omega) = 0$, for example. Then $x_n \overset{a.s.}{\to} x$, as was to be proved.

Uniform convergence of random functions

A sequence of random variables can converge almost surely, and we have just given sufficient conditions for this. But we shall also need convergence of a sequence of random functions $\{x_n(t); t \in T\}$, where $T = [a,b]$ is a closed bounded interval.

Definition A.6. *A sequence of functions* $\{x_n(t); a \leq t \leq b\}$ *converges uniformly to the function* $\{x(t); a \leq t \leq b\}$ *if*

$$\max_{a \leq t \leq b} |x_n(t) - x(t)| \to 0, \quad \text{as } n \to \infty,$$

that is, if x_n *lies close to the limiting function* x *in the entire interval* $[a,b]$ *for all sufficiently large* n.

It is a basic result in real analysis that if a sequence of continuous functions converges uniformly in a closed and bounded interval, then the limiting function is also continuous. This fact was used in the proof of almost sure sample function continuity of a random function, page 43.

Condition (A.12) can be restated to deal with almost sure uniform convergence of random functions: if there exist two sequences of positive numbers δ_n and ε_n such that $\sum_{n=1}^{\infty} \delta_n < \infty$ and $\sum_{n=1}^{\infty} \varepsilon_n < \infty$, and such that

$$P(\max_{a \leq t \leq b} |x_{n+1}(t) - x_n(t)| > \delta_n) < \varepsilon_n, \tag{A.13}$$

then there exists a random function $x(t); a \leq t \leq b$, such that $x_n(t) \overset{a.s.}{\to} x(t)$ uniformly for $t \in [a,b]$.

A.4.4 *Criteria for convergence in quadratic mean*

Some of the representation theorems for stationary processes express a process as a complex stochastic integral, defined as a limit in quadratic mean of approximating sums of complex-valued random variables. To define a quadratic mean integral, or other limit of that kind, one needs simple convergence criteria for when $x_n \stackrel{q.m.}{\to} x$ for a sequence of random variables with $E(|x_n|^2) < \infty$.

The Cauchy convergence criterion for convergence in quadratic mean states that a necessary and sufficient condition for the existence of a (possibly complex) random variable x such that $x_n \stackrel{q.m.}{\to} x$ is that

$$E(|x_m - x_n|^2) \to 0, \tag{A.14}$$

as n and m tend to infinity, independently of each other. (In mathematical language, this is the completeness of the space $\mathscr{L}^2$.)

The limit x has $E(|x|^2) = \lim E(|x_n|^2) < \infty$, and $E(x_n) \to E(x)$. If there are two convergent sequences, $x_n \stackrel{q.m.}{\to} x$ and $y_n \stackrel{q.m.}{\to} y$, then

$$E(x_n \overline{y_n}) \to E(x\overline{y}). \tag{A.15}$$

To show quadratic mean convergence of stochastic integrals, the following criterion is useful:

the Loève criterion: the sequence x_n converges in quadratic mean if and only if

$$E(x_m \overline{x_n}) \text{ has a finite limit } c, \tag{A.16}$$

when m and n tend to infinity independently of each other.

The *if* part follows from the Cauchy criterion and the observation that

$$\begin{aligned} E(|x_m - x_n|^2) &= E(x_m \overline{x_m}) - E(x_m \overline{x_n}) - E(x_n \overline{x_m}) + E(x_n \overline{x_n}) \\ &\to c - c - c + c = 0. \end{aligned}$$

The *only if* part follows from

$$E(x_m \overline{x_n}) \to E(x\overline{x}) = E(|x|^2).$$

A.4.5 Criteria for convergence in probability

Both almost sure convergence and convergence in quadratic mean imply convergence in probability. Further, if $x_n \stackrel{P}{\to} x$, then there exists a subsequence $n_k \to \infty$ as $k \to \infty$, such that $x_{n_k} \stackrel{a.s.}{\to} x$.

To prove this, we use criterion (A.10). Take any sequence $\varepsilon_k > 0$ such that

$$\sum_{k=1}^{\infty} \varepsilon_k < \infty.$$

If $x_n \xrightarrow{P} x$, take any $\delta > 0$ and consider $P(|x_n - x| > \delta) \to 0$ as $n \to \infty$. The meaning of the convergence is that for each ε_k there is an N_{ε_k} such that

$$P(|x_n - x| > \delta) < \varepsilon_k,$$

for all $n \geq N_{\varepsilon_k}$. In particular, with $n_k = N_{\varepsilon_k}$, one has

$$\sum_{k=1}^{\infty} P(|x_{n_k} - x| > \delta) < \sum_{k=1}^{\infty} \varepsilon_k,$$

which is finite by construction. The sufficient criterion (A.10) gives the desired almost sure convergence of the subsequence x_{n_k}.

A.5 Characteristic functions

A.5.1 The characterstic function

The spectral distribution of a stationary process and the covariance function form a Fourier transform pair. This relation corresponds to the relation between a probability distribution and its characteristic function. The "proof" of Bochner's theorem, Theorem 3.3, relied heavily on this correspondence, and it therefore left out some important results about convergence. Naturally, both aspects are part of the general mathematical Fourier theory. The left-out details are summarized here together with the basic properties of characteristic functions. For more details and proofs of the following theorems, see, for example [19, 58].

Definition A.7. *The characteristic function for a random variable x with distribution function $F_x(a) = P(x \leq a)$ is defined as*

$$\varphi(s) = E(e^{isx}) = \int_{-\infty}^{\infty} e^{isu}\, \mathrm{d}F(u).$$

The characteristic function of a p-variate random variable $\mathbf{x} = (x_1, \ldots, x_p)$ is defined as

$$\varphi(s_1, \ldots, s_p) = E\left(e^{i(s_1 x_1 + \ldots + s_p x_p)}\right).$$

The characteristic function is a complex-valued function and it *characterizes* the distribution in the sense that there are not two different distributions that have the same characteristic function. Note, however, that two different characteristic functions may coincide on some finite interval, e.g., on $[-1,1]$.

If the distribution of x is symmetric around 0 then the characteristic function is real valued. Every characteristic function $\varphi(s)$ is also the covariance function for a stationary process in continuous time, possibly complex, and every continuous covariance function $r(t)$ is also a characteristic function if normalized to $\varphi(s) = r(s)/r(0)$.

The following theorem is the basic ingredient behind Theorem 3.3.

Theorem A.7. *If F_n, $n = 1,2,\ldots$, is a sequence of distribution functions with characteristic functions $\varphi_n(s)$, and*

a) $\lim_{n\to\infty} \varphi_n(s) = \varphi(s)$ *exists for all s, and*

b) $\varphi(s)$ *is continuous at $s = 0$,*

then there is a distribution function F such that $F_n(a) \to F(a)$ for all a where F is continuous. Furthermore, $\varphi(s)$ is the characteristic function of F.

The following theorem clarifies the relation between a distribution function and its moments, and the characteristic function.

Theorem A.8. *If the random variable x has finite moments up to order k, i.e., $E(|x|^k) < \infty$, then its characteristic function can be expanded*

$$\varphi(s) = \sum_{j=0}^{k-1} \frac{(is)^j}{j!} E(x^j) + \frac{(is)^k}{k!} (E(x^k) + \delta(s)),$$

where $\delta(s)$ is a function bounded as $|\delta(s)| \leq 3E(|x|^k)$ for all s; [19, Prop. 8.44].

It is an important consequence of the theorem that the moments of a distribution are determined from the derivatives at $s = 0$ of the characteristic function. The converse is not true, however. There are examples of characteristic functions that coincide in an interval around $s = 0$, and have the same series of moments of all orders, but belong to two different distributions. As a consequence, we see that two different covariance functions may well be equal up to some finite time lag t, but differ for larger s-values.

The characteristic function and convolution

If x and y are independent random variables with characteristic functions $\varphi_x(s)$ and $\varphi_y(s)$, then the characteristic function of the sum $x+y$ is the product

$$\varphi_{x+y}(s) = E(e^{i(x+y)s}) = E(e^{ixs})\,E(e^{iys}) = \varphi_x(s)\,\varphi_y(s). \tag{A.17}$$

Note that if the distributions of x and y are absolutely continuous with probability densities f_x and f_y, the probability density of $x+y$ is the convolution

$$f_{x+y}(u) = \int_v f_x(v)\,f_y(u-v)\,\mathrm{d}v.$$

Convolution of distributions corresponds to multiplication of the characteristic functions, just as for other Fourier transform pairs.

Characteristic function of normal variables

The characteristic function of a univariate normal variable x with $E(x) = m$, $V(x) = \sigma^2$, is

$$\varphi(s) = e^{ims-\sigma^2 s^2/2}.$$

A multivariate normal variable $\mathbf{x} = (x_1,\dots,x_p)$ with mean $E(\mathbf{x}) = \mathbf{m}$ and covariance matrix $\mathbf{\Sigma}$ has the characteristic function $\varphi(\mathbf{s}) = e^{i\mathbf{ms}'-\frac{1}{2}\mathbf{s\Sigma s}'}$.

A.5.2 *The inversion theorems*

We present here a proof of statement (a) in the inversion theorem for covariance functions, Theorem 3.4. The proof is a slight reformulation of that in [19].

Theorem A.9. *a) If $\varphi(s)$ is the characteristic function of the statistical distribution function $F(x)$, and $\int_{-\infty}^{\infty} |\varphi(s)|\,ds < \infty$, then $F(x)$ is absolutely continuous with a bounded density $f(x)$, and*

$$f(x) = \frac{1}{2\pi}\int_{-\infty}^{\infty} e^{-ixs}\varphi(s)\,ds. \qquad \text{(A.18)}$$

b) The integrability condition $\int |\varphi(s)|\,ds < \infty$ is sufficient for the inversion formula but not necessary. In fact, (A.18) *holds for each x where f is continuous with left and right derivatives; see [5] for an even weaker condition.*

Proof. We prove part a). Assume that we have found a continuous random variable Y with distribution function $G(y)$, density $g(y)$, and characteristic function $\psi(s)$ such that (A.18) holds for that particular case. We are going to show that (A.18) holds in general.

First, observe that (A.18) also holds for the random variable εY with characteristic function $\psi_\varepsilon(s)$ and density $g_\varepsilon(y) = \frac{1}{2\pi}\int e^{-iys}\psi_\varepsilon(s)\,ds$. Write

$$\frac{1}{2\pi}\int \varphi(s)\,\psi_\varepsilon(s)\,e^{-isy}\,ds = \frac{1}{2\pi}\int_s\int_x e^{isx}e^{-isy}\psi_\varepsilon(s)\,dF(x)\,ds, \qquad \text{(A.19)}$$

and change the order of integration, to obtain

$$\frac{1}{2\pi}\int_s \varphi(s)\psi_\varepsilon(s)\,e^{-isy}\,ds = \int_x g_\varepsilon(y-x)\,dF(x) = h_\varepsilon(y), \text{ say.} \qquad \text{(A.20)}$$

The convolution in the right hand side is the density of a continuous distribution for a sum of two independent variables: X, with distribution F, and εY with density g_ε. The product, $\varphi(s)\psi_\varepsilon(s)$, is its characteristic function, so (A.18) holds for $X+\varepsilon Y$.

As $\varepsilon \to 0$ the characteristic function $\psi_\varepsilon(s) = \psi(\varepsilon s)$ tends to one everywhere. From the assumption on $\varphi(s)$ we can then conclude, by dominated convergence, that the left hand side of (A.20) has a limit,

$$h(y) = \frac{1}{2\pi}\int_s \varphi(s)e^{-isy}\,ds,$$

which is also the limit of $h_\varepsilon(y)$.

Next, we must show that $h(y)$ is the density for the distribution function F. Take $a < b$ to be continuity points of $F(x)$. Then, since $X + \varepsilon Y \xrightarrow{\mathcal{L}} X$,

$$P(a < X \leq b) = \lim_{\varepsilon \to 0} P(a < X + \varepsilon Y \leq b)$$
$$= \lim_{\varepsilon \to 0} \int_a^b h_\varepsilon(y)\,\mathrm{d}y = \int_a^b h(y)\,\mathrm{d}y,$$

by dominated convergence. Thus F has density $h(y)$, which satifies (A.18).

It remains to find a suitable $G(y)$. Common choices are the normal distribution or the two-sided exponential distribution. □

The same method can be used to prove the general inversion theorem, Theorem 3.4(b); see [19, Sect. 8.10].

A.5.3 Plancherel's and Parseval's relations

Let $\widehat{f}(y) = \int e^{iyx} f(x)\,\mathrm{d}x$, with $f(x) = \frac{1}{2\pi}\int e^{-iyx}\widehat{f}(y)\,\mathrm{d}y$, define a Fourier transform pair. When $\phi_1, \widehat{\phi_1}$ and $\phi_2, \widehat{\phi_2}$ are Fourier transform pairs with ϕ_1 and ϕ_2 both integrable and squared integrable, the Plancherel's and Parseval's relations hold, respectively:[6]

$$\int_{-\infty}^{\infty} \phi_1(t)\,\overline{\phi_2(t)}\,\mathrm{d}t = \frac{1}{2\pi}\int_{-\infty}^{\infty} \widehat{\phi_1}(\omega)\,\overline{\widehat{\phi_2}(\omega)}\,\mathrm{d}\omega, \tag{A.21}$$

$$\int_{-\infty}^{\infty} |\phi(t)|^2\,\mathrm{d}t = \frac{1}{2\pi}\int_{-\infty}^{\infty} |\widehat{\phi}(\omega)|^2\,\mathrm{d}\omega. \tag{A.22}$$

The Plancherel relation has the following interesting consequence: If $r(t)$ is a covariance function with spectral density $f(\omega)$, it is always possible to find a Fourier pair $\phi(t), \widehat{\phi}(\omega)$, such that $|\widehat{\phi}(\omega)|^2 = f(\omega)$. Then $\widehat{\phi}(\omega)e^{-i\omega s}$ is the transform of $\phi(t+s)$. Relation (A.21) with $\phi_1(t) = \phi(t)$ and $\phi_2(t) = \phi(t+s)$ gives

$$r(s) = \frac{1}{2\pi}\int_{-\infty}^{\infty} e^{i\omega s} f(\omega)\,\mathrm{d}\omega = \int_{-\infty}^{\infty} \phi(t)\overline{\phi(t+s)}\mathrm{d}t,$$

and we have shown that every covariance function can be expressed in this form; cf. Campbell's formula, (4.45).

---

[6] Marc-Antoine Parseval des Chênes (1755 - 1836), French mathematician, presented the relation in 1799, about ten years before Fourier published his work on Fourier series.

A.6 Hilbert space and random variables

A.6.1 Hilbert space and scalar products

A Hilbert space is a set of elements, which can be added and multiplied by complex numbers, and for which there is defined a *scalar (inner) product.* The distance between elements and convergence of sequences is defined by means of the scalar product.

For reading on Hilbert spaces and on metric spaces, see, e.g., the book by Royden [105].

Definition A.8. *A general Hilbert space $\mathcal{H}$ over the complex numbers $\mathbb{C}$ is a set of elements, usually called points or vectors, with the following properties:*

1. *The operations addition and subtraction are defined, and there exists a unique "zero" element $\mathbf{0} \in \mathcal{H}$ and to each $x \in \mathcal{H}$ there is a unique inverse $-x$: $x+y = y+x \in \mathcal{H}$, $x+\mathbf{0} = x$, $x+(-x) = \mathbf{0}$.*
2. *Multiplication with complex scalar is defined (usually written $cx = c \cdot x$): $c \cdot x \in \mathcal{H}$, $0 \cdot x = \mathbf{0}$, $1 \cdot x = x$.*
3. *A scalar (inner) product (x,y) is defined such that:*

$$
\begin{aligned}
&(x,y) = \overline{(y,x)} \in \mathbb{C}, \\
&(ax+by,z) = a(x,z)+b(y,z), \text{ for any complex } a \text{ and } b, \\
&(x,x) \geq 0, \text{ with equality if and only if } x = \mathbf{0}.
\end{aligned}
$$

4. *A norm $\|x\|$ and a distance $d(x,y) = \|x-y\|$ are defined, and convergence has the standard meaning: if $x \in \mathcal{H}$ then $\|x\| = (x,x)^{1/2}$, and if $x_n, x \in \mathcal{H}$, then $\lim_{n\to\infty} x_n = x$ if and only if $\|x_n - x\| \to 0$.*
5. *The space is complete in the sense that if $x_n \in \mathcal{H}$ and $\|x_m - x_n\| \to 0$ as $m,n \to \infty$, then there is a point $x \in \mathcal{H}$ such that $\lim_{n\hookrightarrow\infty} x_n = x$.*

A sequence x_n is called a Cauchy sequence if $\|x_n - x_m\| \to 0$ as n and m go to infinity independently of each other. In a Hilbert space any Cauchy sequence has a finite limit, $\|\lim x_n\| < \infty$.

Remark A.4. *If $\mathcal{H}$ is a space that satisfies (1-3) in the definition, then it can be completed and made a Hilbert space that satisfies also (5).*

We list some further properties of Hilbert spaces and scalar products, which will be seen to have parallel meanings for random variables:

Schwarz' inequality: $|(x,y)| \leq \|x\| \cdot \|y\|$ with equality if and only if $(y,x)x = (x,x)y$,

Triangle inequality: $\|x+y\| \leq \|x\| + \|y\|$,

Continuity: if $x_n \to x$ and $y_n \to y$ then $(x_n, y_n) \to (x,y)$,

Pythagorean theorem: if x and y are orthogonal, i.e., $(x,y) = 0$, then

$$\|x+y\|^2 = \|x\|^2 + \|y\|^2.$$

Linear subspaces

Let $L = \{x_j \in \mathscr{H}; j = 1,2,\ldots\}$ be a set of elements in a Hilbert space $\mathscr{H}$, and let

$$M_0 = \left\{a_1x_1 + \ldots + a_kx_k; k = 1,2,\ldots; a_j \in \mathbb{C}\right\}$$

be the family of all finite linear combinations of elements in L. Then

$$\mathscr{M} = \overline{M_0} = \mathscr{S}(L) = \left\{x \in \mathscr{H}; x = \lim_{n\to\infty} x_n \text{ for some } x_n \in M_0\right\}$$

is called *the subspace of $\mathscr{H}$ spanned by L*. It consists of all elements in $\mathscr{H}$ which are linear combinations of elements in L or are limits of such linear combinations. It is a *subspace* in the sense that it is closed under addition, multiplication by scalar, and passage to a limit.

Projections in Hilbert space

Two elements in a Hilbert space are called *orthogonal*, written $x \perp y$, if $(x,y) = 0$. Two subsets L_1 and L_2 are said to be orthogonal, $L_1 \perp L_2$, if all elements $x \in L_1$ are orthogonal to all elements $y \in L_2$. Similarly, two subspaces $\mathscr{M}_1$ and $\mathscr{M}_2$ are orthogonal, $\mathscr{M}_1 \perp \mathscr{M}_2$, if all elements in $\mathscr{M}_1$ are orthogonal to all elements in $\mathscr{M}_2$. The reader should check that if $L_1 \perp L_2$, then $\mathscr{S}(L_1) \perp \mathscr{S}(L_2)$.

For a sequence of subspaces, $\mathscr{M}_1, \ldots, \mathscr{M}_k$ of $\mathscr{H}$, write

$$V = \mathscr{M}_1 \oplus \ldots \oplus \mathscr{M}_k$$

for the *vector sum* of $\mathscr{M}_1, \ldots, \mathscr{M}_k$, which is the set of all vectors $x_1 + \ldots + x_k$, where $x_j \in \mathscr{M}_j$, for $j = 1, \ldots, k$.

Theorem A.10 (Projection). *Let $\mathscr{M}$ be a closed subspace of a Hilbert space $\mathscr{H}$, and let x be a point in $\mathscr{H}$ not in $\mathscr{M}$. Then x can be written in exactly one way as a sum*

$$x = y + z$$

with $y \in \mathscr{M}$ and $z = (x - y) \perp \mathscr{M}$. Furthermore, y is the point in $\mathscr{M}$ that is closest to x,

$$d(x,y) = \min_{w \in \mathscr{M}} d(x,w),$$

and equality holds if and only if $w = y$.

The most common use of the projection theorem is to approximate a point x in a general Hilbert space by a linear combination, or a limit thereof, of a finite or infinite number of certain elements in $\mathscr{H}$.

Separable spaces and orthogonal/orthonormal bases

A Hilbert space $\mathscr{H}$ is called *separable* if it contains a countable set of elements $x_1, x_2, \ldots$ such that the subspace spanned by all the x_j is equal to $\mathscr{H}$. If the x-variables are *linearly independent*, i.e., there is no non-trivial linear combination equal to $\mathbf{0}$, $a_1 x_1 + \ldots + a_n x_n = \mathbf{0}$, it is possible to find *orthogonal* elements $z_1, z_2, \ldots$, such that

$$\begin{aligned} z_1 &= c_{11} x_1, \\ z_2 &= c_{21} x_1 + c_{22} x_2, \\ &\ldots \\ z_n &= c_{n1} x_1 + c_{n2} x_2 + \ldots + c_{nn} x_n, \\ &\ldots \end{aligned}$$

This is the *Gram-Schmidt orthogonalization process*. We chose to normalize the z_k to unit length, so

$$(z_j, z_k) = \delta_{jk} = \begin{cases} 1, & \text{for } j = k, \\ 0, & \text{for } j \neq k. \end{cases}$$

The sequence $z_1, z_2, \ldots$ is called a *complete orthonormal basis* for the Hilbert space $\mathscr{H}$. It is a basis, i.e., every element in $\mathscr{H}$ can be written as a linear combination of z_k-elements or as a limit of such combinations, and it is orthonormal by construction. It is furthermore complete, i.e., there is no element $u \in \mathscr{H}$ such that

$$\|u\| > 0, \quad (u, z_j) = 0, \quad \text{for all } j.$$

Suppose now that $\{z_k\}$ is a complete basis for the subspace $\mathscr{M}$. (In particular, this means that for every $u \in \mathscr{M}$, $u \perp z_k$, for all k implies that $u = 0$.) Take any $y \in \mathscr{M}$, and define $c_k = (y, z_k)$. Then, by the orthonormality,

$$0 \le E\left(|y - \sum_0^n c_k z_k|^2\right) = \ldots = E(|y|^2) - \sum_0^n |c_k|^2,$$

so $\sum_0^n |c_k|^2 \le \|y\|^2$ for all n, and hence $\sum_0^\infty |c_k|^2 \le \|y\|^2$. This means that $\sum_0^\infty c_k z_k$ exists as a limit in quadratic mean, and also that

$$y - \sum_0^\infty c_k z_k \perp z_n$$

for all n. But since $\{z_k\}$ is a complete family, $y = \sum_0^\infty c_k z_k = \sum_0^\infty (y, z_k)\, z_k$.

A.6.2 Stochastic processes and Hilbert spaces

A Hilbert space is a set of elements which can be added and multiplied by complex numbers, and for which there is defined an *inner product*. The inner product in a Hilbert space has the same mathematical properties as the covariance between two random variables with mean zero, and therefore it is natural to think of random variables as elements in a Hilbert space.

We shall consider a very special Hilbert space, namely the space of all random variables x on a probability space $(\Omega, \mathscr{F}, P)$, which have zero mean and finite variance.

Theorem A.11. *If $(\Omega, \mathscr{F}, P)$ is a probability space, then*

$$\mathscr{H} = \left\{ \begin{array}{l} \text{random variables } x \text{ on } (\Omega, \mathscr{F}, P), \\ \text{such that } E(x) = 0, E(|x|^2) < \infty \end{array} \right\}$$

with the scalar product

$$(x, y) = E(x\overline{y})$$

is a Hilbert space; it will be denoted $\mathscr{H}(\Omega)$.

First, it is clear that $(x,y) = E(x\overline{y})$ has the properties of a scalar product; check that. It is also clear that we can add random variables with mean zero and finite variance to obtain new random variables with the same properties. Also, $\|x\| = \sqrt{E(|x|^2)}$, which means that if $\|x\| = 0$, then $P(x=0) = 1$, so random variables which are zero with probability one are, in this context, defined to be equal to the zero element $\mathbf{0}$.

Convergence in the norm $\|\cdot\|$ is equal to convergence in quadratic mean of random variables, and if a sequence of random variables x_n is a Cauchy sequence, i.e., $\|x_m - x_n\| \to 0$ as $m,n \to \infty$, then we know that it converges to a random variable x with finite mean, which means that $\mathscr{H}(\Omega)$ is complete. Therefore it has all the properties of a Hilbert space.

A stochastic process as a curve in $\mathscr{H}(\Omega)$

A random variable with mean zero and finite variance is a point in the Hilbert space $\mathscr{H}(\Omega)$. Two equivalent random variables x and y are represented by the same point in $\mathscr{H}(\Omega)$, since they are equal with probability one, $P(x=y) = 1$, and hence $\|x-y\|^2 = E(|x-y|^2) = 0$.

A stochastic process is a family of random variables, and thus a stochastic process $\{x(t), t \in \mathbb{R}\}$ with one-dimensional parameter t is a *curve* in $\mathscr{H}(\Omega)$ (not to be confused by a sample function). Further, convergence in the norm $\|x\| = \sqrt{E(|x|^2)}$ is equivalent to convergence in quadratic mean. In other words, if a stochastic process is continuous in quadratic mean, then the corresponding curve in $\mathscr{H}(\Omega)$ is continuous.

The generated subspace

A set of points in a Hilbert space generates the subspace of all finite linear combinations and their limits. If $\{x(t); t \in T\}$ is a stochastic process, write

$$\mathscr{H}(x) = \mathscr{S}(x(s); s \in T)$$

for the subspace spanned by $x(\cdot)$. Also, for a process $\{x(t), t \in \mathbb{R}\}$, define

$$\mathscr{H}(x,t) = \mathscr{S}(x(s); s \le t)$$

as the subspace spanned by all variables observed up till time t. It contains all variables which can be constructed by linear operations on the available observations. Examples of random variables in $\mathscr{H}(x,t)$ are

$$\frac{x(t) + x(t-1) + \ldots + x(t-n+1)}{n}, \quad \int_{-\infty}^{t} e^{-(t-u)} x(u)\, du,$$

and $x'_-(t) + 3x''_-(t)$, where $x'_-(t), x''_-(t)$ denote left derivatives.

A.6.3 Projection and the linear prediction problem

The projection theorem in Hilbert spaces (Theorem A.10) states that if $\mathscr{M}$ is a closed linear subspace of a Hilbert space $\mathscr{H}$, and x is a point in $\mathscr{H}$ not in $\mathscr{M}$, then there is a unique element y in $\mathscr{M}$ closest to x, and then $z = x - y$ is orthogonal to $\mathscr{M}$.

Formulated in statistical terms, if x is a random variable and $y_1, \dots, y_n$ is a finite set of random variables, then there is a unique element $\widehat{x}$, expressed as a linear combination $\widehat{x} = c_1 y_1 + \ldots + c_n y_n$, that is closest to x in the $\|\cdot\|$-norm, i.e., such that

$$\left\|x - \sum c_j y_j\right\|^2 = E(|x - \sum c_j y_j|^2)$$

is minimal. This linear combination is characterized by the requirement that the residual $x - \sum c_j y_j$ is orthogonal to, i.e., uncorrelated with, all the y_j-variables. This is the least squares solution to the common *linear regression problem*. Expressed in terms of covariances, the coefficients in the optimal predictor $\widehat{x} = c_1 y_1 + \ldots + c_n y_n$ satisfy the linear equation system

$$C(x, y_j) = c_1 C(y_1, y_j) + \ldots + c_n C(y_n, y_j), \quad j = 1, \dots, n, \tag{A.23}$$

which follows from $C(x - \sum_k c_k y_k, y_j) = 0$.

Note that the projection theorem says that the random variable $y = \widehat{x}$ is unique, in the sense that if $\widetilde{y}$ is another random variable that minimizes the prediction error, i.e., $E(|x - \widehat{x}|^2) = E(|x - \widetilde{x}|^2)$ then $E(|\widehat{x} - \widetilde{x}|^2) = 0$ and $P(\widehat{x} = \widetilde{x}) = 1$. This does not mean that the coefficients in the linear combination $\sum c_j y_j$ are unique; if the variables $y_1, \dots, y_n$ are linearly dependent, then many combinations produce the same best predictor.

Example A.4 (MA(1)-process)**.** *Taking an* MA(1)*-process, i.e., from uncorrelated variables* $e(t), t = \dots, -1, 0, 1, 2, \dots$, *with* $E(e(t)) = 0$, $V(e(t)) = 1$, *we construct*

$$x(t) = e(t) + b_1 e(t-1).$$

If $|b_1| < 1$, *the process can be inverted and* $e(t)$ *retrieved from* $x(s), s \leq t$:

$$\begin{aligned} e(t) &= x(t) - b_1 e(t-1) = x(t) - b_1 (x(t-1) - b_1 e(t-2)) \\ &= \sum_{k=0}^{n} (-b_1)^k x(t-k) + (-b_1)^{n+1} e(t-n-1) = y_n(t) + z_n(t), \text{ say.} \end{aligned}$$

Here, $y_n(t) \in \mathscr{S}(x(s);\, s = t-n, \dots, t) \subseteq \mathscr{S}(x(s); s \leq t) = \mathscr{H}(x, t)$, *while*

$$\|z_n(t)\| = |b_1|^{n+1} \to 0$$

as $n \to \infty$. Thus $e(t) - y_n(t) \to 0$ and we have that

$$e(t) = \sum_{k=0}^{\infty} (-b_1)^k x(t-k) = \lim_{n\to\infty} \sum_{k=0}^{n} (-b_1)^k x(t-k) \in \mathscr{H}(x,t)$$

if $|b_1| < 1$. The representation of $e(t)$ as a limit of finite linear combinations of $x(t-k)$-values is explicit and obvious.

For $|b_1| = 1$ it is less obvious that $e(t) \in \mathscr{H}(x,t)$, but it is still possible to represent $e(t)$ as a limit. For example, if $b_1 = -1$, $x(t) = e(t) - e(t-1)$, and $z_n(t) = e(t-n-1)$ does not converge to anything. But in any case,

$$e(t) = \sum_{k=0}^{n} x(t-k) + e(t-n-1),$$

and so, since the left hand side does not depend on n,

$$\begin{aligned} e(t) &= \frac{1}{N}\sum_{n=1}^{N} e(t) = \frac{1}{N}\sum_{n=1}^{N}\sum_{k=0}^{n} x(t-k) + \frac{1}{N}\sum_{n=1}^{N} e(t-n-1) \\ &= \sum_{k=0}^{N}\left(1 - \frac{k}{N}\right) x(t-k) + \frac{1}{N}\sum_{n=1}^{N} e(t-n-1) = y_N(t) + z_N(t). \end{aligned}$$

Now, $z_N(t) = \frac{1}{N}\sum_{n=1}^{N} e(t-n-1) = e(t) - y_N(t) \to 0$ by the law of large numbers, since all $e(t)$ are uncorrelated with $E(e(t)) = 0$ and $V(e(t)) = 1$. We have shown that $e(t)$ is in fact the limit of a finite linear combination of $x(s)$-variables, i.e., $e(t) \in \mathscr{H}(x,t)$. Note that $\mathscr{H}(e,t) = \mathscr{H}(x,t)$.

Example A.5. *We can extend the previous example to have $\mathscr{H}(x,t) \subset \mathscr{H}(e,t)$ with strict inclusion. Take a series of variables $e^*(t)$ and a random variable u with $E(u) = 0$ and $V(u) < \infty$, everything uncorrelated, and set*

$$e(t) = u + e^*(t).$$

Then $x(t) = e(t) - e(t-1) = e^*(t) - e^*(t-1)$, and $\mathscr{H}(e,t) = \mathscr{H}(u) \oplus \mathscr{H}(e^*,t)$ with $\mathscr{H}(u)$ and $\mathscr{H}(e^*,t)$ orthogonal, and $\mathscr{H}(e,t) \supseteq \mathscr{H}(e^*,t) = \mathscr{H}(x,t)$.[7]

Exercises

A:1. Prove Chebyshev's and Markov's inequalities:

$$P(|x - E(x)| > \lambda) \le \frac{V(x)}{\lambda^2}, \quad \text{and} \quad P(|x| > \lambda) \le \frac{E(|x|^h)}{\lambda^h}.$$

[7]For the definition of the direct sum $\oplus$, see Appendix A.6, page 302.

A:2. Prove the Borel-Cantelli lemma:

a) If A_k are events in a probability space $(\Omega, \mathscr{F}, P)$, then $\sum_k P(A_k) < \infty$ implies that $P(A_k \text{ infinitely often}) = 0$.

b) If the events A_k are independent, then $\sum_k P(A_k) = \infty$ implies that $P(A_k \text{ infinitely often}) = 1$.

A:3. Let x_n and y_n, $n = 1, 2, \ldots$, be two sequences of integer valued random variables such that $x_n \leq y_n$ and $\lim_{n\to\infty} E(x_n) = \lim_{n\to\infty} E(y_n) = \theta$,

$$\lim_{n\to\infty} P(x_n = k) = p_k, \quad \sum_k p_k = 1, \quad \sum_k k p_k = \theta.$$

Prove that $\lim_{n\to\infty} P(y_n = k) = p_k$.

A:4. Let $x_1, x_2, \ldots$ be random variables with values in a countable set $\mathscr{F}$, and suppose there are real constants a_k such that

$$\sum_{k=1}^{\infty} P(x_k \neq a_k) < \infty, \qquad \sum_{k=1}^{\infty} a_k < \infty.$$

Prove that $x = \sum_{k=1}^{\infty} x_k$ has a discrete distribution, i.e., there exists a countable set D such that $P(x \in D) = 1$. Hint: Use the Borel-Cantelli Lemma.

Show by example that it is possible for independent random variables x_k to have a sum $\sum_{k=1}^{\infty} x_k$ with a continuous distribution, although all x_k are discrete variables with a common value space – obviously they can not be identically distributed.

A:5. Let $x_1, x_2, \ldots$ be independent identically distributed random variables. Show that $E(|x_k|) < \infty$ if and only if $P(|x_k| > k \text{ infinitely often}) = 0$.

A:6. Suppose the random sequences x_n and x_n' have the same distribution. Prove that if $x_n \overset{a.s.}{\to} x$ then there exists a random variable x' such that $x_n' \overset{a.s.}{\to} x'$.

A:7. Let e_k be a sequence of independent normal variables with mean 0 and variance 1. Prove that, for $|a| < 1$, the sum $\sum_{k=0}^{N} a^k e_{n-k}$ converges in each of the convergence modes as $N \to \infty$, and thus $x_n = \sum_{k=0}^{\infty} a^k e_{n-k}$ is well defined.

A:8. A statistical distribution is uniquely determined by its characteristic function $\varphi(s)$. Use this fact to show, that if $(x_1, \ldots, x_n)$ is an n-dimensional random variable and its characteristic function factorizes,

$$\varphi(s_1, \ldots, s_n) = \prod_1^n \varphi_k(s_k),$$

then the variables are independent.

Appendix B

Spectral simulation of random processes

B.1 The Fast Fourier Transform, FFT

A stationary process $\{x(t), t \in \mathbb{R}\}$ with continuous spectrum $f(\omega)$ can be efficiently simulated by Fourier methods from its spectral representation. One then has to discretize the continuous spectrum and use the approximation (3.44) from Section 3.4.1.

Fourier simulation is most effectively performed with the help of the Cooley and Tukey Fast Fourier Transform (FFT), or rather the inverse transform. This algorithm transforms a sequence of real or complex numbers $Z(0), Z(1), \ldots, Z(N-1)$ into its (inverse) discrete Fourier transform

$$z(n) = \sum_{k=0}^{N-1} Z(k) \exp(i2\pi kn/N), \tag{B.1}$$

for $n = 0, 1, \ldots, N-1$, where the integer N is a power of 2, $N = 2^m$. In the literature, there are as many ways to write the Fourier sum as there are combinatorial possibilities, with or without a factor N in the denominator and with or without a minus sign in the exponential function. Almost every mathematical computer software toolbox contains efficient algorithms to perform the FFT according to (B.1); see [29] for the main reference to the Fast Fourier Transform. Read also the history behind it in [28].

The basis for the use of (B.1) to generate a sample sequence lies in the representation of a stationary process as an approximating sum of harmonic functions with random phase and amplitude; see (3.44) and the alternative form (3.45). The $Z(k)$ will then be chosen as complex random variables with absolute value and argument equal to the desired amplitude and phase. When using the formula for simulation purposes, there are however a number of details that need attention, concerning the relation between the sampling interval and the frequency resolution, as well as the aliasing problem.

Before we describe the steps in the simulation we repeat the basic facts about processes with discrete spectrum, and the special problems that arise when sampling a continuous time process.

B.2 Random phase and amplitude

To see how (B.1) can be used to generate a sample function we consider first the special stationary process (3.25) with discrete spectrum in Section 3.4.1, or the normalized form (3.45). Including the spectral jump at zero frequency it has the form

$$x(t) = \rho_0 + \sum_{k=1}^{\infty} \rho_k \cos(\omega_k t + \phi_k). \tag{B.2}$$

Here ρ_0 is a random level shift, while $\{\rho_k\}$ are the amplitudes and $\{\phi_k\}$ the phases of the different harmonic components of $x(t)$. The frequencies $\omega_k > 0$ can be any set of fixed positive frequencies.

If we define

$$\begin{aligned} Z(0) &= \rho_0, \\ Z(k) &= \rho_k \exp(i\phi_k), \quad \text{for } k = 1, 2, \ldots \end{aligned}$$

it is easy to see that $x(t)$ in (B.2) is the real part of a complex sum, so if we write $y(t)$ for the imaginary part, then

$$x(t) + iy(t) = \sum_{k=0}^{\infty} Z(k) \exp(i\omega_k t). \tag{B.3}$$

We repeat the fundamental properties of this representation.

> If amplitudes and phases in (B.2) are independent and the phases ϕ_k are uniformly distributed over $[0, 2\pi)$, then $\{x(t), t \in \mathbb{R}\}$ is stationary and has a discrete spectral distribution with mass $\sigma_k^2 = \frac{1}{2}E(\rho_k^2)$ and $\sigma_0^2 = E(\rho_0^2)$ at the frequencies $\omega_k > 0$, and $\omega_0 = 0$, respectively. Further, the complex variables $Z(k) = \rho_k \exp(i\phi_k) = \sigma_k(U_k + iV_k)$ have the desired properties if the real and imaginary parts are independent standardized Gaussian random variables, with $E(U_k) = E(V_k) = 0$ and variance $V(U_k) = V(V_k) = 1$.

It is possible to approximate every spectral distribution by a discrete spectrum. The corresponding process is then an approximation of the original process.

Choosing the spectrum: The sampling and aliasing effect, Section 3.2.5, needs attention. If a stationary process $\{x(t), t \in \mathbb{R}\}$ with continuous two-sided spectral density $f_x(\omega)$ is sampled with a sampling interval d, the sequence $\{x(nd), n = 0, \pm 1, \ldots\}$ has a spectral density $f_x^{(d)}(\omega)$ that can be restricted to any interval of length $2\pi/d$, for example the interval $(-\pi/d, \pi/d]$. There it can be written as a folding of the original spectral density,

$$f_x^{(d)}(\omega) = \sum_{j=-\infty}^{\infty} f_x\left(\omega + \frac{2\pi j}{d}\right), \quad \text{for } -\pi/d < \omega \leq \pi/d.$$

The corresponding one-sided spectral density $g_x^{(d)}(\omega)$ can then be defined on $[0, \pi/d]$ as

$$f_x^{(d)}(\omega) + f_x^{(d)}(-\omega).$$

For reasons that will become clear later we prefer to define it instead on $[0, 2\pi/d)$ by

$$g_x^{(d)}(\omega) = \sum_{j=-\infty}^{\infty} f_x\left(\omega + \frac{2\pi j}{d}\right), \quad \text{for } 0 \leq \omega < 2\pi/d. \tag{B.4}$$

B.3 Simulation scheme

In view of (B.1) and (B.3) we would like to generate a finite part of the sum in (B.3) to get $z(n)$ and then take the real part to get $x(t)$ for $t = nd$, $n = 0, 1, \ldots, N-1$. To see the analogy clearly we repeat the expressions:

$$z(n) = \sum_{k=0}^{N-1} Z(k) \exp(i2\pi kn/N) \tag{B.5}$$

$$x(t) = \Re \sum_{k=0}^{\infty} Z(k) \exp(i\omega_k t). \tag{B.6}$$

Here is the scheme to follow:

We have: A real one-sided spectral density $g_x(\omega)$, for $\omega \geq 0$, for a stationary process $\{x(t), t \in \mathbb{R}\}$.

We want: A discrete time sample $x(nd)$, $n = 0, 1, \ldots, N-1$ of $\{x(t), t \in \mathbb{R}\}$ of size $N = 2^m$, with sampling interval d, equally spaced over the time interval $[0, T)$ with $T = Nd$.

Means: Generate random variables $Z(k) = \sigma_k(U_k + iV_k)$, $k = 0, 1, \ldots, N-1$, with independent, standard Gaussian variables U_k and V_k, and take, using the Fast Fourier Transform,

$$z(n) = \sum_{k=0}^{N-1} Z(k) \exp(i2\pi kn/N),$$

according to (B.1). Then set

$$x(nd) = \Re z(n), \quad n = 0, 1, \ldots, N-1.$$

This will give the desired realization.

B.4 Difficulties and details

The Fourier simulation scheme raises a number of questions which have to be dealt with before it can be implemented. Here we shall comment on the important issues. For further comments, see [82, Section 5.6].

Frequency spacing: We have requested N time points regularly spaced in $[0, T)$ in steps of d, and we want to use the special sum (B.1). This will impose a restriction both on the frequency spacing and on the maximum frequency $\omega_{\max}$ that can be represented. Comparing (B.5) and (B.6), bearing in mind that $t = nd$, we find that only frequencies that are of the form

$$\omega_k = \frac{2\pi k}{Nd} = \frac{2\pi k}{T} \quad \text{for} \quad k = 0, 1, \ldots, N-1,$$

appear in the simulation, and further that the highest frequency in the sum is $\frac{2\pi(N-1)}{dN}$, just barely below

$$\omega_{\max} = \frac{2\pi}{d}.$$

Discretization of spectrum: The continuous spectrum with density $g_x(\omega)$ has to be replaced by a discrete spectrum with mass only at the frequencies $\omega_k = \frac{2\pi k}{Nd}$ that enter into the sum (B.5). The mass at ω_k should be equal to

$$\sigma_k^2 = \frac{2\pi}{Nd} g_x^{(d)}(\omega_k), \quad k = 0, 1, \ldots, N-1. \tag{B.7}$$

Generation of the $\mathbf{Z(n)}$: Generate independent random variables

$$Z(k) = \sigma_k (U_k + iV_k)$$

with U_k and V_k from a normal distribution with mean zero and variance 1, for instance by the Box-Müller technique,

$$U_k = \cos(2\pi R_1)\sqrt{-2\ln R_2},$$
$$V_k = \sin(2\pi R_1)\sqrt{-2\ln R_2},$$

where R_1 and R_2 are independent random numbers uniformly distributed in $(0,1]$.

Aliasing: The restricted frequency range in (B.5) implies that the generated $x(nd)$ will have variance $\sum_{k=0}^{N-1}\sigma_k^2$, where each σ_k^2 is an infinite sum:

$$\sigma_k^2 = \frac{2\pi}{Nd}\sum_{j=-\infty}^{\infty} f_x(\omega_k + \frac{2\pi j}{d}).$$

In practice one has to truncate the infinite series and use

$$\sigma_k^2 = \frac{2\pi}{Nd}\sum_{j=-J}^{J} f_x(\omega_k + \frac{2\pi j}{d}), \quad k = 0,1,\ldots,N-1, \tag{B.8}$$

where J is taken large enough. If $f_x(\omega) \approx 0$ for $\omega \geq \omega_{\max}$ one can take $J = 0$.

Simulation of the envelope: The Fourier simulation will not only yield a realization of $x(nd) = \Re z(n)$ but also of its Hilbert transform $y(nd) = \Im z(n)$. Therefore we can get the envelope as a byproduct,

$$\sqrt{x(nd)^2 + y(nd)^2}.$$

Thus, generation of $2N$ Gaussian random numbers U_k, V_k, $k = 0,1,\ldots,N-1$, will result in $2N$ useful data points. If the aim is to generate only the $x(nd)$-series, one could restrict the sum (B.5) to only $n = 0,1,\ldots,N/2-1$ and thus generate only N Gaussian variates.

B.5 Summary

In order to simulate a sample sequence of a stationary process $\{x(t), t \in \mathbb{R}\}$ with spectral density $f_x(\omega)$ over a finite time interval one should do the following:

1. Choose the desired time interval $[0,T)$.
2. Choose a sampling interval d or the number of sample points $N = 2^m$. This will give a sequence of N process values $x(nd)$, $k = 0,1,\ldots,N-1$.

3. Calculate and truncate the real discretized spectrum

$$\sigma_k^2 = \frac{2\pi}{Nd} \sum_{j=-J}^{J} f_x(\omega_k + \frac{2\pi j}{d}), \quad k = 0, 1, \ldots, N-1,$$

   and take J so large that $f_x(\omega) \approx 0$ for $\omega > 2\pi(J+1)/d$.

4. Generate independent standard normal variables

$$U_k, V_k, \quad \text{for } k = 0, 1, \ldots, N-1,$$

   with mean zero and variance 1.

5. Set $Z(k) = \sigma_k(U_k + iV_k)$ and calculate the (inverse) Fourier transform

$$z(n) = \sum_{k=0}^{N-1} Z(k) \exp(i2\pi kn/N).$$

6. Take the real part,

$$x(nd) = \Re z(n), \quad n = 0, 1, \ldots, N-1;$$

   this is the desired sequence.

7. To generate the envelope, take the imaginary part

$$y(nd) = \Im z(n), \quad n = 0, 1, \ldots, N-1;$$

   the envelope is then

$$\sqrt{x(nd)^2 + y(nd)^2}.$$

Commonly used spectra

$$r(t) = \int_{-\infty}^{\infty} e^{i\omega t} f(\omega)\,\mathrm{d}\omega, \quad f(\omega) = \frac{1}{2\pi}\int_{-\infty}^{\infty} e^{-i\omega t} r(t)\,\mathrm{d}t$$

…nce $r(t)$	Spectral density $f(\omega)$	Comment	Page
…at	$\frac{1}{2}(\delta_a(\omega)+\delta_{-a}(\omega))$	discrete spectrum	84
…$t)$	$1/(2\pi)$	"white noise"	137
… $= \frac{\sin at}{at}$	$(2a)^{-1}\mathrm{rect}\,(\omega/a)$	"rect" is a box over $[-a,a]$	105
…(at)	$(2a)^{-1}\mathrm{tri}\,(\omega/(2a))$	"tri" is triangular over $[-2a,2a]$	113
…$(1-a\lvert t\rvert,0)$	$(2\pi a)^{-1}\,\mathrm{sinc}^2(\omega/(2a))$		
…$\frac{a^2t^2}{2})$	$\frac{1}{a\sqrt{2\pi}}\exp(-\frac{\omega^2}{2a^2})$	"Gaussian"	154
…$a\lvert t\rvert)$	$\frac{1}{\pi}\frac{a}{a^2+\omega^2}$	Ornstein-Uhlenbeck	103
…$^2t^2$	$\frac{1}{2a}e^{-\lvert\omega\rvert/a}$	"inverse O-U"	154
…$K_\nu(at)$	$\frac{1}{2\pi}\frac{A}{(a^2+\omega^2)^{\nu+1/2}}$	Matérn covariance $A=\frac{\Gamma(\nu+1/2)}{\Gamma(\nu)}\sqrt{4\pi}\,a^{2\nu}$	149

Covariance $r(t)$	Spectral density $f(\omega)$	Comment	Page
relative damping $\zeta = \sqrt{(1+a)/2}$	$\frac{A}{\omega^4+2a\omega_0^2\omega^2+\omega_0^4}$	damped linear oscillator $2\alpha^2 = \omega_0^2(1+a),\ a > -1$	138
$^{-\alpha\|t\|}(\cos\beta t + \frac{\alpha}{\beta}\sin\beta\|t\|)$	$\frac{A}{(\omega^2-\omega_0^2)^2+4\alpha^2\omega^2}$	under-damped, $-1 < a < 1$ $A = \frac{2\alpha\omega_0^2}{\pi},\ \beta = \omega_0\sqrt{(1-a)/2}$	
$e^{-\alpha\|t\|}(1+\alpha\|t\|)$	$\frac{A}{(\omega^2+\alpha^2)^2}$	critically damped, $a = 1$ $A = 2\alpha^3/\pi$	
$e^{-\alpha\|t\|}\frac{a_1 e^{\beta\|t\|} - a_2 e^{-\beta\|t\|}}{2\beta}$	$\frac{A}{\omega^4+2a\omega_0^2\omega^2+\omega_0^4}$	over-damped, $a > 1$ $A = \frac{2\alpha\omega_0^2}{\pi},\ \beta = \omega_0\sqrt{(a-1)/2}$ $a_1 = \alpha+\beta_1,\ a_2 = \alpha-\beta_1$	
$e^{-a\|t\|}\cos(b\|t\|-\psi)$	$\frac{c\omega^2+d^2}{\pi((\omega^2+a^2-b^2)^2+4a^2b^2)}$	$c = a\cos\psi - b\sin\psi$ $d = (a^2+b^2)(a\cos\psi + b\sin\psi)$	138
wave spectra, $\omega > 0$	$PM = \frac{\alpha}{\omega^5}e^{-1.25(\omega_0/\omega)^4}$	Pierson-Moskowitz	231
	$PM \times \gamma^{\exp(-(1-\omega/\omega_0)^2/2\sigma_m)}$	JONSWAP	231

Appendix D

Solutions and hints to selected exercises

1 Some probability and process background

1:1. Let $[x]$ denote the integer part of x, and take, for $\omega \in \Omega$,

$$x_n(\omega) = \begin{cases} 0 & \text{if } [2^{n-1}\omega] \text{ is even,} \\ 1 & \text{if } [2^{n-1}\omega] \text{ is odd.} \end{cases}$$

Then $\mathbf{y} = (x_1, x_2, \ldots)$ is as desired. The sum, $x = \sum_{k=1}^{\infty} x_k/2^k$, is uniformly distributed over $[0,1]$: its distribution function $F(u) = u$ for each x that has a finite dyadic expansion, $u = \sum_{k=1}^{N} \varepsilon_k/2^k$, with $\varepsilon_k = 0$ or 1. Since any distribution function is non-decreasing, $F(u) = u$ for all $u \in [0,1]$.

1:2. Write $\mathscr{A}$, $\mathscr{A}_o$, and $\mathscr{A}_c$ for the half-open, open, and closed intervals, respectively. The Borel set $\mathscr{B}$ is the smallest σ-field that contains all half-open intervals. Open and closed intervals are unions and intersections of half-open interval: $(a,b) = \cup_{n=1}^{\infty}(a, b-1/n]$, and $[a,b] = \cap_{n=1}^{\infty}(a-1/n, b]$. On the other hand, a half-open interval is an intersection, or a union, of open and closed intervals: $(a,b] = \cap_{n=1}^{\infty}(a, b+1/n)$, and $(a,b] = \cup_{n=1}^{\infty}[a+1/n, b]$. Therefore $\mathscr{A}$, $\mathscr{A}_o$, and $\mathscr{A}_c$ generate the same σ-field $\mathscr{B}$.

1:4. $\mathscr{A}$ is not a field. To be a field, a family has to be closed under finite unions, and $\mathscr{A}$ is not. Let $A = \{1,3,5,\ldots\}$ contain the odd numbers, so

$$\theta_n^A = n^{-1} |A \cap \{1,2,3,\ldots,n\}| \to 1/2,$$

as $n \to \infty$. Then construct $B = (b_1, b_2, \ldots, b_n, \ldots)$ in the following way. Start by letting $b_1 = 2, b_2 = 4, \ldots, b_k = 2k$ be the even numbers up to $b_{N_1/2} = N_1$; then $\theta_{N_1}^{A \cup B} = 1$. Continue to fill B with odd numbers $b_{N_1/2+1} = N_1 + 3, \ldots, b_k = 2k+1$, until $\theta_{N_2}^{A \cup B} < 2/3$, say. Again, shift to even numbers until $\theta_{N_3}^{A \cup B} > 3/4$, say. Alternating with even and odd numbers in B prevents $\theta_N^{A \cup B}$ to converge, although both A and B have asymptotic density $1/2$.

1:8. We show that the set C_Q of functions, which are continuous over the rational numbers, is in the σ-field generated by the intervals $\mathscr{I}$. First take one rational number q_0, and consider the functions which are continuous at q_0 over $\mathbb{Q}$, $C_{q_0} = \{x \in \mathbb{R}^T; \lim_{q\to q_0} x(q) = x(q_0)\}$. Write C_{q_0} as union/intersection:

$$C_{q_0} = \cap_{n=1}^{\infty} \cup_{N=1}^{\infty} \cap_{q;|q-q_0|<1/N} \{x; |x(q) - x(q_0| < 1/n\}.$$

This is the usual formulation of continuity: $x(q)$ arbitrarily close to $x(q_0)$ if only q is sufficiently close to q_0, although we have restricted q to the rational numbers. Now, for each $q \in \mathbb{Q}$, $\{x; |x(q) - x(q_0)| < 1/n\}$ is generated by intervals, and hence is in $\mathscr{B}_T$, and therefore $C_{q_0} \in \mathscr{B}_T$. Taking the intersection over countably many q_0, we still get $C_Q = \cap_{q_0} C_{q_0} \in \mathscr{B}_T$.

1:9. Compute the covariances for the increments.

1:10. For the first inequality:

$$\begin{aligned} E(N) - \frac{1}{2}E(N(N-1)) &= \sum_{k=1}^{\infty} kP(N=k) - \frac{1}{2}\sum_{k=2}^{\infty} k(k-1)P(N=k) \\ &= \sum_{k=1}^{\infty} P(N=k) + \sum_{k=2}^{\infty} (k-1)P(N=k) - \frac{1}{2}\sum_{k=2}^{\infty} k(k-1)P(N=k) \\ &= P(N>0) - \frac{1}{2}\sum_{k=2}^{\infty} (k-1)(k-2)P(N=k) \leq P(N>0) \leq E(N). \end{aligned}$$

The general inequality follows from the following identity, proved by induction over n:

$$\begin{aligned} \frac{1}{k!}\sum_{i=0}^{n} (-1)^i \frac{1}{i!}\alpha_{(k+i)} =& P(N=k) \\ &+ (-1)^n \sum_{i=k+n+1}^{\infty} P(N=i)\binom{i}{k}\binom{i-k-1}{n}. \end{aligned}$$

Assume this holds for n (obviously it holds for $n=1$). Then

$$\begin{aligned} \frac{1}{k!}\sum_{i=0}^{n+1} (-1)^i \frac{1}{i!}\alpha_{(k+i)} &= P(N=k) \\ &+ (-1)^n \sum_{i=k+n+1}^{\infty} P(N=i)\binom{i}{k}\binom{i-k-1}{n} \\ &+ (-1)^{n+1}\frac{(k+n+1)!}{k!(n+1)!} \sum_{i=k+n+1}^{\infty} P(N=i)\binom{i}{k+n+1} \end{aligned}$$

$$=P(N=k)+(-1)^{n+1}\sum_{i=k+n+1}^{\infty}P(N=i)$$
$$\times\left\{\binom{k+n+1}{k}\binom{i}{k+n+1}-\binom{i}{n}\binom{i-k-1}{n}\right\}$$
$$=P(N=k)+(-1)^{n+1}\sum_{i=k+n+2}^{\infty}P(N=i)\binom{i}{k}\binom{i-k-1}{n+1}.$$

2 Sample function properties

2:3. With $Q=\{r_j\}_{j=1}^{\infty}$, the event $N_j=\{x(r_j)\neq y(r_j)\}$ has $P(N_j)=0$. Since both $\{x(t),\, t\in\mathbb{R}\}$ and $\{y(t),\, t\in\mathbb{R}\}$ have, with probability one, continuous sample paths,

$$\begin{aligned}P(x(t)=y(t),\forall t\in R)&=P\left(\cap_{n=1}^{\infty}\{x(r_j)=y(r_j),\, j=1,2,\ldots,n\}\right)\\&=\lim_{n\to\infty}P(\{x(r_j)=y(r_j),\, j=1,2,\ldots,n\})\\&=\lim_{n\to\infty}P(\Omega\setminus\cup_1^n N_j)=1.\end{aligned}$$

2:4. Expanding $e^{-|t|}=1-|t|+t^2/2-|t|^3/6+O(t^4)$, one has

$$\begin{aligned}e^{-|t|}(1+a|t|+bt^2)=1&+(a-1)|t|+(b-a+1/2)t^2\\&+(a/2-b-1/6)|t|^3+O(t^4).\end{aligned}$$

In order that the process should be twice differentiable, both the $|t|$ and the $|t|^3$-term have to disappear, giving $a=1$ and $b=1/3$.

2:7. Set $X_n(k):=\left|w\left(\frac{k+1}{2^n}\right)-\left(\frac{k}{2^n}\right)\right|=\frac{\sigma}{2^{n/2}}|U_{k,n}|$, where $U_{k,n}\overset{\mathcal{L}}{=}U$ are standard normal variables. Thus, $E(X_n(k))=\frac{\sigma}{2^{n/2}}E(|U|)$, $V(X_n(k))=\frac{\sigma^2}{2^n}V(|U|)$, where $E(|U|)$ $(=\sqrt{2/\pi})$ and $V(|U|)$ $(=(1/2-1/\pi))$ are finite constants, whose exact values are of no interest to us here. Thus

$$E(Y_n)=2^n E(X_n(1))=C_1 2^{n/2},\quad V(Y_n)=2^n V(X_n(1))=C_2,$$

independent of n. Thus, by Markov's inequality,

$$\begin{aligned}P(Y_n<n)&=1-P\left(|Y_n-E(Y_n)|>\left(\frac{n-E(Y_n)}{D(Y_n)}\right)D(Y_n)\right)\\&\leq\left(\frac{D(Y_n)}{n-E(Y_n)}\right)^2\leq C\frac{1}{2^n}.\end{aligned}$$

Therefore $\sum_{k=1}^{\infty}P(Y_n<n)<\infty$, and, by the first Borel-Cantelli lemma, $Y_n\geq n$ for infinitely many n.

2:9. The covariance function is $\sigma^2(\min(s,t) - st/T)$.

2:13. The process $x(t) = \cos(2\pi t + \phi)$ with random $\phi \in U(0, 2\pi)$ has a tangent maximum equal to 1. Calculate the distribution function of $x(0)$:

$$\begin{aligned}
1 - F_{x(0)}(x) &= P(\cos(2\pi t + \phi) > x) = P(\cos\phi > x) \\
&= P(\cos\phi > x \mid \phi \in [0,\pi)) \cdot P(\phi \in [0,\pi)) \\
&+ P(\cos\phi > x \mid \phi \in [\pi, 2\pi)) \cdot P(\phi \in [\pi, 2\pi)) \\
&= 2P(\phi < \arccos x) = \frac{\arccos x}{\pi}.
\end{aligned}$$

The density $f_{x(t)}(x) = \frac{1}{\pi\sqrt{1-x^2}}$ is unbounded.

2:17. Take U as standard normal and choose $V = U$ with probability $1/2$ and $V = -U$ with probability $1/2$.

3 Spectral and other representations

3:1. The condition is $|b| \leq a/2$.

3:2. The product $r_1 r_2$ is a covariance function (for the product of two independent processes) with spectral density $f_1 * f_2$. The convolution $r_1 * r_2$ is a covariance function if f_1 and f_2 are squared integrable.

3:9. We first calculate the moment $E(u^2 v^2)$ for bivariate normal variables u, v. It can be found from the coefficient of $s^2 t^2/4!$ in the series expansion of the moment generating function (or the characteristic function),

$$\begin{aligned}
E(e^{su+tv}) &= 1 + sE(u) + tE(v) + \frac{s^2 E(u^2) + 2st E(uv) + t^2 E(v^2)}{2!} \\
&+ \ldots + \frac{\ldots + 6 s^2 t^2 E(u^2 v^2) + \ldots}{4!} + \ldots.
\end{aligned}$$

If u, v are bivariate Gaussian variables, with mean 0, variance 1, and correlation coefficient ρ, then the moment generating function is

$$E(e^{su+tv}) = e^{\frac{1}{2}(s^2 + 2\rho st + t^2)} = \ldots + \frac{\ldots + 2s^2t^2 + 4\rho^2 s^2 t^2 + \ldots}{8} + \ldots,$$

giving $E(u^2 v^2) = 1 + 2\rho^2$. Thus, $r_y(\tau) = C(x^2(0), x^2(\tau))$ is equal to $E(x^2(0)x^2(\tau)) - r_x^2(0) = 2r_x^2(\tau)$. The second part follows immediately from the fact that multiplication of two covariance functions (here $r_x(\tau)$ and $r_x(\tau)$) corresponds to convolution of the corresponding spectral densities. Note the analogy between convolution of probability densities and multiplication of the characteristic functions.

3:10. Differentiation of a stationary process implies multiplication of the spectral density $f(\omega)$ by $|i\omega|^2$. Thus, since $u(t) = 2x(t)x'(t) = \frac{d}{dt}x^2(t)$,

$$f_u(\omega) = \omega^2 f_{x^2}(\omega) = 2\omega^2 \int_\mu f_x(\mu) f_x(\omega - \mu)\, \mathrm{d}\mu.$$

An alternative, more cumbersome solution goes via the spectral representation $x(t) = \int e^{i\lambda t}\, \mathrm{d}Z(\lambda)$, and $x'(t) = \int (i\mu)e^{i\mu t}\, \mathrm{d}Z(\mu)$:

$$\begin{aligned} r_u(t) &= E(u(t)\overline{u(0)}) \\ &= 4 \iiiint \lambda\, \mu e^{-i(\lambda+\mu)t} E\big(\mathrm{d}Z(\lambda)\,\mathrm{d}Z(\mu)\,\overline{\mathrm{d}Z(\lambda')}\,\overline{\mathrm{d}Z(\mu')}\big). \end{aligned}$$

Here, most of the cross-expectations vanish; the only non-zero contributions are

$$E\big(\mathrm{d}Z(\lambda)\,\mathrm{d}Z(\mu)\,\overline{\mathrm{d}Z(\lambda')}\,\overline{\mathrm{d}Z(\mu')}\big) = \begin{cases} \mathrm{d}F(\lambda)\,\mathrm{d}F(\lambda'), & \mu = -\lambda, \mu' = -\lambda', \\ \mathrm{d}F(\lambda)\,\mathrm{d}F(\mu), & \lambda' = \lambda, \mu' = \mu, \\ \mathrm{d}F(\lambda)\,\mathrm{d}F(\mu), & \mu' = \lambda, \lambda' = \mu. \end{cases}$$

The quadruple integral in $r_u(t)$ therefore becomes

$$\begin{aligned} r_u(t) &= 4 \iint \lambda\lambda' \mathrm{d}F(\lambda)\,\mathrm{d}F(\lambda') + 4 \iint \mu^2 e^{-i(\lambda+\mu)t}\, \mathrm{d}F(\lambda)\,\mathrm{d}F(\mu) \\ &\qquad + 4 \iint \lambda\mu\, e^{-i(\lambda+\mu)t}\, \mathrm{d}F(\lambda)\,\mathrm{d}F(\mu) \\ &= 0 + 4 \iint e^{-i(\lambda+\mu)t} \mu(\lambda+\mu) f_x(\lambda)\, f_x(\mu)\, \mathrm{d}\lambda\, \mathrm{d}\mu \\ &= 4 \int e^{i\omega t} \omega \{ \int \omega' f_x(\omega - \omega') f_x(\omega')\mathrm{d}\omega' \}\, \mathrm{d}\omega \\ &= \int e^{i\omega t} \{ 2\omega^2 f_x(\omega - \omega') f_x(\omega')\, \mathrm{d}\omega' \}\, \mathrm{d}\omega, \end{aligned}$$

with $\lambda + \mu = -\omega$, $\mu = -\omega'$.

3:11. Start by the spectral representation of the innovations e_t:

$$e_t = \int e^{i\omega t}\, \mathrm{d}Z_e(\omega).$$

Then, from the definitions of x_t and y_t, we obtain

$$\int e^{i\omega t}\, \mathrm{d}Z_x(\omega) - \theta \int e^{i\omega(t-1)}\, \mathrm{d}Z_x(\omega) = \int e^{i\omega t}\, \mathrm{d}Z_e(\omega),$$

$$\int_{-\pi}^{\pi} e^{i\omega t}\, \mathrm{d}Z_y(\omega) = \int e^{i\omega t}\, \mathrm{d}Z_e(\omega) + \psi \int e^{i\omega(t-1)}\, \mathrm{d}Z_e(\omega),$$

which gives

$$dZ_e(\omega) = (1 - \theta e^{-i\omega})\, dZ_x(\omega),$$
$$dZ_y(\omega) = (1 + \psi e^{-i\omega})\, dZ_e(\omega),$$
$$dF_{xy} = E(dZ_x(\omega) \cdot \overline{dZ_y(\omega)}) = \frac{1+\psi e^{i\omega}}{1-\theta e^{-i\omega}}\, dF_e(\omega),$$

where $dF_e(\omega) = \frac{1}{2\pi}\, d\omega$. The cross-spectral density is $\frac{1}{2\pi}\frac{1+\psi e^{i\omega}}{1-\theta e^{-i\omega}}$.

3:12. First note that $m_y = a_1 + b_1 m_x$ and $m_x = a_2 - b_2 m_y$ (assuming $E(u_n) = E(v_n) = 0$); then subtract $m_x = (a_2 - b_2 a_1)/(1 + b_1 b_2)$, $m_y = (a_1 + b_1 a_2)/(1 + b_1 b_2)$, and consider instead $x_n - m_x$, $y_n - m_y$. They obey the same equations as x_n, y_n, but with $a_1 = a_2 = 0$. With $u_n = \int_{-\pi}^{\pi} e^{i\lambda n}\, dZ_u(\lambda)$, $v_n = \int_{-\pi}^{\pi} e^{i\lambda n}\, dZ_v(\lambda)$, we therefore have (with $E(|u_n|^2) = E(|v_n|^2) = 1$),

$$\int e^{i\lambda n}\, dZ_y(\lambda) = b_1 \int e^{i\lambda(n-1)}\, dZ_x(\lambda) + \int e^{i\lambda n}\, dZ_u(\lambda)$$
$$= b_1 e^{-i\lambda} \int e^{i\lambda n}\, dZ_x(\lambda) + \int e^{i\lambda n}\, dZ_u(\lambda),$$
$$\int e^{i\lambda n}\, dZ_x(\lambda) = b_2 \int e^{i\lambda n}\, dZ_y(\lambda) + \int e^{i\lambda n}\, dZ_v(\lambda).$$

Solving for dZ_x and dZ_y, we get

$$dZ_x(\lambda) = \frac{(1 + b_1 b_2 e^{i\lambda})(dZ_v(\lambda) - b_2 dZ_u(\lambda))}{1 + b_1^2 b_2^2 + 2 b_1 b_2 \cos\lambda},$$

$$dZ_y(\lambda) = \frac{(1 + b_1 b_2 e^{i\lambda})(dZ_u(\lambda) + b_1 e^{-i\lambda} dZ_v(\lambda))}{1 + b_1^2 b_2^2 + 2 b_1 b_2 \cos\lambda}.$$

Finally, we get the spectral densities as

$$f_x(\lambda) = \frac{\sigma^2}{2\pi} \frac{1 + b_2^2}{1 + b_1^2 b_2^2 + 2 b_1 b_2 \cos\lambda},$$
$$f_y(\lambda) = \frac{\sigma^2}{2\pi} \frac{1 + b_1^2}{1 + b_1^2 b_2^2 + 2 b_1 b_2 \cos\lambda},$$
$$f_{xy}(\lambda) = \frac{\sigma^2}{2\pi} \frac{b_1 e^{i\lambda} - b_2}{1 + b_1^2 b_2^2 + 2 b_1 b_2 \cos\lambda}.$$

3:15. The covariance function is $r_y(t) = \frac{1}{2} r_x(t) \cos\Gamma t$, and the spectral density is $f_y(\omega) = (f_x(\omega + \Gamma) + f_x(\omega - \Gamma))/4$. The increments of the spectral process is $dZ_y(\omega) = (e^{i\phi} dZ_x(\omega - \Gamma) + e^{-i\phi} dZ_x(\omega + \Gamma))/2$.

3:19. For $0 < \kappa < 1$.

3:21. $\lim_{h \downarrow 0} h^{-1} P(N([0,h] = 2) = 1/2 \neq 0$.

4 Linear filters – general properties

4:1. b) $r_y(t) = r_x(t)^2 + r_x(t-T)r_x(t+T)$,
$f_y(\omega) = (f_x * f_x)(\omega) + \frac{1}{4}\int e^{ivT/2} f_x(\omega+v/2) f_x(\omega-v/2)\,\mathrm{d}v$.

4:2. Linear interpolation is equivalent to a convolution of the discrete covariance with the "tent function," $h(u) = \max(1-|u|,0)$, with Fourier transform $(1/\pi^2)\,\mathrm{sinc}^2\omega \geq 0$. Thus, the interpolation has a non-negative, integrable, and symmetric Fourier transform.

4:4. $r_Y(s,t) = (e^{-(t-s)} - e^{-(t+s)})/2$ for $0 < s < t$.

4:5. $h(u) = \delta_0(u) + e^{-u} I_{u\geq 0}$, $g(\omega) = 1 + \frac{1}{1+i\omega}$, $f_y(\omega) = \frac{2}{\pi(1+\omega^2)}$, $r_y(t) = 2e^{-|t|}$. Hint: Find f_y first.

4:7. $\lim_{T\to\infty} Tv(T) = f(0)$.

4:8. Try $f(t) = e^{\beta t}$.

5 Linear filters – special topics

5:2. Hint: $|r(s,t)|^2 \leq r(s,s)r(t,t)$.

5:3. Solve Exercise 2.9.

5:4. Use $E((A-\widehat{A}(t))x(s)) = 0$ for $0 \leq s \leq t$ to get $\widehat{A}(t) = x(t)/(1+t)$.

6 Ergodic theory and mixing

6:1. $Tx = 2x \bmod 1$ is measurable, since $f(x) = ax+b$ with constants a, b takes Borel sets into Borel sets. It is measure preserving since

$$\begin{aligned}
&P(T^{-1}A) = P(\{x \in [0,1); 2x \in A\}) \\
&= P(\{x \in [0,1/2); 2x \in A\}) + P(\{x \in [1/2,1); 2x-1 \in A\}) \\
&= \int_0^{1/2} I_{\{2x\in A\}}\,\mathrm{d}x + \int_{1/2}^1 I_{\{2x-1\in A\}}\,\mathrm{d}x \\
&= \int_0^{1/2} I_{\{x\in A/2\}}\,\mathrm{d}x + \int_{1/2}^1 I_{\{x\in (A+1)/2\}}\,\mathrm{d}x = \frac{1}{2}P(A) + \frac{1}{2}P(A) = P(A).
\end{aligned}$$

6:2. With $x_n(\omega) = x(T^{n-1}\omega)$, we shall show that $\{x_n\}$ are independent $Bin(1,1/2)$-variables. Take a finite selection of x_n's, for example (x_1,x_3,x_6) and a sequence of 0/1's, for example $(0,1,1)$. Then, with $\omega = \sum_{k=1}^\infty \varepsilon_k 2^{-k}$,

$$\begin{aligned}
P((x_1,x_3,x_6) = (0,1,1)) &= P(\varepsilon_1 = 0, \varepsilon_3 = 1, \varepsilon_6 = 1) \\
&= P(\varepsilon_1 = 0)P(\varepsilon_3 = 1)P(\varepsilon_6 = 1).
\end{aligned}$$

6:4. To construct a two-sided sequence, $\mathbf{x} = \{x_n, n \in Z\}$, we have to produce a consistent family of *finite-dimensional* distributions, in which the half-sequence $\mathbf{x}^+ = \{x_n, n \geq 0\}$ has the given distribution. So take a finite sequence of times, $\mathbf{t} = (t_1 < \ldots < t_n)$, with $t_1 < 0$. In order that the new sequence be strictly stationary it is necessary and sufficient that $x_\mathbf{t} = (x_{t_1}, \ldots, x_{t_n})$ has the same distribution as $x_{\mathbf{t}-t_1} = (x_{t_1 - t_1}, \ldots, x_{t_n - t_1})$. But the distribution of $x_{\mathbf{t}-t_1}$ is already defined by the given finite-dimensional distributions. So just let $x_\mathbf{t}$ have that distribution.

6:5. For $x_{n+1} = 4x_n(1-x_n)$, take $y_n = 2x_n - 1$, so $y_{n+1} = 1 - 2y_n^2$. Use the trigonometric identity, $\cos 2\phi = 1 - 2\sin^2\phi$, and let ϕ, be a random variable with uniform distribution in $(-\pi/2, \pi/2)$. Then with

$$y_0 = \sin\phi, \quad x_0 = (1+\sin\phi)/2,$$
$$y_1 = 1 - 2y_0^2 = \cos 2\phi, \quad x_1 = (1+\cos 2\phi)/2,$$

one has

$$P(x_0 \leq x) = P(\sin\phi \leq 2x-1) = \frac{\arcsin(2x-1)}{\pi},$$
$$P(x_1 \leq x) = P(\cos 2\phi \leq 2x-1) = P(\sin\phi \leq 2x-1) = \frac{\arcsin(2x-1)}{\pi};$$

the arguments can be continued and whatever n-dimensional set $(x_{k_1}, \ldots, x_{k_n})$ we take, as long as the first element x_{k_1} has the same distribution as x_0, the distribution is shift-invariant.

6:6. Show that, for every $\varepsilon > 0$, the sum $\sum_n P(|x_n|/n > \varepsilon) < \infty$, and use the Borel-Cantelli lemma. But

$$\infty > E(|x_n|) = \int_0^\infty P(|x_n| > u)\,du > \varepsilon \sum_{k=1}^\infty P(|x_n| > k\varepsilon) = \varepsilon \sum_{n=1}^\infty P(|x_n| > n\varepsilon),$$

since x_n is (supposed to be) stationary.

6:8. Take $\Omega = \{0,1\}$ with $P(0) = P(1) = 1/2$. Then $Tx = 1-x$ is measure preserving. The invariant sets are $\emptyset$ and Ω and they have trivial probability. Hence T is ergodic. $T^2x = x$ is also measure preserving, but there are now non-trivial invariant sets, namely $\{0\}$ and $\{1\}$.

6:13. Take $\Omega = [0,1] \times [0,1]$ and start with two independent uniformly distributed variables x_0 and y_0 in $[0,1]$. Define the transformation

$$x_{n+1} = x_n + \theta_1 \mod 1,$$
$$y_{n+1} = y_n + \theta_2 \mod 1,$$

for irrational θ_1 and θ_2. This is ergodic if θ_1/θ_2 is irrational, not otherwise.

7 Vector processes and random fields

7:1. These are consequences of (4.7).

7:4. Compute the cross-covariance between the two processes.

7:9. It is a product of two covariance functions; which?

8 Level crossings and excursions

8:1. (a) The probability $P(x_n < 0 < y_n)$ is the volume under the density function in the second quadrant $x < 0 < y$. By a linear transformation to independent variables with mean zero and unit variance, this quadrant is transformed into a sector, with probability equal to the relative size of the opening angle. The transformation to independent Gaussian variables with unit variance is $w_n = \frac{x_n - \rho_n y_n}{1-\rho^2}$, $v_n = y_n$.
Suppose $\rho_n > 0$. Then the boundary lines $x_n < 0, y_n = 0$ and $x_n = 0, y_n > 0$ are transformed into $w_n < 0, v_n = 0$, and $w_n = -\rho_n v_n/(1-\rho_n^2)$. The opening angle ϕ of the sector satisfies $\cos\phi = \rho_n$. Thus $\phi = \arccos \rho_n$ and $P(x_n < 0 < y_n) = \frac{\arccos \rho_n}{2\pi}$.
(b) First note that also $u_n = x_n + y_n$ and $z_n = y_n - x_n$ are independent Gaussian with mean 0 and variances $V(u_n) = 2(1+\rho_n)$ and $V(z_n) = 2(1-\rho_n)$, respectively, and hence with density

$$f(u,z) = f_{u_n,z_n}(u,z) = c\, e^{-u^2/4(1+\rho_n)}\, e^{-z^2/4(1-\rho_n)}.$$

The conditional densities of $x_n + y_n | x_n < 0 < y_n$ and $y_n - x_n | x_n < 0 < y_n$ are

$$\begin{aligned} f_{u_n|x_n<0<y_n}(u) &= c\int_{z>|u|} f(u,z)\,\mathrm{d}z \\ &= c_1\, e^{-u^2/4(1+\rho_n)} \left\{1 - \Phi\Big(\frac{|u|}{\sqrt{2(1-\rho_n)}}\Big)\right\}, \\ f_{z_n|x_n<0<y_n}(u) &= c\int_{|u|<z} f(u,z)\,\mathrm{d}u \\ &= c_2\, e^{-z^2/4(1-\rho_n)} \left\{2\Phi\Big(\frac{z}{\sqrt{2(1+\rho_n)}}\Big) - 1\right\}. \end{aligned}$$

(c) Finally,

$$\begin{aligned} f_{c_n u_n|x_n<0<y_n}(u) &= c_{1,n}\, e^{-\frac{u^2(1-\rho_n)}{2(1+\rho_n)}} \{1 - \Phi(|u|)\} \to \sqrt{\pi/2}\,\{1-\Phi(|u|)\}, \\ f_{c_n z_n|x_n<0<y_n}(u) &= c_{2,n}\, e^{-\frac{z^2}{2}} \left\{2\Phi\Big(z\sqrt{\frac{1-\rho_n}{1+\rho_n}}\Big) - 1\right\} \\ &\to z e^{-z^2/2}, \quad z > 0. \end{aligned}$$

8:2. E.g., $x(t) = \cos(t + \phi)$.

8:4. With $c_n = 1/\sqrt{2(1-\rho_n)} = \frac{n}{\sqrt{\lambda_2}}(1 + o(1))$, we get

$$\zeta_n = \frac{x(1/n) - x(0)}{1/n} = c_n z_n (1 + o(1)).$$

According to the solution to Exercise 8.1(c), the conditional density, given $x_n < 0 < y_n$, is in the limit, as $n \to \infty$, a Rayleigh density. The conclusion is that the derivative of a Gaussian process at the times of upcrossings of the mean level has a Rayleigh distribution. This is what we found by other means in Theorem 8.8.

A Some probability theory

A:3. The variable $\varepsilon_n = y_n - x_n \geq 0$, and $E(\varepsilon_n) = E(y_n) - E(x_n) \to 0$. Thus $P(\varepsilon_n \geq 1) \to 0$ as $n \to \infty$. But $P(x_n = k) \leq P(y_n = k) + P(\varepsilon_n \geq 1) \leq P(x_n = k) + 2P(\varepsilon_n \geq 1)$.

A:4. Since $\sum_{k=1}^{\infty} P(x_k \neq a_k) < \infty$, the Borel-Cantelli lemma implies that, with probability one, only a finite number of x_k differ from the corresponding a_k; after a while $x_k = a_k$. This means that there is a random N such that

$$x = \sum_{k=1}^{\infty} x_k = \sum_{k=1}^{N} x_k + \sum_{k=N+1}^{\infty} x_k = \sum_{k=1}^{N} x_k + \sum_{k=N+1}^{\infty} a_k.$$

Here the second term,

$$\sum_{k=N+1}^{\infty} a_k = \sum_{k=1}^{\infty} a_k - \sum_{k=1}^{N} a_k = a - \sum_{k=1}^{N} a_k = b_N,$$

can take only a countable number of values, one for each N. The first term, $\sum_{k=1}^{N} x_k$, is a sum of finitely many x_k, each taking values in the countable set $\mathscr{F}$. The possible values form a countable set, and hence the sum can take only a countable number of different values. The union over N is countable. For an example of a continuous $\sum x_k$, take $\sum \varepsilon_k / 2^n$ from Exercise 1.1.

A:5. With $F(c) = P(|x_k| \leq c)$, use that $1 - F([c]+1) \leq 1 - F(c) \leq 1 - F([c])$ and $E(|x|) = \int_0^{\infty} P(|x| > c)\,\mathrm{d}c$.

Bibliography

[1] Åberg, S. (2007); Wave intensities and slopes in Lagrangian seas. *Adv. Appl. Probab.*, **39**, 1020–1035. Cited on p. 273.

[2] Adler, R. (1981): *The geometry of random fields.* John Wiley & Sons, Chichester. Cited on p. 217, 220.

[3] Adler, R.J. and Taylor, J.E. (2007): *Random fields and geometry.* Springer, New York. Cited on p. 60, 217.

[4] Andrews, D.W.K. (1984): Non-strong mixing autoregressive processes. *J. Appl. Probab.*, **21**, 930–934. Cited on p. 205.

[5] Apostol, T.M. (1974): *Mathematical analysis.* 2nd ed., Addison-Wesley, Reading. Cited on p. 299.

[6] Azaïs, J.-M. and Wschebor, M. (2009): *Level sets and extrema of random processes and fields.* John Wiley & Sons, Hoboken. Cited on p. 26, 217, 219, 236, 238, 244, 272, 272.

[7] Bartlett, M.S. (1963): The spectral analysis of point processes. *J. Roy. Stat. Soc. B (Methodological)*, **25**, 264–296. Cited on p. 105.

[8] Belyaev, Yu.K. (1959): Analytic random processes. *Theor. Prob. Appl.*, **4**, 402–409. Cited on p. 47, 51, 197.

[9] Belyaev, Yu.K. (1960): Properties of sample functions of a stationary Gaussian process. *Theor. Prob. Appl.*, **5**, 117–120. Cited on p. 47, 51.

[10] Belyaev, Yu.K. (1961): Continuity and Hölder's conditions for sample functions of stationary Gaussian processes. *Proceedings of the Fourth Berkeley Symposium on Mathematical Statistics and Probability*, **2**, 22–33. Cited on p. 48.

[11] Belyaev, Yu.K. (1968): On the number of exits across the boundary of a region by a vector stochastic process. *Theor. Prob. Appl.*, **13**, 320–324. Cited on p. 268.

[12] Benedetto, J.J. (1992): Irregular sampling and frames. In *Wavelets: A tutorial in theory and applications*, Ed: C.K. Chui, pp. 445–507. Academic Press, San Diego. Cited on p. 163.

[13] Beran, J. (1994): *Statistics for long-memory processes.* Chapman &

Hall, New York. Cited on p. 142–144.

[14] Billingsley, P. (1999): *Convergence of probability measures.* 2nd ed., John Wiley & Sons, New York. Cited on p. 50, 54.

[15] Billingsley, P. (1995): *Probability and measure.* 3rd ed., John Wiley & Sons, New York. Cited on p. 275, 283.

[16] Bochner, S. (1933): Monotone Funktionen, Stieltjessche Integrale und harmonische Analyse. *Math. Ann.*, **108**, 378–410. Cited on p. 74.

[17] Bolin, D. and Lindgren, F. (2011): Spatial models generated by nested partial differential equations, with an application to global ozone mapping. *Ann. Appl. Stat.*, **5**, 523–550. Cited on p. 228, 233.

[18] Bradley, R.C. (2005): Basic properties of strong mixing conditions. A survey and some open questions. *Probability Surveys*, **2**, 107–144. Cited on p. 202.

[19] Breiman, L. (1968): *Probability.* Addison-Wesley, Reading. Reprinted 1992 in SIAM Classics in Applied Mathematics, SIAM, Philadelphia. Cited on p. xviii, 80, 106, 126, 252, 275, 296–300.

[20] Brémaud, P. and Massoulié, L. (2002): Power spectra of general shot noises and Hawkes point processes with a random excitation. *Adv. Appl. Probab.*, **34**, 205–222. Cited on p. 141.

[21] Brillinger, D. (1993): The digital rainbow: Some history and applications of numerical spectrum analysis. *Can. J. Stat.*, **21**, 1–19. Cited on p. 97.

[22] Brockwell, P.J. and Davis, R.A. (1991): *Time series: theory and tethods*, 2nd ed., Springer-Verlag, New York. Cited on p. 82, 101, 134, 152, 210, 215.

[23] Bulinskaya, E.V. (1961): On the mean number of crossings of a level by a stationary Gaussian process. *Theor. Prob. Appl.*, **6**, 435–438. Cited on p. 58, 236, 262, 273.

[24] Campbell, N. (1909): The study of discontinuous phenomena. *Proc. Cambr. Phil. Soc.*, **17**, 117–136. Cited on p. 141.

[25] Capiński, M. and Kopp, E. (2004): *Measure, Integral and Probability.* 2nd ed., Springer, London. Cited on p. 275.

[26] Chung, K.L. (2001): *A course in probability theory.* 3rd ed., Academic Press, San Diego. Cited on p. 77, 80, 291.

[27] Chung, K.L. and Williams, R.J. (1990): *Introduction to stochastic integration.* 2nd ed., Birkhäuser, Boston. Cited on p. 61.

[28] Cooley, J.W. (1992): How the FFT gained acceptance. *IEEE Signal Proc. Mag.*, **9**, 10–13. Cited on p. 309.

[29] Cooley, J.W. and Tukey, J.W. (1965): An algorithm for the machine calculation of complex Fourier series. *Math. Comp.*, **19**, 297–301. Cited on p. 309.

[30] Cramér, H. (1939): On the representation of a function by certain Fourier integrals. *Trans. Amer. Math. Soc.*, **46**, 191–201. Cited on p. 74.

[31] Cramér, H. (1942): On harmonic analysis in certain function spaces. *Arkiv Mat. Astron. Fysik*, **28B**, no. 12. Cited on p. 25, 85.

[32] Cramér, H. (1945): *Mathematical methods of statistics*. Princeton University Press, Princeton. Cited on p. 28.

[33] Cramér, H. (1946): On the theory of stochastic processes. *Proc. Tenth Scand. Congr. of Math.*, Copenhagen, pp. 28–39. Cited on p. 25.

[34] Cramér, H. (1962): Décompositions orthogonales de certains processus stochastiques. *Ann. Fac. Sci. Clermont*, **11**, 15–21. Cited on p. 196.

[35] Cramér, H. and Leadbetter, M.R. (1967): *Stationary and related stochastic processes*. John Wiley & Sons, New York. Reprinted by Dover Publications, 2004. Cited on p. xvii, 41, 47–54, 66, 88, 106, 159, 173, 236, 240, 244, 247, 253.

[36] Cressie, N. and Wikle, C.K. (2011): *Statistics for spatio-temporal data*. John Wiley & Sons, Hoboken. Cited on p. 217, 228.

[37] Daley, D.J. and Vere-Jones, D. (2003): *An introduction to the theory of point processes*, Vol I and II, 2nd ed., Springer, New York. Cited on p. 17, 105, 108, 111.

[38] Dedecker, J., Doukhan, P., Lang, G., et al. (2007): *Weak dependence: With examples and applications*, Springer Lecture Notes in Statistics, No. 190. Cited on p. 202.

[39] Diggle, P. (2003): *Statistical analysis of spatial point patterns*. 2nd ed., Edward Arnold, London. Cited on p. 17, 105.

[40] Dobrushin, R.L. (1960): Properties of sample functions of a stationary Gaussian process. *Theor. Probab. Appl.*, **5**, 120–122. Cited on p. 48.

[41] Doob, J.L. (1953): *Stochastic processes*. John Wiley & Sons, New York. Reprinted 1990 in Wiley Classics Library. Cited on p. 40, 198–200.

[42] Doukhan, P. and Louhichi, S. (1999): A new weak dependence condi-

tion and applications to moment inequalities. *Stoch. Proc. Appl.*, **84**, 313–342. Cited on p. 205.

[43] Doukhan, P. and Louhichi, S. (2001): Functional estimation of a density under a new weak dependence condition. *Scand. J. Stat.*, **28**, 325–341. Cited on p. 205.

[44] Durbin, J. (1985): The first-passage density of a continuous Gaussian process to a general boundary. *J. Appl. Probab.*, **22**, 99–122. Cited on p. 261.

[45] Durrett, R. (1991): *Probability: Theory and examples*. 10th ed., Cambridge University Press, Cambridge. Cited on p. 80, 173.

[46] Einstein, A. (1905): *Investigations on the theory of Brownian movement*. Reprinted 1956 by Dover Publications, New York. Cited on p. 23.

[47] Eyer, L. and Bartholdi, P. (1999): Variable stars: Which Nyquist frequency? *Astron. Astrophys. Sup.*, **135**, 1–3. Cited on p. 101.

[48] Feichtinger, H.G. and Gröchenig, K. (1994): Theory and practice of irregular sampling. In *Wavelets: Mathematics and applications*, pp. 305–363. Stud. Adv. Math., CRC, Boca Raton. Cited on p. 163.

[49] Fernique, X. (1964): Continuité des processus Gaussiens. *C.R. Acad. Sci. Paris*, **258**, 6058–6060. Cited on p. 60.

[50] Fernique, X. (1974): Régularité des trajectoires des fonctions aléatoires Gaussiennes. *Lecture Notes in Mathematics*, **480**, Springer-Verlag, New York. Cited on p. 220.

[51] Fisher, R.A. (1929): Tests of significance in harmonic analysis. *P. R. Soc. Lond. A-Conta.*, **125**, 54–59. Cited on p. 97.

[52] Garsia, A.M. (1965): A simple proof of Eberhard Hopf's maximal ergodic theorem. *J. Math. and Mech.*, **14**, 381–382. Cited on p. 184.

[53] Genz, A. (1992): Numerical computation of multivariate normal probabilities. *J. Comput. Graph. Stat.*, **1**, 141–149. Cited on p. 262.

[54] Genz, A. and Kwong, K.-S. (2000): Numerical evaluation of singular multivariate normal distributions. *J. Stat. Comput. Sim.*, **68**, 1–21. Cited on p. 262.

[55] Gradshteyn, I.S. and Ryzhik, I.M. (2007): *Table of integrals, series, and products*. 7th ed., Academic Press, New York. Cited on p. 146, 224, 226.

[56] Grenander, U. (1950): Stochastic processes and statistical inference.

Ark. Mat. **1**, 195-277. Cited on p. 28, 170, 193.

[57] Grimmett, G.R. and Strizaker, D.R. (2001): *Probability and random processes*. 3rd ed., Oxford University Press, Oxford. Cited on p. 275.

[58] Gut, A. (2009): *An intermediate course in probability*. 2nd ed., Springer, Dordrecht. Cited on p. 275, 296.

[59] Hahn, S.L. (1996): *Hilbert transforms in signal processing*. Artech House, Norwood. Cited on p. 157.

[60] Hawkes, A.G. (1971): Spectra of some self-exciting and mutually exciting point processes. *Biometrika*, **58**, 83–90. Cited on p. 116.

[61] Hurst, H.E. (1951): Long-term storage capacity of reservoirs. *Trans. ASCE*, **116**, 770–799. Cited on p. 144.

[62] Ibragimov, I.A. and Linnik, Yu.V. (1971): *Independent and stationary sequences of random variables*. Wolters-Noordhoff, Groningen. Cited on p. 202–205.

[63] Ibragimov, I.A. and Rozanov, Y.A. (1978): *Gaussian random processes*. Springer-Verlag, New York. Cited on p. 198, 204.

[64] Isserlis, L. (1918): On a formula for the product-moment correlation of any order of a normal frequency distribution in any number of variables. *Biometrika*, **12**, 134–139. Cited on p. 152.

[65] Jordan, D.W. and Smith, P. (2007): *Nonlinear ordinary differential equations*. 4th ed., Oxford University Press, Oxford. Cited on p. 133, 134, 138.

[66] Kac, M. and Slepian, D. (1959): Large excursions of Gaussian processes. *Ann. Math. Stat.*, **30**, 1215–1228. Cited on p. 253.

[67] Kadane, J.B. and O'Hagan, A. (1995): Using finitely additive probability: Uniform distributions on the natural numbers. *J. Am. Stat. Assoc.*, **90**, 626–631. Cited on p. 277.

[68] Karhunen, K. (1947): Über lineare Methoden in der Wahrscheinlichkeitsrechnung. *Ann. Acad. Sci. Fennicae, Ser. A*, **37**, 1–79. Cited on p. 166.

[69] Karlin, S. and Taylor, H.M. (1981): *A second course in stochastic processes*. Academic Press, New York. Cited on p. 275.

[70] Kinsman, B. (1965): *Wind waves – their generation and propagation on the ocean surface*. Dover Editions 1984, 2002. Dover, Mineola. Cited on p. 232.

[71] Kolmogorov, A. (1933): *Grundbegriffe der Wahrscheinlichkeitsrech-*

nung. Springer-Verlag, Berlin. Cited on p. 6, 285.

[72] Lamperti, J. (1962): Semi-stable stochastic processes. *T. Am. Math. Soc.*, **104**, 62–78. Cited on p. 126.

[73] Lasota, A. and Mackey, M.C. (1994): *Chaos, fractals, and noise; stochastic aspects of dynamics*. 2nd ed., Springer-Verlag, New York. Cited on p. 173, 180.

[74] Leadbetter, M.R. (1974): On extreme values in stationary sequences. *Z. Wahrscheinlichkeit*, **28**, 289–303. Cited on p. 205, 246.

[75] Leadbetter, M.R., Lindgren, G. and Rootzén, H. (1983): *Extremes and related properties of random sequences and processes*. Springer-Verlag, New York. Cited on p. 205, 235, 237, 244–248, 254.

[76] Lindgren, F., Rue, H. and Lindström, J. (2011): An explicit link between Gaussian fields and Gaussian Markov random fields: The stochastic partial differential equation approach (with discussion). *J. R. Statist. Soc. B*, **73**, 423–498. Cited on p. 217, 228.

[77] Lindgren, G. (1970): Some properties of a normal process near a local maximum. *Ann. Math. Stat.*, **41**, 1870–1883. Cited on p. 254, 260.

[78] Lindgren, G. (1972): Local maxima of Gaussian fields. *Ark. Mat.*, **10**, 195–218. Cited on p. 273.

[79] Lindgren, G. (1975): Prediction from a random time point. *Ann. Probab.*, **3**, 412–423. Cited on p. 253.

[80] Lindgren, G. (1980): Model processes in nonlinear prediction with application to detection and alarm. *Ann. Probab.*, **8**, 775–792. Cited on p. 268.

[81] Lindgren, G. (2012): A detailed statistical representation of optical vortices in random wave fields. *J. Opt.*, **14**, 035704. Cited on p. 273.

[82] Lindgren, G., Rootzén, H. and Sandsten, M. (2014): *Stationary stochastic processes for scientists and engineers*. Chapman & Hall/CRC, Boca Raton. Cited on p. xviii, 13, 101, 152, 215, 312.

[83] Lindgren, G. and Rychlik, I. (1991): Slepian models and regression approximations in crossing and extreme value theory. *Int. Stat. Rev.*, **59**, 195–225. Cited on p. 254, 263.

[84] Loève, M. (1948): Fonctions aléatoire du second ordre, appendix to P. Lévy, *Processus stochastiques et mouvement Brownien*. Gautier-Villard, Paris. Cited on p. 85.

[85] Loève, M. (1962): *Probability theory*. Van Nostrand Reinhold, Prince-

ton. Cited on p. 166.

[86] Longuet-Higgins, M.S. (1957): The statistical analysis of a random moving surface. *P. Tr. R. Soc. S-A*, **249**, 321–387. Cited on p. 27, 272.

[87] Mandelbrot, B. (1965): Une classe de processus stochastiques homothetiques a soi; application a loi climatologique de H.E. Hurst. *Comptes Rendus (Paris)*, **260**, 3274–3277. Cited on p. 144.

[88] Maruyama, G. (1949): The harmonic analysis of stationary stochastic processes. *Mem. Fac. Sci. Kyusyu Univ.* **A4**, 45–106. Cited on p. 193.

[89] Matérn, B. (1960): Spatial variation. *Meddelande från statens skogsforskningsinstitut*, **39(5)**. Cited on p. 226.

[90] Matheron, G. (1973): The intrinsic random functions and their applications. *Adv. Appl. Probab.*, **5**, 439–468. Cited on p. 143, 146, 217.

[91] Mercadier, C. (2005): MAGP toolbox. Available at http://math.univ-lyon1.fr/~mercadier/MAGP/. Cited on p. 262.

[92] Micheson, A.A. and Stratton, S.W. (1898): A new harmonic analyser. *American Journal of Science*, **5**, 1–13. Cited on p. 97

[93] Miller, K.S. (1974): *Complex stochastic processes.* Addison-Wesley, Reading. Cited on p. 74.

[94] Mittal, Y. and Ylvisaker, D. (1975): Limit distributions for the maxima of stationary Gaussian processes. *Stoch. Proc. Appl.*, **3**, 1–18. Cited on p. 246.

[95] Øksendal, B. (1998): *Stochastic differential equations, an introduction with applications.* 5th ed., Springer, Berlin. Cited on p. 136.

[96] Percival, D.B. and Walden, A.T. (1993): *Spectral analysis for physical applications.* Cambridge University Press, Cambridge. Cited on p. 101.

[97] Perrin, J.B. (1913): *Les atomes*, Félix Alcan, Paris. Cited on p. 24

[98] Petersen, K. (1983): *Ergodic theory.* Cambridge University Press, Cambridge. Cited on p. 173.

[99] von Plato, J. (1994): *Creating modern probability*. Cambridge University Press, Cambridge. Cited on p. 180.

[100] Rényi, A. (1967): Remarks on the Poisson process. *Stud. Sci. Math. Hung.*, **5**, 119–123. Also in Lecture Notes in Mathematics 1967, **31**, 280–286. Cited on p. 108.

[101] Rice, S.O. (1944, 1945): Mathematical analysis of random noise. *Bell Syst. Tech. J.*, **23**, 282–332, and **24**, 46–156. Reprinted in: Wax, N. (1954): *Selected papers on noise and stochastic processes.* Dover

Publ., New York. Cited on p. 25, 26, 69, 102, 141, 237.

[102] Rice, S.O. (1963): Noise in FM-receivers. In: *Time Series Analysis*, Ed: M. Rosenblatt, Ch. 25, pp. 395–422. Wiley, New York. Cited on p. 26.

[103] Rosenblatt, M. (1956): A central limit theorem and a strong mixing condition. *P. Natl Acad. Sci. USA*, **42**, 43–47. Cited on p. 202.

[104] Rosiński, J. and Zak, T. (1997): The equivalence of ergodicity and weak mixing for infinitely divisible processes. *J. Theor. Probab.*, **10**, 73–86. Cited on p. 203.

[105] Royden, H.L. (1988): *Real Analysis*. 3rd ed., Prentice Hall, Englewood Cliffs. Cited on p. 198, 301.

[106] Rozanov, Yu.A. (1982): *Markov random fields*. Springer-Verlag, New York. Cited on p. 228.

[107] Rue, H. and Held, L. (2005): *Gaussian Markov random fields: Theory and applications*. Chapman & Hall, Boca Raton. Cited on p. 147, 217.

[108] Rychlik, I. (1987): A note on Durbin's formula for the first-passage density. *Stat. Probab. Lett.*, **5**, 425–428. Cited on p. 261.

[109] Rychlik, I. (2000): On some reliability applications of Rice's formula for the intensity of level crossings. *Extremes*, **3**, 331–348. Cited on p. 237, 239.

[110] Samorodnitsky, G. (2007): *Long range dependence*. now Publishers Inc, Delft. Cited on p. 142–144.

[111] Sato, K.-I. (1999): *Lévy processes and infinitely divisible distributions*. Cambridge University Press, Cambridge. Cited on p. 126.

[112] Schottky, W. (1912): Über spontane Stromschwankungen in verschiedenen Elektrizitätsleitern. *Annalen der Physik*, **362**, 541–567. Cited on p. 141.

[113] Schuster, A. (1898): On the investigation of hidden periodicities with application to a supposed 26 day period of meteorological phenomena. *Terrestrial Magnetism and Atmospheric Electricity*, **3**, 13–41. Cited on p. 97.

[114] Slepian, D. (1963): On the zeros of Gaussian noise. In: *Time Series Analysis*, Ed: M. Rosenblatt, pp. 104–115. Wiley, New York. Cited on p. 235, 254.

[115] Slutsky, E. (1937): Alcune propozitione sulla teoria della funzioni aleatorie. *Giorn. Ist. Ital. Attuari*, **8**, 193–199. Cited on p. 41, 60.

[116] St Denis, M. and Pierson, W.J. (1954): On the motion of ships in con-

fused seas. *Transactions, Soc. Naval Architects and Marine Engineers*, **61** (1953), pp. 280–357. Cited on p. 26, 272.

[117] Stein, M.L. (1999): *Interpolation of spatial data; Some theory for kriging*. Springer, New York. Cited on p. 145, 217, 227, 234.

[118] van Trees, H.L. (1968): *Detection, estimation, and modulation theory. Part I*. John Wiley & Sons, New York. Cited on p. 28, 170.

[119] The WAFO group (2011): WAFO – a MATLAB Toolbox for Analysis of Random Waves and Loads; Tutorial for WAFO version 2.5. March 2011.
Available at http://www.maths.lth.se/matstat/wafo/. Cited on p. 262, 264, 266.

[120] Whittle, P. (1954): On stationary processes in the plane. *Biometrika*, **41**, 434–449. Cited on p. 227.

[121] Whittle, P. (1963): Stochastic processes in several dimensions. *Bull. Intern. Statist. Inst.*, **40:2**, 974–994. Cited on p. 227.

[122] Williams, D. (1991): *Probability with martingales*. Cambridge University Press, Cambridge. Cited on p. 283, 291.

[123] Wold, H. (1954): *A study in the analysis of time series*. 2nd ed., Almquist and Wiksell, Stockholm. Cited on p. 196.

[124] Wong, E. and Hajek, B. (1985): *Stochastic processes in engineering systems*. Springer-Verlag, New York. Cited on p. 168, 172.

[125] Yaglom, A.M. (1962): *An introduction to the theory of stationary random functions*. Prentice Hall, Englewood Cliffs. Cited on p. xvii, 65, 88, 103, 198.

[126] Yaglom, A.M. (1987): *Correlation theory of stationary and related random functions, Vol I and II*. Springer-Verlag, New York. Cited on p. xviii, 217, 218.

Index